The Inner Bird

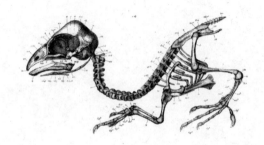

GARY W. KAISER

The Inner Bird

Anatomy and Evolution

UBCPress · Vancouver · Toronto

15 14 13 12 11 10 09 08 07   5 4 3 2 1

Printed in Canada on ancient-forest free paper (100% postconsumer recycled)
that is processed chlorine- and acid-free, with vegetable-based inks.

Library and Archives Canada Cataloguing in Publication

Kaiser, Gary W.
The inner bird: anatomy and evolution / Gary W. Kaiser.

Includes bibliographical references and index.
ISBN 978-0-7748-1343-3

1. Birds – Anatomy. 2. Birds – Evolution. I. Title.

QL697.K24 2007   571.3'18   C2007-900240-4

Canadä
UBC Press gratefully acknowledges the financial support
for our publishing program of the Government of Canada
through the Book Publishing Industry Development Program (BPIDP),
and of the Canada Council for the Arts,
and the British Columbia Arts Council.

Printed and bound in Canada by Friesens
Set in Arrus and Aurora by Artegraphica Design Co. Ltd.
Text designer: Irma Rodriguez
Project editor: Francis Chow

UBC Press
The University of British Columbia
2029 West Mall
Vancouver, BC V6T 1Z2
604-822-5959 / Fax: 604-822-6083
www.ubcpress.ca

For Dr. Alice Wilson,
Dr. Donald C. Maddox, Mr. Herb Groh,
and all the other adult mentors
of the Macoun Field Club,
the Ottawa Field-Naturalists' program
for young people

Contents

Part 3: How Does a Bird Fly?

Appendices

Figures and Tables

Tables

Acknowledgments

Any project of this size is dependent on the help and guidance of friends and colleagues. Robert Bender, Peter Clarkie, Ann Duffy, Michael McNall, Jean-François Savard, and Colleen Wilson read early versions of the manuscript and helped to make it more readable. David Broster, Brian Stafford, and Kyle Elliott caught some important errors in physics and technical terminology that I hope have not crept back in. I owe a particularly deep debt to John G. Robertson, Vincent Nealis, Richard Hebda, and three anonymous reviewers for their help in overcoming some problems with technical language and with the underlying structure of the book. The book would never have been completed without strong editorial support from Jean Wilson, Holly Keller, and Francis Chow.

I wish to express my gratitude to Dr. Hou Lianhai of the Wenya Museum, People's Republic of China, who took the time to point out some of the peculiar features of *Confuciusornis* and other early birds in his collection. I also wish to thank the National Museum of Canada, the Royal Ontario Museum, the Royal British Columbia Museum, the Cowan Museum at the University of British Columbia, the Museum of Zoology at the State University of Louisiana, and the Darwin Research Station, Galapagos Islands, for providing access to valuable skeletal material. Juan Parra provided additional important material for the Hoatzin.

Brett vander Kist and I prepared most of the figures except as noted. We concentrated our efforts on images that are not readily available from other sources. Consequently, there are few illustrations of whole birds or dinosaurs. There are huge numbers of informative images of dinosaurs and ancient birds on websites such as the Dinosauricon (http://dino.lm.com/pages/).

Access to live animals has also been critically important, and I want to thank Mauricio Alvarez, Ivan Jimenez, and Klaus Schutze for leading bird-banding projects in Colombia; Cecilia Lougheed-Betancourt for the opportunity to band albatrosses on the Galapagos Islands; and Juan-Carlos Riveros S. for the projects he organized in coastal Peru. Many of these same people worked on the Marbled Murrelet in North America. They joined Pascal Dehoux, Jesse Grigg, Carlos-Andres Botero, Daniel Cadena, Gustavo Londoño, and many other students and volunteers in early efforts to capture the Marbled Murrelet, and played a critical role in uncovering the secrets of its nesting habitat and developing an understanding of its flight capabilities.

While working on this book, I was privileged to be accepted as a research associate at the Royal British Columbia Museum.

Introduction

Does the world really need another bird book? Ever since Aristotle declared the study of birds to be a useful activity some 2,300 years ago, hundreds of enthusiasts have taken him at his word and written about the experience. You might think from the numbers crowding the world's bookshelves that there are more than enough books to satisfy any conceivable demand, but most deal only with what birds look like. They are guides to the identification of the thousands of different kinds of birds, used as basic reference tools by amateur ornithologists or "birders." There are field guides for every place on earth, but birds are so diverse and widespread that the books have no room for more than a picture and a brief note. This book is not about identification; it deals with the internal structures, how they work, and what they mean to the bird.

The bird portrayed in the field guide is the outward form of an animal whose specializations and sophistications are largely internal. There is far more to a bird than colourful feathers and a cheery song. These are trivial compared with the array of sophisticated design elements in a body plan that has ensured survival since the time of the dinosaurs. Birds have cleverly solved many difficult problems in their long history and found ways to invade most of the world's environments. They may or may not be dinosaurs in some sense,[1] but they are definitely not mammals, and as successful co-inhabitants of the globe, they have much to teach us mammals about alternative ways to exist. To properly appreciate their solutions to life's challenges, we must look at the animal beneath the feathers – the inner bird.

A study of the internal workings of a bird might seem an esoteric and obscure subject, but if you picked up this book you were already in the "bird section" of

---

[1] For the purpose of clarity, I use an arbitrary definition for birds and dinosaurs based on their ability to fly. The former is a feathered animal that flies, whereas the latter is a non-flying creature that may or may not have feathers. Some representatives of both groups are tiny (less than 100 g) and others mid-sized (about 100 kg), but only dinosaurs produced giants of 100 tonnes or more. All dinosaurs are extinct.

the bookstore or library and probably have a collection of field guides at home. You may even have a basic textbook on ornithology and have noticed that its illustrations and examples have changed little since you were in school. In fact, they have changed little since your grandparents' day. This widespread use of outdated material is misleading and weakens the ability of both professional and amateur ornithologists to apply new discoveries made in other sciences. This book attempts to show how modern discoveries in paleontology, aerodynamics, molecular genetics, and other sciences are affecting our fundamental understanding of the bird and how it lives. Increasingly, the results of modern research are challenging traditional interpretations of avian anatomy.

A hundred years ago, anatomy was the single most important aspect of ornithology, and fully half the chapters in an ornithology text might be devoted to the skeleton. Today's texts squeeze a brief summary of the anatomical information into the introductory chapters and the skeleton is lucky to get more than a page or two. Such extreme condensation has required a great deal of generalization and simplification. Over the years, unnoticed errors and subtle misrepresentations of fact have crept into the accounts, and significant pieces of information have been omitted. More importantly, such highly condensed summaries give the impression that avian anatomy is simple or merely a variation on mammalian anatomy. This is not the case.

A bird's skeleton is not just a lightweight and rather rigid assembly of hollow bones; it may, in fact, be heavier than the skeleton of a similarly sized rodent and it is usually just as flexible. Sometimes the vertebrae of the back are indeed fused into a rigid block, but the extent and occurrence of such fusion is difficult to predict and may vary within a family. A great many birds have independent vertebrae and a flexible backbone. Only the hip region is always fused into a rigid block known as the synsacrum. Similarly, the claim that birds have hollow bones is an oversimplification. Some bird bones may be filled with nothing more than air but, just as in a rabbit or other small mammal, some bones are empty while others are filled with marrow. Even the claim that birds have air sacs extending into the limb bones is not universally true. The sandpipers and their relatives probably descended from birds whose bones were open to the respiratory system, but for unknown reasons their bones are now sealed off.

Some familiar anatomical features show surprising variability. You might expect the wishbone or furcula to be a universal characteristic of birds, but it varies greatly in shape from group to group and occasionally disappears. It is usually described as a thin springlike bone that flexes during flight. In birds with very energetic flight, however, such as hawks, it is a large and robust unit that is too stiff to flex. Owls may look like hawks and also hunt on the wing, but many of them have done away with their wishbone. For whatever reason, it is just not necessary to their flight technique and they make do with a pair of clavicles connected by a long loop of ligament. Similar startling differences in the design of other bones have been successful and have contributed to the great diversity of birds throughout the modern world. New information about the function of skeletal features and other internal structures has not gone unnoticed by the research community. It appears regularly in the technical literature but only a little filters down into more general publications.

The widespread emphasis on the external features and habits of birds also extends to the professional literature. Even at the university level, textbooks devote most of their space to avian ecology, behaviour, conservation, and similar topics. Consequently, it can be surprisingly difficult to find useful information on the internal features of birds. General textbooks offer some basic information on physiologically significant organ systems, but it is very difficult to get beyond the most elementary questions about avian anatomy and the skeleton is often dismissed in a page or two. There are veterinarian guides to the domestic chicken that go into great depth, but you must be prepared to navigate through a very large and very dense professional literature. It is easy to get lost in specialized technical jargon.

The chicken is by far the most abundant and most important bird on the planet, and avian anatomy has traditionally relied on it as a conveniently available example of a bird. There was a time when whole chickens were regularly sold in grocery stores and most people had a passing familiarity with their internal anatomy. The world has moved on. In most places, chicken meat is sold as anonymous, sterile lumps in plastic display packs that intentionally obscure any suggestion that the contents came from a living creature. The disjointed bits are not even useful as elementary introductions to avian musculature. Some textbook editors seem to have noticed this social change and have begun to use illustrations from a wider variety of birds.

Although chickens have great cultural significance, they have always been a poor choice as a typical bird. In the wild, jungle fowl spend most of their time walking on the ground and rarely take flight. Pigeons, starlings, or sparrows might be a better choice today. They at least are arboreal animals that hop about on the ground, behaving in the same way as 90% of the other birds. As unwelcome pests, victims of urban hazards, and common offerings from house cats, they also make a convenient subject for casual anatomical investigation.

In spite of the difficulties in learning more about avian anatomy, a flood of exciting new fossils suggests that the time is ripe for such study. Paleontologists have been busily reconstructing ancient life from the remains of birds and birdlike dinosaurs that seem to turn up almost daily. The most exciting finds are often accompanied by the claim that scientists will be forced to rethink the origin of birds. The more dramatic journalistic claims are foolish but not all are, and the fossil record of birds and their ancestors has proven particularly challenging to the scientific community. Before 1980, no one suspected the existence of undescribed subclasses of fossil birds with very distinctive anatomical features. We now know that the Cretaceous fauna included Confuciusornithes and Enantiornithes as well as modern birds (Neornithes). Other recently discovered fossils have demonstrated that many lineages of dinosaur had feathers and some small forms, such as *Microraptor,* may even have achieved a basic form of gliding flight. Without a basic knowledge of avian anatomy, it is difficult to assess the significance of such new information or understand why some some creatures are classified as birds and others as dinosaurs.

New fossils have not been the only challenge to traditional ornithology. Advances in biochemistry and molecular genetics have begun to chip away at the view of living birds as members of a monolithic group. Anatomical studies have long been used to place ostriches and their relatives in a discrete group within Class Aves. The use of the name Paleognathae implied that they were perhaps more primitive than other birds, but it took biomolecular analyses to confirm that they actually represent an early branch from the main lineage. Similar analyses have shown that the galliform and anseriform birds (chickens and ducks) represent another, major branch. The remaining groups of birds are still lumped together informally as the Plethaves because analyses of anatomical features have persistently failed to produce a convincing family tree. Molecular techniques have also failed to produce a family tree, and their results frequently contradict one another. Nonetheless, they have succeeded in undermining confidence in long-standing ideas about the relationships among the major groups of birds. One recent genetic analysis has called for a division of the Plethaves into two large groups, Metaves and Coronaves. Metaves happens to contain many groups that have been most difficult to classify, and it may represent survivors of an early radiation of modern birds. Unlike some other results of molecular studies, it is not one that was anticipated by students of anatomical features, and researchers are scrambling to find further evidence one way or another.

Such a fundamental division within the Class Aves would have profound significance for our understanding of the group's history, but so far it is based only on variation in a single fibrinogen gene. Nonetheless, such results make it clear that students of both fossil and living birds need to take a fresh look at avian anatomy. One of the objectives of this book is to introduce some of the appropriate background material and suggest some new interpretations for familiar structures.

In many ways, this book is a personal statement about birds and it reflects events in my career as a working ornithologist. Like most of my contemporaries, I am a practical avian ecologist with little academic training in the study of birds. I have spent most of my career dealing with questions of human impact on the environment and never needed to call on biomolecular genetics or paleontology. I dissected the occasional cadaver, searching for ingested lead pellets or other evidence of human-induced mortality, but my work never required more than a basic understanding of avian anatomy.

When I first looked for work as a biologist, I was, if anything, a specialist in insect sensory physiology. After a brief stint bouncing chicken carcasses off aircraft for the Bird Hazards to Aircraft Committee, I began a 33-year career as a Migratory Birds population biologist for the Canadian Wildlife Service and spent a great deal of time counting ducks from low-flying aircraft. In the 1970s, continental management of bird populations depended heavily on information from bird bands, and conservation agencies invested in numerous large-scale capture operations. Consequently, over the years I handled thousands of ducks, geese, auks, and sandpipers, as well as loons, grebes, cormorants, pelicans, cranes, herons, petrels, gulls, and terns. References to some of these birds appear throughout the book because I know them best. Fortunately, my experience was not limited to North America. Over the years, I have been lucky enough to help colleagues with field studies of penguins, boobies, frigatebirds, shearwaters, albatrosses, and other marine birds. I have also done a little work with small forest birds in the tropics. For many years I conducted courses in field ornithology and mist-netted a variety of exotic resident and migrant species in Colombia, Ecuador, Peru, Sri Lanka, Borneo, and the Philippines.

Although I have banded all the North American ducks and most of the world's sandpipers, it was handling tiny tropical species in Colombia that really opened my eyes to the structural diversity of birds. During a series of winter field courses for students from the Universidad de los Andes, I handled everything from huge and furious oropendolas to tiny and docile sword-billed hummingbirds. They were exciting and bizarre creatures but none stands out in my mind more than a giant White-collared Swift caught at 3,300 m in the Sierra Nevada of Santa Marta. It lay stiffly in my hand, more like a model airplane than a living creature. Its body felt heavy and densely muscled, surprisingly reminiscent of Marbled Murrelets that we had been catching in British Columbia during the previous summer. It would be 20 years before I realized just how few high-speed aerial specialists there are among birds and why the swift's internal architecture was so different from that of other high-speed fliers, such as a murrelet or a falcon.

I never touched another giant swift but I still work with the Marbled Murrelet whenever I get the chance. The murrelet was largely ignored by conservation agencies until it was declared a threatened species in the early 1990s. Suddenly there was a demand for information on its nest sites and feeding areas. It proved to have such unusual habits, however, that the only way to study it effectively was either through radio telemetry or tracking on radar. Nonetheless, it soon became one of the most intensively studied species in North America and its biology is discussed regularly at seabird conferences and on dozens of websites. Although few of this book's readers will have seen one of these birds, it is now one of the most thoroughly documented species in North America.

The Marbled Murrelet is a secretive, non-colonial seabird that can be studied only one individual at a time, and working with it changed my perception of birds. It forced me to think of them as discrete organisms instead of masses of animals in populations or communities. The murrelet turned out to be one of the fastest birds in the air, and individuals regularly commute more than a hundred kilometres between the nest and feeding areas at sea. Other auks nest near sea level but the murrelet will lay its egg more than a kilometre up one of the coastal mountains and more than 50 km inland. It is famous for finding the right branch in a forest of branches, in complete darkness. For me, its unusual flight capabilities and spe-cial adaptations for life underwater have become useful benchmarks in assessing the performance and structure of other species.

Another intensely documented species has also caught my imagination, although I have never been able to handle a living example. It is the Hoatzin (more or less rhymes with Watson), a tropical species that is about as far anatomically from a murrelet as it is possible to get. It is a large, loose-limbed, and slow-flying chicken-like bird that lives along streams in the Neotropics. It is perhaps best known for the peculiar fingers on the wings of its young. They are suggestively reptilian but are not this bird's only unusual feature. In spite of intensive investigation, it has repeatedly defied classification, confounding both comparative anatomists and biomolecular geneticists.

In 1891, William Kitchen Parker made Victorian naturalists familiar with the peculiar features of the Hoatzin by publishing detailed drawings of the skeleton of a fully developed embryo [155]. Its fingers suggested the intriguing possibility of an evolutionary connection to primitive fossil birds like the *Archaeopteryx*, but its specialized digestive system suggested something much more advanced. It is the avian equivalent of a ruminant, and no other bird has such a hugely enlarged crop for the fermentation of vegetable matter.

Surprisingly, the crop suggests a link to yet another group of primitive birds. In 1981, Cyril Walker described a whole new subclass of fossil birds that he called "Enantiornithes" [221]. These "opposite birds" were a highly successful and widely distributed group until they became extinct with the dinosaurs. The pecto-ral skeleton of the species that gave the group its name, *Enantiornis*, shared many structural features with the modern Hoatzin, implying that the two birds might also have shared a very similar lifestyle. Could the Hoatzin be an unrecognized living fossil? It is highly unlikely but a Victorian amateur would have found the subject much easier to investigate than his or her modern counterparts. There are no illustrations of the Hoatzin skeleton in modern reference texts or on specialized ornithological websites. A recent study of the Hoatzin's skull is available only in Portuguese.

The Hoatzin and the Marbled Murrelet are among the most specialized birds and are somewhat unfamiliar to most ornithologists. Other species might seem more representative of their class, but these two have been so intensively studied that they have become more useful examples than they first appear. Both are dis-cussed on the Internet and in many books. This book also uses other auks and specialized types of birds, such as loons, grebes, frigatebirds, and albatrosses, to illustrate some of its points. It mostly avoids discussions of songbirds, chickens, or pigeons.

The book is divided into three parts, each of which addresses one of the elementary questions that have been the basis of ornithology since the time of Aristotle: "What is a bird?" "What kind of bird is it?" and "How does it fly?"

Aristotle answered the first question, to almost everyone's satisfaction, some 2,300 years ago. In the last 20 years, however, paleontologists have rekindled interest in it by finding one fossil after another that bridges the gap between birds and dinosaurs. The phrase "birds are dinosaurs" has become a journalistic cliché but it is patently foolish from an ornithologist's point of view, if only because we see living birds every day whereas dinosaurs have been dead for 65 million years.[2] For most avian ecologists, this is far too long a period for dinosaurs to be very relevant in the study of living species. The distinction between birds and dinosaurs is of more interest to evolutionary theorists investigating the origin of birds and their method of flight.

Since the late John Ostrom found the large predatory dinosaur *Deinonychus* and described its birdlike characteristics, there has been increasing acceptance of the idea that birds arose from within a group of advanced carnivorous dinosaurs known as theropods. In this century, a civil court jury would certainly decide in favour of a dinosaur origin "on the balance of probabilities," but there are still enough questions that a criminal court jury might have difficulty deciding "beyond a reasonable doubt."

Ironically, it is Aristotle's comparative method that lies at the root of any ongoing difficulty in distinguishing between tiny birds and giant dinosaurs. For centuries, biologists have compared one specimen with another and used similarity as an indicator of relationship. Unfortunately for those who want simple answers, similarity does not imply causality.

There is an evolutionary process called convergence that strongly limits the usefulness of similarity and is particularly widespread in birds. When otherwise unrelated animals occupy similar habitats and share similar lifestyles, natural selection is likely to increase the similarity of their appearance and they are said to converge. The process is different from mimicry, where an innocuous species gains protection by looking like a dangerous or toxic neighbour. Convergence may occur between animals that live in opposite hemispheres or, like the Hoatzin and *Enantiornis* mentioned above, that are separated by great expanses of time. The degree of convergence can be stunning, and in some cases it has confounded ornithologists for centuries. Only recently have we come to understand that American vultures are a kind of stork even though they look very much like African vultures, which are related to eagles.

Sometimes the true relationships between animals are so thoroughly hidden that they can be resolved only by the use of DNA and gene sequences. It turns out, however, that there is even convergence of DNA, and biomolecular analysts

---

2 The subject of dinosaurs and ancient extinctions brings up the issue of geological time scales. I have placed a bird-oriented version in Appendix 2. Readers need only be aware that dinosaurs became extinct at the end of the Cretaceous Era about 65 million years ago, while we live in the Tertiary, which began at that time. Little additional information is needed to understand this book.

must be careful of their own comparisons. Most importantly, DNA cannot be applied to fossils and paleontologists must depend on anatomical details winkled out from confusing fragments of bone. In the last few years, distinguishing feathered dinosaurs from the early reptile-like birds has become one of their greatest challenges.

The question "What is a bird?" is investigated in the first two chapters in terms of the internal structure and architecture of living species. Chapters 3 and 4 look at the same question in terms of a bird's similarities and differences with regard to the animals commonly known as dinosaurs.

The question "What kind of bird is it?" is usually answered with a name, but like a person's name, a bird's name makes sense only in a broader context of family relationships and connections to larger groups. Nonetheless, the name is central to the popular hobby of birdwatching and the basis for the vast number of field guides. For centuries, the question of names was the most important issue in ornithology. After Christopher Columbus and other explorers introduced the rest of the world to the Europeans, huge numbers of unknown species were brought back to Europe. For over 400 years, the description of newly discovered birds absorbed the careers of most ornithologists. By the mid-20th century, the great majority of the world's unknown species had been catalogued and ornithologists, like most other biologists, began to turn their attention to other kinds of study. Many chose to link ornithological studies to experimental forms of science. Today there are more professional ornithologists than ever, but only a handful work on issues of bird identification. Most specialize in bird-oriented components of other fields, such as avian ecology, avian demography, avian physiology, or avian genetics.

This book does not deal with individual species of birds except as representatives of general types. Usually these types have such distinctive and specialized characteristics that they have been placed in distinct families or groups of families known as orders. Generally ornithologists and biomolecular geneticists have had some success at working out the relationships between species and clustering them in large groups, but agreement on the relationships among those larger groups has eluded them. The history of the difficulties in determining ranks for the various groups and the relationships among the families of birds is discussed in Chapters 5 and 6. Many problems remain unresolved; consequently, ornithologists have failed to produce a generally accepted family tree or phylogeny for the birds. Without such a tree there can be no evolutionary story.

Most ornithologists agree that flight capability is the key to understanding avian evolution, and that "How does a bird fly?" lies at the heart of ornithology. This question has had a much greater impact on our culture and history than any issue of bird identity. For centuries our ancestors watched birds and dreamed of achieving flight themselves. Without a basic knowledge of the composition of air, however, it remained a fantasy until science advanced and pioneer aviators solved the problems of generating sufficient mechanical power. Even then, they built aircraft that looked like box kites and few thought that avian flight merited serious scientific investigation.

In the mid-20th century, when aeronautical engineers began to design truly sophisticated aircraft, they also began to reconsider the performance of birds as flying animals and to make systematic investigations of avian flight. It proved to be a rewarding field of investigation, and avian aerodynamics made many important contributions to aircraft design. In return, aeronautical engineers provided ornithologists with new perspectives on the energetics of bird flight and developed mathematical models of avian performance in the air. These models of energy expenditure have enabled ecologists to determine the cost of the various aspects of the bird's life cycle, such as parental investment in reproduction. Energetic factors have key components in conservation plans and assessments of the bird's ability to exploit modified environments. A host of more esoteric questions arising from the mathematical models remain the subject of intense investigation, and new ideas about flight are reported regularly in the *Journal of Experimental Biology* and other scientific publications.

Feathers are central to avian flight but remain the single most difficult element to model or investigate effectively. Their importance is indicated by the energy invested by the bird in preening and sophisticated physiological processes that keep the plumage in good condition. Feathers are so intimately linked to the flight of birds that the discovery of feathered dinosaurs was initially met with controversy and disbelief. None of those dinosaurs show adaptations necessary for flapping flight, but the same can be said about *Archaeopteryx* and some other primitive species. Birds appear to have taken a very long time to develop basic aerial skills. It may have been an expensive or risky process, and some early lineages appear to have abandoned the attempt. The role of feathers and some early experiments in avian design are discussed in Chapters 7 and 8.

The evidence is sparse but it is beginning to look as though birds were surprisingly diverse and abundant in the Cretaceous, and at least some modern groups were around long before dinosaurs died out. Chapter 8 looks at the three major evolutionary experiments undertaken by birds in the Cretaceous: the Confuciusornithes, the Enantiornithes, and the Neornithes, or modern birds. You can tell from the tongue-twisting names that the first two of these very important groups are fairly recent discoveries. They have yet to be given a catchy tag by amateur dinosaur enthusiasts. To help the text flow, I refer to the first group as Confucius birds and the second as ball-shouldered birds.

Many of the Enantiornithes are almost indistinguishable from modern birds except for details of their skeletal anatomy. Their shoulder includes a "ball" that fits into a socket at the base of the shoulder blade – hence, "ball-shouldered birds." Neornithes have a socket in the shoulder that receives a "ball" on the shoulder blade. I suppose they could be referred to as "socket-shouldered birds," but "modern birds" is more meaningful. The Enantiornithes were not recognized as a group until 1981 but they may have been more widespread and successful than the modern birds in the Cretaceous. Why did they become extinct while modern birds survived? The answer may have to wait until we unravel the mysteries surrounding the events[3] that exterminated the dinosaurs and many other life-forms at the end of the Mesozoic.

Chapters 9 and 10 look at the influence of flight on the lifestyles of modern birds. Chapter 9 examines the adaptations of a dozen groups usually described as small forest birds. They include about 85% of all species, but in spite of their numbers, they are remarkably uniform in their basic body plans and flight technique. Ornithologists use details of skull structure and internal anatomy to distinguish the families and orders, but in the field the most striking differences are often limited to superficial matters of plumage colour and feather shape.

Although the great majority of small forest birds are capable of flight, nearly all are pedestrians that earn their living by foraging on the ground or some other substrate. Many fly only when pressed to escape predators or to commute between a roost and a feeding area. Only a small minority of birds have learned to forage on the wing or undertake long migrations, but even these aerial specialists are little different from their more sedentary relatives. Except for the extreme specializations of hummingbirds and swifts, most aerial specialists among the small forest birds show only modest variation in the length and shape of the wing.

Less than 15% of all bird species have developed the specialized flight techniques needed to actually earn a living on the wing. In spite of this small number, they are represented in about half the major lineages or orders of birds. Smaller types, such as hummingbirds, swifts, swallows, nightjars, and the smaller owls, have retained a close relationship with forest habitats. They may forage in open areas but typically return to the forest's shelter to roost or nest. Larger species have more difficulty manoeuvring among the branches and often inhabit open habitats on a full-time basis. Some are terrestrial while others are marine or aquatic. The terrestrial types are typically found on grasslands and estuaries, where they exploit thermal updrafts to soar long distances at only a small energy cost. The thermals enable them to investigate sparsely distributed foraging opportunities across vast areas. A few close relatives of those birds of open country, such as the accipiter hawks and forest-falcons, have returned to the forest, where they use high-speed flight to pursue smaller forest species. Others have learned to swim and make use of the many small lakes and wetlands scattered across the continents.

Only a few groups have been able to move out onto the open oceans, but their spectacular success is one of the major achievements of birds and was never

---

3   "K-T event" is the usual paleontological shorthand for occurrences that led to the extinction of the dinosaurs and many other lineages at the end of the Cretaceous (K) and before the beginning of the Tertiary (T).

accomplished by dinosaurs. The oceans offer a wealth of opportunities but are a challenging environment, and birds have been forced to undertake an array of novel adaptations in order to survive. Food is often abundant at sea but is usually patchily distributed, so marine birds have been forced to find ways to search over huge areas. Once they locate prey, they must be able to catch it even if it disappears beneath the surface. As a result, birds have learned to dive deeper and faster than their prey.

The most successful groups of seabirds exhibit one of three basic strategies for life at sea. The petrels depend on exceptionally efficient flight to search wide areas without exhausting their energy supply. Penguins have exchanged the ability to fly for bodies capable of storing dense masses of fat that they use to fuel long swimming voyages at sea. Auks look a little like very small penguins but have retained the ability to fly and are neither as large nor as heavy (except for some extinct flightless forms). Like penguins, they use their wings for underwater propulsion. In the air, energetically expensive flight depletes energy reserves much more rapidly than the more efficient flight of petrels and albatrosses. Auks have successfully based their lifestyle on the advantages offered by the combination of high-speed flight and the regular availability of energy-rich prey in the sea.

For many years, ornithologists were unimpressed by the flight capabilities of auks, but advances in avian aerodynamic theory have forced them to change that opinion. Auks were once seen as a mere evolutionary waypoint on the path to becoming as flightless as penguins. Just as modern analyses of DNA suggest that seabirds represent advanced lineages, however, modern aerodynamic theory suggests that their flight is a highly sophisticated adaptation for a specialized existence. The auk's choice of speed and energetic extravagance is just as specialized a flight technique as the grace and efficiency of petrels. In each case, the aerial technique has made demands on the owner's internal structure while facilitating reproductive strategies that are denied to all other types of birds.

Where possible, I have illustrated the ideas in this book with material from my own research and experience, but the great majority of illustrations are based on the publications of other ornithologists. By the mid-20th century, the scientific literature had become so vast that no one library could contain it all, and so cumbersome that it impeded research. For better or worse, the Internet has changed all that. Much of the 19th- and 20th-century technical literature for North America is now available, without charge, through SORA, the <u>S</u>earchable <u>O</u>rnithological <u>R</u>esearch <u>A</u>rchive (http://www.elibrary.unm.edu/sora/index.php). Other science journals have websites where you can browse through the titles of current and past issues. Most will allow you to read the abstract for free and give you the opportunity to purchase a copy of a particularly interesting paper (usually online as a PDF file) for between US$10 and US$30.

The volume of paleontological literature has also become very large. Fortunately, much of it is accessible on the Internet, where it has become a common subject in various dinosaur chat groups. Hard copies of paleontological papers are not as readily available as the bird literature, but SAPE, the Society of Avian Paleontology and Evolution (http://www2.nrm.se/ve/birds/sape/sape001.html.en), offers access to the texts of many recent papers.

Since 1998, dramatic reports of new fossils in newspapers and science magazines have helped rekindle public curiosity about the origin of birds. Debate on this issue has flared intermittently since the discovery of *Archaeopteryx* in 1861 and encouraged the production of such books as *The origin and evolution of birds* (1996) by Alan Feduccia [60], *The rise of birds* (1997) by Sankar Chatterjee [24], *The mistaken extinction* (1998) by Lowell Dingus and Timothy Rowe [50], and *Dinosaurs of the air* (2002) by Gregory Paul [157]. Paul's book is particularly useful because it includes illustrations of many of the important characteristics that have been used to identify the various lineages of dinosaurs and Mesozoic birds. The most recent round of debates also led to the publication of four large compilations of research papers: Michael J. Benton, Mikhail A. Shishkin, David M. Unwin, and Evgenii N. Kurochkin edited *The age of dinosaurs in Russia and Mongolia* (2000); J. Gauthier and L.F. Gall edited *New perspectives on the origin and early evolution of birds* (2001); Luis M. Chiappe and Lawrence M. Witmer edited *Mesozoic birds: Above the heads of dinosaurs* (2002); and Philip Currie, Eva Koppelhus, Martin Shugar, and Joanna Wright edited *Feathered dragons: Studies in the transition from dinosaurs to birds* (2004). Individual papers from these books are cited at various points in the text, and the books are all recommended as a source of technical detail and information on closely related topics.

Throughout this book, I have tried to help the reader find additional information either on the Internet or in the technical literature by giving the names of the researchers whose work forms the basis of my discussion and the correct scientific name for the organism and for its component parts. The specialized terminology does not appear in typical dictionaries so I have included a descriptive glossary among the appendices at the end. Nonetheless, the book is a natural history and not a technical treatise. I have had to compromise between readability and appropriate citation. I have tried to give credit to the scientists who had the most important ideas and did the most innovative work, but the subject is complex and I apologize if I have failed to give full credit to every deserving worker. I hope that there are enough citations that readers can explore the literature further and keep abreast of new developments.

This book touches on many highly specialized fields and there are bound to be mistakes. For those you have my sincere apologies. There will also be claims that you cannot believe – you are not supposed to believe them! Examine them for yourself, in other books, in museums, or even on the Internet.

PART 1

What Is a Bird?

The Bird beneath the Feathers

*How adaptations for flight
allow a bird to earn a living on the ground*

There are only four groups of terrestrial vertebrates: amphibians, reptiles, mammals, and birds. The first three usually walk, crawl, run, and jump using a basic four-legged architecture, although a few have been modified for swimming or even for flying. These tetrapods differ in body covering and physiology but they are similar enough in structure that it is easy to imagine them as distantly related cousins. A family relationship between these three and the birds is much more difficult to imagine. Like mammals, birds are advanced, warm-blooded vertebrates. They share some characteristics with specialized mammals, such as the human's ability to walk on two legs or the bat's ability to fly, but it is clear that they have little else in common with that group. Birds may still walk on the ground but are otherwise completely specialized as flying creatures and are clothed in a unique type of body covering that contributes to structural integrity and participates in aerial locomotion.

The body covering or plumage also completely obscures the bird's internal activity and has greatly impeded our attempts to understand it as a functioning animal. The plumage has even made it difficult for writers to find ways of describing bird movement. They always seem to be referring to some sort of solid object that just happens to be moving through space. When writers describe the movement of mammals, whether human athletes or fast horses, they turn to well-worn clichés like "grace in motion." The rippling muscles slide smoothly over one another in ways that capture the eye and the imagination. It does not matter whether the mammal under consideration is a mouse or a hippopotamus because we recognize that their underlying body plans are similar to our own and know that all the component parts will act in a predictable manner. It has been much more difficult to understand the bird. For centuries, we knew little more about a bird's movement than that it was a mystery that we could not imitate but that seemed to be based on the flapping of wings.

Other aspects of avian locomotion added to the early observers' inability to understand the flapping wing. Dense plumage seems to enable the bird to change shape while disguising the movement of the body or limbs. The albatross resting on the sea might be carved from a block of wood until it transforms itself and drifts away effortlessly on hugely extended wings. The squat duck takes off more energetically but it also stretches out its neck and adopts a more streamlined conformation for high-speed flight. Even the backyard sparrow transforms itself from a momentarily tense visitor to a vibrating projectile as it dives into protective shrubbery.

The changes in shape are largely the responsibility of the feathers but feathers do much more than define the bird's external shape. Feathers are found on no other living animal,[1] and, more than any other feature, they give meaning to the word "bird." They protect the fragile body, give it the power of flight, insulate it from the cold, advertise its species, and display its readiness for reproduction. Feathers are also a key element in an architectural strategy that exposes as little as possible of the bird's living body to the outer world. Not surprisingly, they are often the only feature we notice about a bird. Like an inept house cat with a disappointing mouthful of feathers, we often get distracted by externalities and miss the meaty bit underneath. By concentrating on appearance, we lose sight of two fundamental characteristics of birds.

First, almost none of the visible parts of a bird are alive. From one end to the other, birds are covered by tough proteins called keratins that are formed as a lifeless excretion of the skin. Keratins come in several forms. Some form the airy layer of feathers that gives the fleshy parts of the body a flexible covering, while other types create a durable surface for working structures such as the beak, legs, and claws. Even the living surface of a bird's eye lies beneath a glassy layer of transparent keratin.

Avian keratin is unique but similar material forms the patches of hair, fur, and claws in mammals and scales in reptiles. Like the scales of reptiles, the feathers are shed at regular intervals, but the eye covering and leg scales remain for life.

The second fundamental characteristic of birds is that they are pedestrians. Few actually earn their living in the air or spend a great deal of time in flight. Wherever they are, the feathers protect them from excess heat or cold and decrease the risk of mechanical injury. Most birds forage by walking about on the ground, looking for food just like their distant reptilian ancestors. Those that live in forests have traded level ground for more vertical substrates such as trunks, branches, leaves, and flowers, but they still forage on foot. Although the great majority of birds fly only to escape from predators or commute between feeding areas and a roost, the benefits must have made it worthwhile for their ancestors to undergo the complex adaptations that led to aerial flight. In modern birds, almost every anatomical feature reflects the demands of aerial locomotion, but we should never lose track of the bird's heritage as a pedestrian animal.

---

[1] The recent confirmation of feathers in some dinosaurs further blurs the distinction between these two groups. In *Dinosaurs of the Air*, Gregory S. Paul sees some of the most advanced dinosaurs as descendants of birds that lost the power of flight [158].

Although the outer cloak of feathers is an important part of the animal, all living functions are the responsibility of the inner bird. The inner bird is a strange goblin-like creature that manipulates its appendages by pulling on long tendons just as the human operator within the muppet Big Bird pulls on a network of internal wires and strings. The puppeteer gets to enjoy an independent existence when he sheds his casing at the end of the workday, but the plumage of the inner bird is part of an integrated whole animal.

Suggesting that we should look at the inner bird as a puppeteer behind a screen of feathers may seem an extreme point of view but it does not overstate the case. The muppet analogy extends beyond the feathered suit to the bird's basic design. In mammals, the limbs are moved by muscles that are distributed over the whole skeleton so that they lie close to the joints they move. There is ample additional room on the limbs for generous blood supplies and even stores of fat. In birds, the muscles are remote from the joints and the limbs often appear thin and sticklike. Even the fleshier wings are little more than nubs, with no more tissue than is needed to carry and manipulate the feathers during flight.[2] The limbs are not one of the places where a bird can store its reserves of fat, and they offer little space for blood vessels or nerves. Because the muscles are collected together in a dense mass and anchored as close as possible to the body's core, they often lie far from their point of action and depend on long tendons to carry out their responsibilities. Tendons from muscles along the breastbone move the wing, while others along the backbone move the tail. Even the muscles that curl the toes are mounted high on the leg.

In mammals, tendons are usually short links of connective tissue that travel in a straight line across a single joint. In birds, tendons are often very long and many cross two or more joints. In some cases, they are responsible for movements that are more complex than the simple bending of a hinge. Perhaps the best-known example involves the supracoracoideus and the pectoral muscles that lie parallel to each other along the keel of the sternum. The large pectoral muscle depresses the wing by pulling directly on the humerus. The smaller supracoracoideus raises the wing because its tendon reverses direction as it passes through a pulley in the shoulder.

Dependence on the forelimbs for flight and the hind limbs for walking has left birds with few options for manipulating prey or other objects that might be worth investigating. Parrots grip food with their feet and raptors kill with their talons, but most other birds are entirely dependent on coordinated interaction of their neck and beak. In effect, the neck has become the bird's equivalent of an arm and the beak is its equivalent of a hand. To meet the demands of this role, the neck must be exceptionally long (Figure 1.1). In mammals, such as the giraffe, this is achieved through the elongation of a small number of vertebrae. The neck of a

---

2 The muscles that put a little protein into "Buffalo wings" are found only in chickens. Other birds have replaced them with a lightweight tendon.

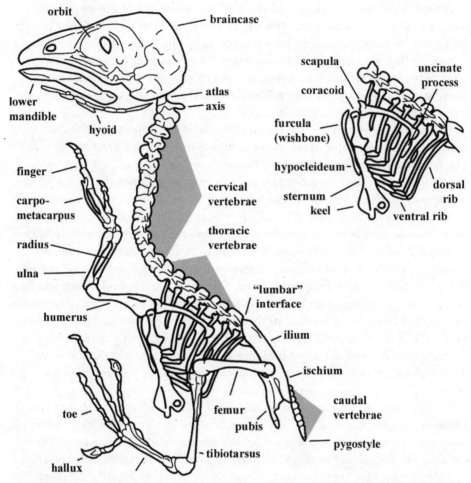

Figure 1.1   Major elements in the avian skeleton. This illustration is derived from W.K. Parker's depiction of a newly hatched Hoatzin [155]. This is not a typical bird but all of the important bones are present. At this stage of development, many are not completely fused and the individual components of some, such as the tarsometatarsus, are visible. In this species, the bones of the fingers are independent and are similar to those in the early ancestors of birds. More typical avian hand bones can be seen in Figure 1.11.

bird, however, is assembled from numerous small components, supported and strengthened by a complex array of muscles, tendons, and ligaments. In many ways, this elaborate structure is reminiscent of the type seen in dinosaurs and other ancient reptiles.

The assignment of manipulative duties to the head and neck has required that birds solve some complex biomechanical problems. Birds often use rapid movements of the head to capture prey, but if the bird's head were heavy, it would generate

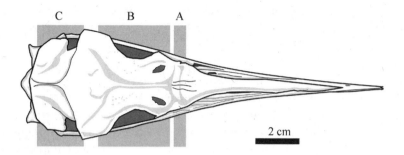

C   B   A

2 cm

Figure 1.2 Dorsal view of the skull of a Pacific Loon. (A) Flexion zone between beak and braincase; (B) troughs over the eyes for the salt glands; (C) troughs to hold unusually large jaw muscles.

considerable momentum that might strain the neck joints or injure the fragile spinal cord. The ancestors of birds solved this problem by reducing the amount of bony material and soft tissues in the head. The long-necked dinosaurs used the same tactic but birds have carried it to an extreme. Bones in a bird's skull are often paper-thin and have lost much of their protective capacity. Even the jaws have become lightweight structures in spite of the variety of services that they are called on to perform. They lost their heavy teeth early in the bird's evolutionary story, and their muscles have been moved closer to the body core. The muscles lie behind the eyes or beneath the skull, near the top of the neck. The brain and eyes have remained exceptionally large and contribute the greatest amount of weight to a bird's head.

The massive jaw muscles found in many mammals are usually attached to a distinctive ridge along the midline of the skull, called the sagittal crest. Most birds have small jaw muscles and a skull that is a smoothly rounded dome without such obvious sites for muscle attachment. Crests and other structures appear, however, among birds with exceptionally heavy beaks or birds that regularly handle strug-gling prey. In heavy-billed ibises and fish-eating cormorants, and in boobies, a re-inforcing crest of bone creates a distinctive collar near the base of the skull and leads to a trough that contains the jaw muscles. Puffins and petrels have a more modest array of ridges in the same general area, but broad troughs mark the spaces occupied by their jaw muscles. Only loons and large grebes have well-developed, mammal-like sagittal crests along the midline of the skull (Figure 1.2).

Even the most robustly jawed birds can only begin to process food in the mouth. Most birds are satisfied with the ability to orient the food for comfortable swallowing, but some species are able to tear large prey into edible chunks or crack unwanted shells off seeds. There are no teeth or elaborate grinding surfaces in the mouth for chewing. Even the saliva produced by some species lacks digestive en-zymes. Both mechanical and chemical reduction is delayed until the food enters the muscular foregut deep within the body. Even there, birds lack a hard grinding surface and, like their distant relatives among the dinosaurs,[3] birds swallow grit or small pebbles to aid mechanical reduction. The grit might not be lighter than a set of teeth but it is stored deep in the body, not in the head.

Long necks, such as those found in all birds, are uncommon in nature because they are vulnerable to attack and radiate valuable body heat. To reduce the risk, birds adopt a protective posture by folding their necks and tucking their heads tightly into their bodies. This pulls the neck's muscles close to the warmth of the central core of the body without limiting the bird's ability to reach distant objects. In flight, the plumage covers the neck and smoothes the angular contours of the shoulders, reducing drag by allowing air to pass smoothly over the body. A few long-necked birds fly with the neck extended. Flamingos and cranes fly so slowly that drag is not a concern, but ducks, geese, and swans achieve higher speeds where drag can significantly increase energetic costs. They are careful to maintain the head in a central position so that it does not increase drag by extending beyond the cross-section of the body.

No characteristics highlight the distinctions between the inner and outer bird more than the structures used by a bird to interact with the world. The continuity of a bird's plumage is broken only by its eyes, beak, and feet, whose shiny, armoured hardness contrasts sharply with the proverbial softness of the feathers. Using only these three external features, the bird has thrived for millions of years while all of its nearest relatives have long since disappeared. Each structure is based on a fragile core of living tissue that protects itself by excreting a tough but lifeless outer case.

Interacting with the External World

Perceiving the External Environment

Eyes and Vision
The eyes of birds are truly vital organs; no bird can fly blind.[4] Although the eyes lie near the base of the beak, they are perhaps the most exposed part of the bird's body. To protect them from normal wear and tear, they are covered by the same tough sclerotic film that lines the lungs and the walls of the hollow bones. This covering is one feature that birds seem to have inherited from reptilian ancestors, but it is different from the transparent scales that cover the eyes of modern reptiles. Snakes and lizards lose their eye coverings in one piece whenever they shed the rest of their skin. Birds keep their eye covering for life, adding to it from within. They also have a third eyelid, the semi-transparent nictitating membrane, which can sweep across the eye to clear it of dust and other irritants.

---

3 Dinosaurs will appear regularly in this story. Paleontologists now say that birds are dinosaurs, but for clarity and simplicity, I use the word "dinosaur" in its traditional and original sense as the name of an extinct lineage of ancient reptiles. A bird is a flying (or secondarily flightless) animal that shares the ancestry of *Archaeopteryx*, the earliest known member of its lineage.

4 Oilbirds and some swifts nest in caves and appear to have developed basic skills in echolocation.

Within the bird's eye, there is a ring of a dozen or so small bony plates set around its pupil. Similar structures were found in dinosaurs and other primitive reptiles. The ring appears in all birds but its individual small plates vary slightly in form and number and have occasionally been used as a characteristic in classification. Birds that forage underwater need to modify the shape of their eyes to accommodate the greater optical density of water. The small plates help to protect their eyes from extreme distortion.

Falcons and other hunting birds have long been famous for their keen eyesight but almost all birds have exceptionally acute vision. They achieve it by having exceptionally large eyes. A bird's eye may be larger than that of a much bigger mammal [16], so that the eye of a 200 g owl is as large as or larger than that of a 100 kg human. The eye of the ostrich is the largest of any terrestrial vertebrate and may approach the upper limits of optical effectiveness in a biological structure. Many factors affect the growth of biological materials and organic tissues cannot provide sufficient clarity and consistency of structure to make a practical lens beyond a certain size. In addition, the single lenses in a large eye may have problems with diffraction that are not noticeable in a smaller eye [121].

There are two clear benefits to a large eye. There is ample room for a large receptor field on the retina and a long distance between the lens and that receptor field. As in a digital camera, acuity or resolution is based on the number of activated receptors. A large eye can hold more receptors and therefore has greater potential acuity. The greater the distance between the lens and the retina, the greater the image on the receptor field. In mammals, the eyes are usually spherical so that the size of receptor field is directly proportional to the diameter of the eye. Not all parts of a spherical eye are important to image quality, however, even though they contribute to its weight. Birds have adopted an eye shaped like a flattened disc, which offers a saving in weight without decreasing the diameter of the receptor field (Figure 1.3).

Even unusual eyes require special accommodation in the bird's head and leave little room for other structures in the skull. The eyeballs are so tightly fitted that their backs press on each other and are separated by only a paper-thin sheet of bone. They squeeze the brain into the back of the skull and restrict the neural connections for the olfactory system to a narrow channel along the midline.

The flattened disc is a particularly suitable shape when it comes to fitting a large eye into the avian skull. In general, bird skulls tend to be long and narrow, but some are more exaggerated than others. For instance, cormorants have a particularly narrow skull that is much less bulbous than that of a typical bird. They accommodate their flattened eye discs by placing them more or less back to back so that their maximum diameter lies parallel to the long axis of the skull. The arrangement places the eyes on the side of the bird's head, where they give a very broad field of vision that helps the bird detect threats over a wide arc and provides a good general view of the surrounding environment. Unfortunately, it significantly reduces opportunities for binocular vision and these birds have only a small wedge

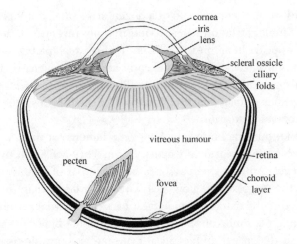

Figure 1.3 The flattened disc of the avian eye in cross-section. The view looks into the right eye. The pectin is an elaborately folded structure with many blood vessels that carries nutrients to the internal tissues of the eye (after G.R. Martin [121]).

of overlap in the visual field, directly in front and above. Such an arrangement is somewhat surprising in an active predator that uses its sight to pursue prey. Perhaps cormorants modify their hunting behaviour to accommodate this shortcoming.

There is evidence for special behaviour in other active predators that depend on effective binocular vision to locate prey in three dimensions. Falcons are high-speed aerial hunters that need the enhanced visual acuity provided by exceptionally large eyes. They have a wider type of skull than a cormorant but must still make do with a much narrower wedge of visual overlap than humans. To get the clearest possible image of a prey item, a predator needs to place the image of its target near the centre of the retina, in the most sensitive part of the eye. In a falcon, that region is not within the area of binocular overlap, so the clearest image can be held in only one eye at a time. If it tried to keep the clearest possible target image straight ahead in its line of flight, it would have to turn its face about 40° to one side. It would then present the side of the head to the oncoming air and increase drag by as much as 50% [210]. To avoid the energetic costs of additional drag, a falcon uses a specialized behaviour that makes the best of its skull design. It attacks obliquely, spiralling in on the prey in a way that keeps the target in the best part of one eye's visual field [211]. Auks are also high-speed predators but they chase fish underwater. Water is a much denser medium than air, making drag an even greater issue. It is not surprising, then, to find that auklets also make use of a spiralling attack.

Predatory birds generally have larger eyes than other species of similar body size [83]. A hawk's eye is about 1.4 times bigger, and an owl's eye about 2.4 times bigger, than that of a grouse. Because a larger eye resolves objects at greater distances, a large predator can see a sparrow at a much greater distance than a sparrow can see another bird of its own size. This kind of increased ability to resolve small objects is particularly useful to birds that attempt to fly through forests at high speed.

Although falcons and hawks are spectacularly effective hunters, their overall design is not very different from that of other birds. In comparison, owls are far more specialized predators and have evolved a suite of unique features. Unusual feather construction gives them nearly silent flight; special facial plumage and modifications to their skull enhances their highly sensitive hearing; and a specialized optical system gives them binocular vision that is exceptionally effective at night. The eyes of other avian predators are merely large versions of the flattened discs found in most birds, but the eyes of the owl have a unique tubular shape (Figure 1.4). Much of the outer disc has been lost, leaving only the central core. Where a ring of small bones, or scleral ossicles, strengthens the flat surface of the eye in most birds, owls have platelets that stand on end to form a tall supporting collar, giving the eye a peculiar columnar shape that looks and functions like a short telescope. It allows the owl's eye to retain a great distance between the lens and the retina while fitting into the limited space in an owl's crowded skull. The narrow tubes lie beside each other on the owl's flat face to give the owl an unusual degree of binocular vision.

Most owls do not fly at high speeds but use their high-resolution eyes to hunt in poor light. Compared with a hawk, their receptor field is rather small for the

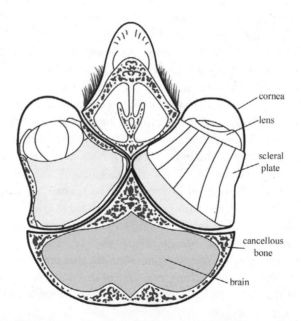

Figure 1.4   The specialized eyes of an owl. The left eye is shown in cross-section (as in Figure 1.3); the right eye is shown with the supporting tube of scleral ossicles in place. Note the relative sizes of the brain and the eyes (after G.R. Martin [121]).

large lens, but the lens concentrates as many photons as possible onto the small retina, providing the owl with an intensely lit image of a small patch of habitat. In combination with excellent binocular vision, this makes the owl an exceptionally efficient predator.

Owls are not the only birds that specialize in nocturnal vision. The tropical Oilbird is a fruit eater that roosts deep in caves where there is no light at all. In the absolute darkness of the caves, it may depend on a primitive form of echolocation and its sense of smell to find its way about, but its eyes are specially constructed to function in extremely low light. "The retina is dominated by small rod receptors arranged in a banked structure that is unique among terrestrial vertebrates. This arrangement achieves a photoreceptor density that is the highest so far recorded (1 million rods/mm^2) in any vertebrate eye" [122]. In fact, the Oilbird has 2.5 times more rods than any other bird; in order to fit, they are arranged in banks like those found in some deep-sea fish. Photons that pass through the first layer of rods appear to be picked up by deeper layers. The structure provides extreme sensitivity at very low light levels. It comes at a cost, however, and the Oilbird may find it difficult to resolve visual information into a detailed image.

The Beak, Taste, and Olfaction

The beak is usually the most prominent of the bird's three exposed structures, and is perhaps the most flexible in evolutionary terms. Its shape offers clues to the bird's food habits and frequently makes a colourful contribution to its owner's repertoire of displays and signals. Probing, tapping, or gaping with the beak is often an important part of courtship behaviour or territorial defence. Nonetheless, it is primarily a feeding apparatus that varies in shape from group to group according to food preferences. Most birds use it as a basic set of pincers or snips, but others may call upon it to function as a chisel, spear, or sieve. Herons, cranes, and storks have simple, elongated beaks for picking up small prey, but the length is important if the bird is to hold struggling prey away from the all-important eyes. From a distance, the long bills of shorebirds look like they might be as hard and sharp as a heron's, but they are sensitive probes. The blunt tip of the beak in a Dunlin is as soft and sensitive as a fingertip and it is well irrigated with blood.[5] It can sense the movement of tiny worms hidden deep in fine mud, and its tip can open, without moving the whole jaw, to pick items from their hiding places. In contrast, the tip of the beak in a goose or parrot is a robust and rigid structure that acts as a hard-edged set of sharp shears that can tear through the toughest vegetable fibres.

Even the most robust beaks include some sophisticated structures and specialized components. Near the tip, there are spaces for tiny taste organs and other sensory receptors that help the bird sort out edible from inedible items. Usually

---

5 Artists who work from dead specimens often portray them with needle-sharp beaks because the soft covering shrinks and collapses as the blood in the tip dries.

these sensors are concentrated along the margin of the upper bill, where there is some risk that the continuously growing sheath of the beak might cover them over. They must be kept clear, and birds frequently trim their beaks by wiping their edges against some hard object. The cuttlefish "bone" given to captive parrots helps them wear away the rapidly growing edge of the beak and keep the sensitive areas exposed.

The finches on the Galapagos Islands have perhaps the most famous beaks in history. The structural variety among the beaks was a key piece of evidence in Darwin's development of the theory of evolution. Recent evidence suggests that they have even more to offer astute observers. The islands undergo long cycles between slightly wetter and slightly drier climates. The seeds that the finches eat get larger under the moist regime and smaller under the dry regime. Apparently, the length of the beak changes within some species so that new generations of birds can handle changes in the size of edible seeds [73].

The hard, smooth surface of the outer beak hides a great deal of internal complexity that is not associated with feeding – at least not directly. The mid section of the upper beak covers and protects any organs of smell that the bird might possess. A number of thin bony plates above the palate often turn the nasal area into a convoluted maze of chambers lined with a thin layer of sensitive tissue that collects odours carried on the inhaled air. The tissue's blood supply also helps to warm air coming into the body and conserve moisture as it is exhaled. If a sense of smell is unusually important to a bird, this central section will also include large bony nostrils. Large nostrils and large eyes in Old World vultures suggest that they have both a keen sense of smell and keen eyesight, but it is not immediately obvious which sense they use to find their prey. The great medieval ornithologist Frederick von Hohenstaufen tackled the question in some of the earliest experiments of modern science, and concluded that they could not find food by scent alone.

The openings created by the nostrils lead directly to the lungs, and one might expect flooding to be a potential problem for birds that dive underwater. Species that plunge into the sea from 30 or 40 m up, such as gannets, boobies, and pelicans, would face a more extreme challenge if they had not done away with the nostrils by sealing them within the sheath of the beak. Needless to say, such birds hunt by sight alone and have little or no sense of smell. Other seabirds, such as petrels, make great use of their sense of smell to find food across stretches of open ocean, and have large, well-developed nostrils. Birdwatchers sometimes exploit this sensitivity by spreading fish oil on the water to attract petrels within binocular range. Petrels do not plunge violently into the water so there is no high-pressure rush of water into the mouth. When they put their head beneath the surface, a small plug of cartilage beneath the tongue seals the inner airways. It is adequate to protect the air passages of shearwaters and other diving petrels that swim to depths of 70 m or more [229]. A similar simple device protects the lungs of auks and loons.

Although modern birds do not have teeth in adult life, the embryos develop one and sometimes two "egg teeth" before they hatch. The egg tooth is usually a small deposit of white bony material that collects near the tip of the beak and is used by the fully grown embryo to crack the egg's shell. If a second egg tooth develops, it appears near the tip of the lower mandible. Typically, an egg tooth will drop off very soon after hatching, but a fledgling Marbled Murrelet may keep its upper tooth for a month or more after heading out to sea. It is quite large and, when it catches the light, the white flash offers an easy way to distinguish recent fledglings from slightly older birds.

In most birds, the horny beak is no more complicated than a pair of chopsticks. Like the set of chopsticks, it is an excellent example of how an elegantly designed but simple tool is capable of many functions. Like that set of chopsticks, however, the beak is useful only because it is manipulated by a complex and sophisticated hand. The "hand" in this case is the avian jaw, an assembly of small bones that is far more complicated than anything found in mammals. Where mammals have two robust bones that work against each other, birds have nine or more small bones that sometimes work together and sometimes in opposition (see Chapter 2). Special hinges allow the upper jaw to rotate against the skull, and the arms of the lower jaw can flex outward near their midpoints. Both actions increase the gape and enable the bird to cope with inconveniently shaped foods. Unlike the bones in other parts of the skeleton, some bones in the anterior part of the jaw are not attached to muscles. They are passive pushrods or levers bound by ligaments to bones further back in the mouth that do have their own muscles. The action of a bird's jaw has more in common with the shuttling of a mechanical loom than the relatively simple grip of the human hand.

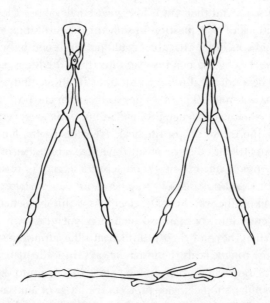

Figure 1.5 The hyoid apparatus of the Chattering Lory, a parrot. Based on an drawing by G.V. Mivart in F.E. Beddard [8].

If a bird's jaw functions as its fingers, the tongue is its thumb. Birds have a very non-mammalian tongue that is useful in the initial stages of handling food. It is neither soft nor fleshy and has its own internal skeleton, called the hyoid apparatus (Figure 1.5). The outer surface of the tongue is sometimes armed with arrays of little spikes that hold food in place or help direct it into the gullet. When a puffin is already holding one fish but wants to catch another, it uses its spiky tongue to hold the first fish against the roof of its mouth until it can chase and catch the next. Sometimes it uses this trick to bring home a dozen small fish to its nestling.

The beak and jaws are useful as a "hand" for the bird because the avian neck is a very capable arm. The structure and function of the neck is discussed with other parts of the skeleton in Chapter 2.

Ears and Hearing

A bird's ears are not marked by external structures. In some owls, facial feathers may reflect sounds towards the ear's opening, but typically there are no external structures, only a narrow canal leading to the inner ear. In spite of the ear's small size, many birds depend on songs and calls for social interaction.

The Avian Head

The concentration of the bird's most important sensory receptors close to its only feeding apparatus has had a profound effect on the design of the head and the organization of its component parts. The avian head is one of the most specialized structures in nature.

As in all vertebrates, the shape of the bird's head reflects the fundamental function of the skull as a case for containing and protecting the brain. In spite of the slanderous implications of the term "bird-brained," the bird's brain is almost as large and well developed as that of a mammal of similar size. In its efforts to contain a large brain, very large eyes, and mobile jaws, the avian skull has gone through a series of evolutionary contortions that are rather similar to those that accommodate the large brain in humans. Our face has rotated "forward" on the skull so that the eyes face front when we stand erect. In strict anatomical terms, our face has rotated ventrally and now lies on the same surface as our belly. This movement allowed our brain to curl and to expand into the spacious dome formed by the skull so that its ventral surface now faces downward. The faces of some birds have also rotated forward, and birds are the only other kind of terrestrial vertebrate with a large spherical head. Their brain has retained its primitive orientation, however.

In a primitive vertebrate, the brain is small and elongated, with a long axis that lies on the same plane as the long axis of the skull. This arrangement offers ample space in the head for eyes and other sensory organs without the need for long neural connections. The bird's brain has retained some of its primitive orientation to the spinal cord in spite of its enlargement. Unfortunately, when the brain of a bird is illustrated in a textbook, it is usually shown in a prone position without the surrounding braincase, and its orientation within the skull of the living animal is not

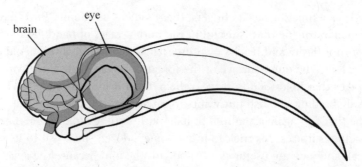

Figure 1.6   Relationship of the brain and eye in the skull of a Yellow-billed Hornbill.

at all obvious. In fact, such illustrations are misleading and unhelpful because the bird's brain typically stands on its end. It is not rotated or folded forward like the human brain.

We have already seen that the bird's eyes are so large that they press against each other at the midline. In many birds, there is no room for the brain to sit between or beneath them. Consequently, the brain sits vertically in the back of the skull, its long axis lying across the long axis of the skull (Figure 1.6). This posture may increase the effectiveness of important neural connections. Vision, the most important sense in birds, depends on the quick integration of vast quantities of visual data. Shorter nerve connections between the sensors in the retina and data processors in the brain speed the bird's response in an emergency or when presented with an opportunity to feed. In the vertical posture, the optical lobes are virtually pressed against the backs of the eyes, so the optic nerve is very short.

The peculiar vertical posture of a typical bird's brain allows it to maintain a simple linear arrangement with the spinal cord that is not much different from the relationship between the brain and spinal cord in the most primitive of vertebrates. In species where the brain's axis lies across that of the beak, only the bones of the skull have changed shape. Those above the eyes have become greatly elongated to allow the beak to rotate forward onto the same surface as the belly. The bones between the ears and around the brainstem have remained more or less unchanged. The most extreme example of these changes appears in the skull of the American Woodcock [32]. If the beak were held horizontally, the back of the skull would be tipped backward about 30° beyond vertical, to the point where the woodcock's brain would be lying on its back. The beak normally points downward in life, however, and the woodcock can enjoy an unequalled view of predators in the sky even as it probes for earthworms in soft soil.

Not all birds have a brain oriented across the beak. As mentioned earlier, cormorants (and some other diving birds) have unusually elongated skulls and their brains and beaks lie on the same plane. Consequently, they look somewhat primitive and often rest with their beaks pointing skyward. In keeping with the simple linear arrangement, the opening for the spinal cord, the foramen magnum, lies

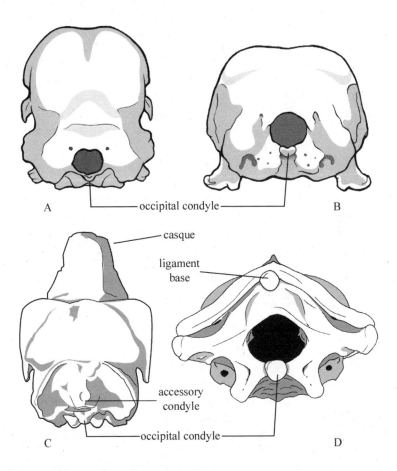

Figure 1.7 Caudal or posterior views of the skulls in four birds, showing the positions of the foramen magnum and occipital condyle. (A) Hoatzin; (B) Black Currasow; (C) Ground Hornbill, with an accessory condyle that articulates with the atlas and supports the weight of a large casque on the skull; and (D) Double-crested Cormorant, with a large boss above the foramen magnum for the attachment of very large neck ligaments (A and B adapted from drawings by M.L. Vidiera Marceliano [218]).

centrally in the base of the skull (Figure 1.7), just above the nub of bone, the occipital condyle, that articulates with the first vertebra of the neck.

In other birds, the expansion of bones in the skull included the forward components of the braincase and it ballooned into a large, smooth dome. Because the bones near the brainstem did not expand, the skull appears to balance across the top of the neck, as it does in humans. The extent of this process has been highly variable; consequently, the foramen magnum and its associated connection to the neck may vary from a position on the "back" of the skull to one that lies beneath it (Figure 1.7). Anatomically speaking, the location of the brainstem has not changed but the bones of the skull have slid into new positions around it. As in humans, the migration of the face onto the ventral surface of the body is the most immediately apparent result of this process.

The compact arrangements of the avian brain and optical systems contrast sharply with the more attenuated arrangements of the olfactory system. Most birds make little use of their sense of smell and it is unlikely to be involved in any sort of emergency response. There is therefore little risk to the bird in having relatively long nerves between the sensitive tissue in the nostrils and the olfactory lobes in the brain. The olfactory impulses from the midsection of the beak must make a relatively long journey to the brain, through an arch of bone on the midline of the skull that passes over the eyes. New Zealand's nocturnal kiwi is one of the few birds that use a keen sense of smell to find its prey. The kiwi's skull is exceptionally long and allows the olfactory tracts to extend far forward towards nostrils at the tip of the beak. The eyes are exceptionally small and leave ample room in the skull for a specialized olfactory system, complete with large and rather mammal-like olfactory lobes.

Moving through the External Environment

Feet, Legs, and Terrestrial Locomotion

Near the tail end of the bird, scaly legs hold the body off the ground. Just as the feathers create the outer bird and mark it as a flying animal, the well-developed legs mark the inner bird as an active pedestrian, descended from a long line of walkers and runners. Legs are a fundamental feature of birds; many species are flightless and some of those are wingless, but no bird is legless.

In the limbs of most vertebrates, long bones of roughly equal length meet at a major hinge formed by the elbow or the knee. In primitive vertebrates and mammals, a large single bone (humerus or femur) is attached to the body and the outer unit consists of paired slender bones (radius and ulna or tibia and fibula). Birds have moved a long way from that basic pattern. Their hind limb includes three large long bones (femur, tibiotarsus, and tarsometatarsus) and the remnant of a fourth (fibula). As a result, birds have two major joints between the hip and the toes (Figure 1.1).

The femur is the leg bone that lies closest to the body. It is deeply buried in a mass of muscles and completely screened from view by a covering of feathers. Many bones found within a bird's body are thin and flexible, but in spite of its position deep in the thigh, the femur is one of the largest and most robust bones in the skeleton. Its strength reflects its importance in transferring the weight of the body to the legs. In a large dinosaur, the femur, like the other bones in the leg, was a simple column; in birds, it is a much more complex structure. It is usually slightly bowed and twisted, suggesting that it is designed to resist an array of conflicting forces. The right-angled ball-and-socket joint that forms the hip generates some of those forces because birds have abandoned the simple upright posture of a dinosaur. When a dinosaur stood erect, its femur, the tibia, and the fibula were held almost vertically; when the bird stands up, in its "at-rest" posture, the femur is suspended at an angle, almost parallel to the ground. This posture is maintained by tension in the thigh muscles, which are attached at several points to the femur, where they generate shearing forces in the bone that can only be accommodated by increased size and strength. The muscles twist the femur against the hip joint and subject it to almost constant strain unless the bird is lying on its belly or floating on water. Because it is held horizontally, the femur contributes little to the bird's length of stride.

Two long bones of roughly equal length make up the externally visible part of the leg. Both are derived from the fusion of several embryonic bones. The one nearer the body includes a bone of the leg (tibia), roughly equivalent to the mammalian shinbone, and a fused collection of small ankle bones (tarsals). It is called a tibiotarsus. The outer segment of the limb is constructed from a mixture of ankle and foot bones (metatarsals) and is called a tarsometatarsus. Frequently, the tarsometatarsus is the only visible portion of the hind limb. The neighbouring tibiotarsus is often hidden by a skirt of long feathers. These specialized lower limb bones have no precise equivalent in a mammalian limb and there are no widely used common names for them.

In very long-legged birds, the tarsometatarsus and tibiotarsus usually have very similar lengths. When the leg bends at the ankle, the equal length of the two long bones allows the body to move straight up and down like a scissors-lift. If the two bones were of different lengths, long-legged birds might have considerable difficulty sitting squarely on the eggs. Instead, they can place the egg between their feet and confidently drop straight onto the nest cup as the leg folds. The normal resting posture of the leg bones may also be of significance before the egg is laid. Because the femur is almost horizontal, the knee lies forward of the main body cavity so that the upper part of the limb can reach around the front of even a very large egg. The space behind the knee allows the egg to be much wider than the hip joint without interfering with the bird's ability to walk. In straight-legged dinosaurs, the eggs could not have been wider than the hip joint. In their case, the egg size was further limited by the passage of the oviduct through a solid ring of bone formed by the pair of pubic bones. In birds, the tips of these bones do not meet.

The great majority of birds are terrestrial pedestrians, so it is not too surprising that many share a similarly shaped tarsometatarsus. It is usually one of the longest bones in the leg but it may be short and stubby, especially in groups that do not do a lot of walking. It is very short in swifts and hummingbirds, which use their limbs only when perching or moving around the nest. They have such tiny feet that they have been placed in the Order Apodiformes, a name that means "footless." The tarsometatarsus is also stubby in the frigatebird, a seabird that not only does not walk but apparently does not swim and spends much of its life in the air. Like the Apodiformes, it uses its feet only when roosting. In waterbirds, such as grebes and loons, that swim actively, the tarsometatarsus is often flattened, as though it were designed to move through water with little resistance (Figure 1.8). One of the

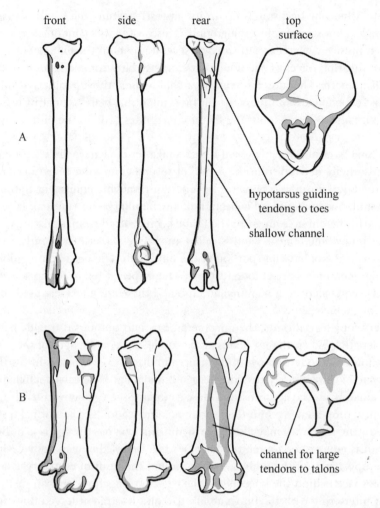

front   side   rear   top
                                          surface

A

hypotarsus guiding
tendons to toes

shallow channel

B

channel for large
tendons to talons

Figure 1.8   Two extremes in the shapes of the tarsometatarsus. In (A), the Yellow-billed Loon, it is a flattened box with modest channels, front and back, for tendons to the toes. In (B), the Northern Hawk Owl, the bone has become a broad, protective trough for the massive tendons that close the talons.

most spectacular modifications occurs in tarsometatarsi of predatory birds. The bone is shaped like a long trough to carry the robust tendons that operate the killing feet (Figure 1.8).

In many familiar land birds, the absence of muscles along the tarsometatarsus makes it look like a particularly fragile and sticklike bone, but it is usually much the same diameter as the neighbouring tibiotarsus. The tibiotarsus, however, is buried within the large mass of muscles that make up the familiar "drumstick" in a chicken or turkey. Except during extreme contortions, the muscle-covered bone is hidden beneath a skirt of long feathers.

The tibiotarsus is accompanied by a thin splint of bone that is all that remains of the fibula. The upper end of this vestige widens to meet the knee, but the lower tip just trails off into gristle without ever reaching the ankle. It is often used as an example of how birds eliminate unnecessary weight in the outer parts of the limbs, where decreased bone mass can help to reduce problems of inertia and momentum during active movement. In dinosaurs, the fibula usually articulated with the ankle bones.

At the outer end of the bird's leg, the toes form the most variable part of the limb. They offer valuable clues to the bird's lifestyle and may vary in number and position as well as shape and size. Typically, the claws and the covering scales are more dramatically modified for special functions than the underlying bones. The toes of raptorial predators are among the most familiar examples. Their tips are armed with long talons and their soles have a thick armour of large, rough scales to secure wriggling prey. These killing feet quickly subdue prey and reduce the risk of injury to the predator's sensitive head. Climbing birds also have curved claws and rough scales on the toes, but they are designed to grip tree bark and the talons are neither as thick nor as sharply curved as those of raptors. Birds that live in the water tend to have short claws; the skin of their feet is soft, especially if the feet are important heat radiators. The thin web between the toes is flexible for swimming but must also be durable enough to withstand the wear and tear of walking on dry land. Grebes are waterbirds that lack webbed feet. Their paddles are formed from expanded scales that grow along the sides of the toes so that each digit acts as a paddle. Similar structures appear in several unrelated groups of birds, such as phalaropes, coots, and fin-foots.

Many seabirds put their webbed feet through real torture tests. The web of skin between their toes must be both soft and tough. Seabirds nest on craggy islets and must be able to walk around on sharp rocks or excavate deep nesting burrows in stony soil. In the breeding season, their claws are worn to blunt nubs, like those of a dog. The young hatch with large, soft feet that are ideal points of attack for the ticks that swarm in many colonies. The webs heal quickly but their surface may bear scars from these parasites throughout the bird's adult life. On the water, most seabirds use their feet only for slow paddling on the surface. Consequently, worn or damaged webs and even serious injury to the foot need not be a serious problem. It is even less of a problem for the many species that use wing-propelled locomotion underwater.

The number and arrangement of toes is one characteristic of birds that varies from group to group (Figure 1.9) but is not a particularly useful tool for classification. There is not a clear relationship between foot structure and other taxonomic characters. The similarity of the toe arrangements may be merely superficial. In the single most common format for toes, three forward and one back, there is a choice of six different arrangements for the underlying plantar tendons that manipulate the digits. In addition, toe arrangements that might be described as rare or specialized tend to be just another unique feature for groups of birds that are

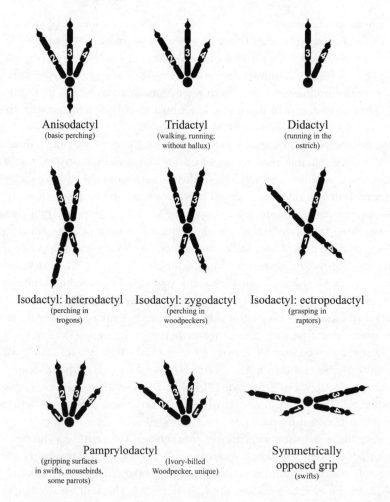

Anisodactyl
(basic perching)

Tridactyl
(walking, running;
without hallux)

Didactyl
(running in the
ostrich)

Isodactyl: heterodactyl
(perching in
trogons)

Isodactyl: zygodactyl
(perching in
woodpeckers)

Isodactyl: ectropodactyl
(grasping in
raptors)

Pamprylodactyl
(gripping surfaces
in swifts, mousebirds,
some parrots)

(Ivory-billed
Woodpecker, unique)

**Symmetrically
opposed grip**
(swifts)

Figure 1.9 The basic arrangements of toes in birds. Most of the isodactyl arrangements appear to be adaptations for perching, but the ectropodactyl form seen in owls and ospreys gives the tips of the talons a very wide spread for grabbing prey. The unusual "opposed grip" seen in swifts is likely a minor modification of the isodactyl type. It is also found in chameleons but is otherwise very rare in nature.

already difficult to classify because they have an abundance of other unique or unusual features.

Two, three, or four toes may fan out from the knuckles of the tarsometatarsus in ways that are better clues to lifestyle than to family relationships. Three-fourths of all birds share the most frequently encountered toe formats, while some closely related families have an assortment of feet. For instance, among the Charadriiformes, gulls, terns, and auks have webbed feet but sandpipers and plovers include species with and without webs. Phalaropes have lobes on their toes like a grebe, whereas others, such as the oystercatcher, have simple toes but have lost the hallux (digit 1) at the back of the foot.

Wings and Aerial Locomotion

Although much of a bird's skeleton reflects the importance of the pedestrians in its ancestry, it is not the ability to walk or even run and swim that has generated interest in birds; it is their ability to fly. Flight is the key characteristic that has enabled birds to occupy every conceivable terrestrial habitat and has played a major role in their successful invasion of the high seas. One might argue that flight has enabled penguins to invade Antarctica, even though they "fly" only underwater.

Except for the penguin's specialized flipper, the modern bird's forelimb is completely adapted for aerial flight. It, and it alone, provides the motive power for aerial locomotion. Unlike the legs or other structures with which the bird interacts with its environment, none of the wing's living structures are visible. The ancestral bird, *Archaeopteryx*, had naked fingers on its wings, but today only the nestlings of the South American Hoatzin have exposed digits. It is a temporary condition and these curious structures are gradually absorbed as the bird matures. In all other birds, the feathers completely cover the fleshy parts of the limb and provide the lift needed for flight. As in the hind limbs, tendons connect to remote muscles that provide the actual locomotory power. Small muscles lie along the wing bones but these are responsible mostly for subtle changes in feather position or wing shape. The muscles that drive the wings' powerful strokes lie deep within the body, along the breastbone.

Although the forelimbs of birds have been extensively modified to serve as wings, the long bones retain the primitive characteristic of being approximately equal in length. Exceptions usually indicate that the bird has a very specialized flight technique. In wing-propelled diving birds, such as auks and penguins, the humerus is almost 30% longer than the radius or ulna. It is also arched and flattened into a shape not found in any other kind of bird. It has been suggested that the unique shape reduces drag underwater, but the reduction in profile is minimal and the purpose of this design remains a matter of speculation. It will come up again in Chapter 10, in the discussion of underwater wing-propelled locomotion.

The wing skeletons in swifts and hummingbirds are also very unusual (Figure 1.10). The humerus is short, stubby, and robust, like the femur of the leg. Deep sculpting on its surface accommodates tendons and attachments for large muscles, but it is less than 20% of the wing skeleton's length. The paired radius and ulna are also robust but otherwise not highly specialized, while the hand is hugely elongated and accounts for nearly 60% of the wing skeleton's length. No other birds possess such an unusual skeletal arrangement. Swallows soar through the air to forage on flying insects and are somewhat convergent with swifts, but their wing skeleton is similar to that of other small songbirds, with the hand less than 40% of the wing's length.

The short, thick humerus in hummingbirds and swifts may be related to their use of very rapid wing beats. Just as the robust avian femur is designed to bear the stress generated by large muscles in the hip, the stubby humerus may be the appropriate design to absorb unusual forces generated by rapid contraction of the

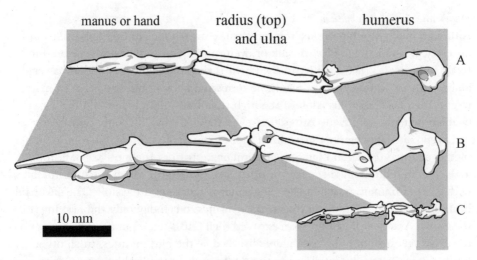

manus or hand   radius (top)   humerus
                       and ulna

A

B

10 mm

C

Figure 1.10 Variations in the proportions of the major skeletal elements in the wings of three aerial specialists among small forest birds. (A) A swallow (*Hirundo pyrrhonota*); (B) a swift (*Aeronautes saxatilis*); and (C) a hummingbird (*Selasophorus rufus*).

flight muscles. It is more difficult to understand why both these birds should have such large hands. Intuitively, you might expect to find smaller bones in the outer part of the wing, where the weight of bone might create issues with momentum and inertia. In fact, the size of the hand is misleading and the total contribution of the skeleton to the wingspan in hummingbirds and swifts is less than 40%. In other birds, it usually ranges from 55% to 60%. The remainder is unsupported feather.

Feathers tend to make a large contribution to the wings of highly aerobatic birds. In the swiftlike swallows, the primaries account for about 60% of the wing-span. Storm-petrels, aptly nicknamed sea swallows, are highly manoeuvrable in the air and have an almost butterfly-like flight. Unsupported feathers make up a bit over half the total wingspan. Even though albatrosses are closely related to storm-petrels, their flight style has called for a completely different architectural strategy. They use greatly elongated wing bones that support 70% of the wingspan to achieve highly controlled and efficient flight on wings that have rather short primary and secondary feathers. Other marine and aquatic birds, such as loons, grebes, and auks, tend to fly at high speeds using a very energetically expensive technique. Their wing skeletons are also rather long. In the case of the auks, which use the wings for underwater locomotion, part of that length is due to elongation of the humerus.

Arm bones are remarkably similar among all kinds of vertebrates, but the hands are often highly specialized. The nature of the avian hand skeleton has been a point of contention in the bird/dinosaur debate. Birds have remnants of only three digits, which studies of the developing embryo suggest are numbers 2, 3, and 4.

Some dinosaurs also have a hand reduced to three digits, but the fossil record appears to provide evidence for the sequential loss of digit 5 and then 4, leaving numbers 1, 2, and 3. If both the avian and the dinosaur sequences have been correctly interpreted, the bird's hand did not descend from a dinosaur's. It is not an argument that is going to go away in the immediate future. It is possible that avian digits 2, 3, and 4 slid over into positions 1, 2, and 3 during the early stages of embryonic development, but recent genetic evidence for this "sleight of hand" has not been conclusive one way or another (see Chapter 4).

The difficulty posed by the sequence of avian digits is fairly typical of a range of long-standing problems that remain unresolved in biology. They tend to be based on the observations of prominent 19th-century figures and have been successfully defended against contradictory evidence ever since. The digit sequence was first put forward by the Victorian anatomist Richard Owen and has survived a series of criticisms by embryologists in both the 19th and 20th centuries.

One of the most significant of these long-standing problems has to do with the classification of birds and the development of an evolutionary story for the group. In spite of the development of sophisticated mathematical techniques and biomolecular genetics, ornithologists have not been able to agree on a family tree for the birds. As a result, the 19th-century classification by Max Fürbringer [62] continues to influence our ideas about avian evolution. It has been updated by Alexander Wetmore and others, but its basic framework remains. Fürbringer placed the loon near the base of his tree because it seemed to have more reptilian characteristics than other birds and was remarkably similar to the toothed fossil *Hesperornis*. He placed the songbirds near the top of the tree because they appeared to have more complex anatomy and more advanced behaviour. Today, loons, grebes, and other waterbirds still appear in the opening pages of field guides.

If that were the only significant consequence of Fürbringer's tree, there would be no problem. Even though he was writing 15 years before the successful flight of the Wright brothers at Kitty Hawk, the loon's position has had a subtle influence on thinking about avian flight. It is quite easy to find derogatory comments about the loon's aerial capabilities in both the technical and popular literature. Throughout the mid-20th century, aeronautical engineers struggled with problems of lifting large loads into the air. As a result, it was only natural that biologists of the time were unimpressed with the loon's need to taxi to gain sufficient speed for take-off. It must be remembered, however, that even though the loon's method may be more labour-intensive and energetically expensive than a sparrow's quick spring into the air, high levels of energy expenditure are not necessarily a primitive trait.

We should look beyond the cost of take-off to the loon's ability to carry its heavy body rapidly across an entire continent. The loon takes advantage of a subtle paradox in aerodynamic theory. If a bird can generate enough thrust to fly very fast, it can increase its payload without a proportional increase in energy expenditure. It is already putting so much energy into thrust that changes in the amount of energy going into lift, to carry additional payload, are relatively small. Although

the loon may have difficulty getting into the air, its large body can carry a great
deal of fuel; once airborne, it can travel long distances very quickly. Modern jet
aircraft use the same strategy by trading high fuel expenditure for a short journey.
We might look at the loon as the avian equivalent of the Boeing 747, which carries its payload at high speeds but consumes huge quantities of fuel, especially on
take-off. No one would argue that the 747 is a type of primitive aircraft.[6]

Victorian notions about avian flight continued to influence ornithology until
1988, when Jeremy Rayner published an important review that made the principles
of aerodynamics accessible to ornithologists [176]. It covered his own work on the
importance of vortices generated by moving wings and the application of fluid dynamics to avian flight [172, 173] as well as the contributions of C.J. Pennycuick,
U.M. Norberg, and other theorists.

The Tail and Aerial Control
Birds often have spectacular fans of tail feathers but beneath this plumage there
is little more than a small nub of flesh. Much of the tail's skeleton lies within the
body, between rearward extensions of the hipbones. Only the bladelike pygostyle
extends into the fleshy nub. In spite of its simple appearance, the tail's internal
components give it a great deal of manoeuvrability and make it an important component of the flight apparatus. Tail feathers vary in length from family to family,
according to variations in their aerodynamic responsibilities. They tend to be longest in birds that require a great deal of aerial manoeuvrability and use the tail as a
rudder. They are unusually short in birds that fly at exceptional speeds or exhibit
unusual levels of aerial efficiency. Many of those short-tailed birds are marine or
aquatic fliers that steer by making subtle changes in wing conformation or by
opening the web between their toes.

The tail of the modern bird is an outstanding example of structural centralization. *Archaeopteryx* and some other ancestral species had long, reptilian tails that
may have been flexible but were not an effective shape for an aerial rudder. Also,
they must have included a considerable weight of muscle and bone. The fleshy
nub of a modern bird's tail contributes to controlled flight even though its movements are controlled by muscles closer to the body's core and it does not contain
any muscles of its own.

Conclusion
Although most paleontologists today agree that birds have evolved from a specific
group of advanced dinosaurs, just a few years ago it was the topic of heated debate
and there are still some unresolved arguments and anomalies. Most ornithologists
have stood aside from these debates, partly because paleontologists deal with a level

---

6 We will revisit the speed-versus-efficiency question in Chapter 10 when we look at its effect on
contrasting lifestyles among seabirds.

of structural detail foreign to most ornithologists and partly because modern birds seem so far removed from their reptilian ancestors. Paleontologists may never convince ornithologists that the origin of birds has practical significance. The relationship of birds to one group of dinosaurs or another has no demonstrable impact on avian conservation, wildlife management, or endangered species issues. Many similarities and differences between birds and dinosaurs are obvious, however. Perhaps an investigation of the design of one would improve our understanding of the construction of the other, if we can avoid getting lost in the technical jargon of unfamiliar sciences.

Further Reading

The topics in this chapter are covered in some way in most ornithology textbooks. Progress in avian anatomy is generally slow, and older technical books that are often readily available in public libraries may contain information that is still current:

Chamberlain, F.W. 1943. Atlas of avian anatomy: Osteology, arthrology, myology. Michigan State College, East Lansing, MI.

Proctor, N.S., and P.J. Lynch. 1993. Manual of ornithology: Avian structure and function. Yale University Press, New Haven, CT.

Reports on research in avian aerodynamics or mathematical models of bird flight frequently appear in the *Journal of Experimental Biology*. A good start can be made by looking for papers in the 1990s by C.J. Pennycuick, J.M.V. Rayner, and U.M. Norberg. Earlier work can be misleading. Rayner puts much of it in perspective in:

Rayner, J.M.V. 1988. Form and function in avian flight. Pages 1-66 *in* R.F. Johnston (ed.). Current ornithology. Vol. 5. Plenum Press, New York.

A Bird Is an Animal with Hollow Bones

2

*Special features of avian
bones and their assembly as a skeleton*

Engineers have become adept at constructing static objects such as buildings and bridges, and their vehicles, from aircraft to automobiles, are sleek and impressive. Their robots and other imitations of organic life, however, always seem rather clumsy and overly power-hungry. Even in science fiction fantasies, robots lurch about and lack the ease of motion that is typical of a living creature. When it comes to combining efficiency, clever design, and the efficient use of power, the hand of nature always seems more skilled. Nowhere is the evidence for this stronger than in the design of bones and their assembly as skeletons.

All terrestrial vertebrates depend on bone to give the body rigidity, to anchor muscles, and to protect vital organs. Gravity and the low buoyancy of solid objects in air have prevented terrestrial animals from using tricks with inflation or fluid pressure that work well underwater. Hydraulic systems give shape to octopus and other complex organisms under water, but even the partially mineralized cartilage found in sharks is inadequate for terrestrial existence.

Bone was not the only possible choice available to primitive animals during the early phases of evolution. Arthropods developed chitin as the basis for their skeleton and applied it to a wide variety of structural problems both in water and on land. Because it comes in many forms and can be rigid, tough, or flexible as needed, chitin forms their entire skeleton and is often sculpted into complex shapes for external decoration or to form elaborate structures in joints. It has some severe limitations, however. Chitin is a non-living, tanned material whose fibres are bound by cross-linked proteins. It cannot be stretched or reabsorbed to allow growth. Its use obliges arthropods to undergo complicated shedding processes that seem to limit their ultimate size, especially on land.

Vertebrates have developed a similar-looking material, called keratin, that is not used within the skeleton but is often supported by bone. It is a tough material that appears as horns, claws, beaks, fur, and, most importantly for birds, feathers. Unlike chitin, it grows continuously from within and is often used for structures that wear away naturally. Both of the most elaborate structures made with keratin occur in birds. One is the beak, which is often large and colourful; the other is the feather, one of the most complex and elaborate non-living structures found in any animal.

Bone has none of the limitations of chitin. It is a physiologically active tissue that can change size and shape to meet the body's needs during the various stages of an animal's life. The assembly of bones as skeletons allows flexible and efficient motion because each individual component can be made stiff enough to resist the local forces of compression, shearing, or bending. Bone was selected as the basic structural material at a very early stage in the history of vertebrates, long before they ventured onto land. As a result, there are some general themes in skeletal architecture that appear consistently across the whole group: the arrangement of the skull and lower jaw; the string of vertebrae in the backbone; and limbs that have a single large bone near the body, a middle element consisting of a pair of bones, and an outer element consisting of a fan of digits. Most of the major innovations among vertebrates involve variation in the shape of these basic structures or changes in their application, rather than the creation of entirely new structures. Nonetheless, there are some examples of novel and specialized bony items scattered through the group. Mammals may have showy but disposable antlers, and female birds have developed physiologically active medullary bone that stores calcium for the reproductive cycle. Although the skeletons of all terrestrial vertebrates are constructed from the same kinds of bony tissues and they share basic structural themes with some very remote common ancestors, the skeletons of birds and mammals are specialized for very different functions and have developed in very different directions.

In terrestrial animals, the skeleton has two basic functions. As in aquatic animals, it helps the body keep its shape during movement. On land, however, it keeps the body off the ground, supporting it against the force of gravity. In mammals, the skeletal design emphasizes strength and stability. Mammals have a somewhat rectilinear architecture, which overcomes gravity simply by resting the body on top of robust, columnar legs. The bones in those legs are heavy and stiff to resist compression from the force of gravity when the animal is at rest and bending during locomotion. Locomotion is achieved by flexion at the hip when muscles pull the whole limb forward and backward beneath the body.

The avian skeleton also lifts its owner off the ground but it is fundamentally different from the mammalian skeleton. Its parts look more like a fragile collection of springs and levers under tension than a building frame. Very large, pedestrian birds such as the ostrich, and some sedentary species such as the Hoatzin, have more mammal-like, robust bones, but species that are active in the air or on water have much more slender and flexible bones. In very active birds, none of the supporting elements is vertical to the ground, and the other pieces are arranged more like a loosely woven wicker basket than some more rigid framework.

While many of the bones in the body are small and flexible, the leg bones are stiff and often exceptionally long. Both birds and humans are bipedal but the resemblance is only superficial. Whereas the human leg consists of two long elements held vertically, the bird's leg has three long elements, none of which is vertical when the bird is at rest.[1] The innermost bones in a bird's leg are analogous to those in a mammal, but the lower element is a novel development. It is the product of a fusion event in the embryo that captures the longer foot bones and some anklebones and turns them into a new long bone. At the outer end of this special bone, there is a fan of very long toes that stabilize the standing bird. Locomotion is achieved by knee flexion when muscles move the lower elements of the leg forward at the knee. There is only a little movement of the leg's uppermost element, the femur.

The forelimb is specially modified as a wing. Like the other special characteristics of the avian skeleton, it reflects the bird's commitment to flight. The bones are large, with ample room for the attachment of muscles, and they are stiff, to resist the stresses generated during flight. As with the leg, the largest and heaviest bones are typically those closest to the body. This distribution of weight reduces the build-up of momentum during rapid movement.

Both the wing bones and many bones in other parts of the skeleton reflect the bird's dependence on extreme centralization of muscles near the body's core. Centralization requires that some bones provide broad surfaces for the attachment of large muscles. It also brings tendons with a variety of unrelated functions together in the joints. Their roles are usually kept separate by special guiding structures on the bones, even though such structures increase the weight of the joint. Because bird tendons frequently cross more than one joint, the guiding structures often give bird bones a highly sculpted surface that is much more elaborate than that on the bones' mammalian equivalents.

Bird skeletons are famous for hollow bones that are said to help the bird meet its need for reduced weight. This perception may not be wholly accurate, however, because any savings in weight are small. A gas-filled tubular bone is only 8%-13% lighter than a comparable marrow-filled bone. The hollow construction may actually be an important response to the kinds of stress placed on a bird's skeleton. S.M. Swartz and other researchers [205] have drawn attention to the problems of torsion, especially that created in the bones of the wing during the downstroke. In their recent analysis, J. Cubo and A. Casinos [38] reported that hollow bones sacrifice bending strength in order to obtain high values of torsional strength.

The hollow cores of avian bones are not merely empty space but are convenient chambers that birds have put to several uses. Most are filled with air and are referred to as "pneumatized" because they connect with air sacs of the respiratory system that extend throughout a bird's body. Although the bones are lined with an impermeable membrane that prevents the exchange of oxygen with adjacent tissues, they may still play important roles in respiration and other physiological processes.

---

[1] The upper leg bone, the femur, is never vertical. In very long-legged birds, the lower bones may look vertical but they are usually held at slight angles. Birds that habitually rest on one leg appear able to lock the supporting leg into a truly vertical posture and hold it there without additional muscle tension.

Gas exchange occurs in the bird's lungs, but the movement of air through the hollow bones may be help the bird radiate the excess heat generated by muscles during flight. Like mammals, birds use some hollow bones for storage. Some long bones hold a special tissue called "medullary bone," which contains a supply of calcium important in the production of eggshells. Other bones are filled with fat-rich marrow, as in mammals.

Elements of the bird's skeleton may perform functions or include components that are very rare or unknown among mammals. For instance, the well-developed hyoid apparatus that supports the tongue (Figure 1.5) has an ancient heritage in primitive vertebrates. Its half-dozen bones and cartilaginous elements appear to be derived from the second and third visceral arches, which supported the gills of early fishes. In various birds, the hyoid and the tongue have been modified for purposes that range from collecting nectar to helping break the shell from nuts. Predatory species have adapted it to the catching and holding of insects or other prey. In other parts of the body, simple bony elements without as historic a heritage as the hyoid appear in soft tissues wherever extra support is needed. These usually appear as simple splints of bone that develop in connective tissue to stiffen complex structures such as the neck or backbone. Connective tissue in more fleshy parts of the body may also develop flexible ribbons of bone along the "phares" that mark the boundaries between adjacent muscles. The thin strips of bone help prevent unwanted deformation when one or both muscles contract.

Maintaining the position and shape of muscles is also one role of more familiar bones, such as the furcula (wishbone) or the scapula (shoulder blade). Both bones also act as springs by automatically returning to their original shape as the adjoining muscles relax. A particularly famous piece of X-ray cinematography shows the wishbone of a starling flexing and relaxing as the bird flies [100]. It has been suggested that the springlike action of the furcula assists with flight, but in most birds this bone is far too thin to exert significant force [5]. In many vigorous fliers, the wishbone is too massive and stiff to flex significantly, and several families have members that lack one without their ability to fly being affected.

Because the avian skeleton must perform specialized functions unlike those of most mammals, it may use otherwise familiar bones in unusual ways or develop structural novelties that make their operation somewhat difficult to understand. Some of these special features hold the keys to both the bird's ancestry and its success in the modern world. This chapter is intended as an introduction to the general features of the avian skeleton that have made birds unique animals.

The vertebrate skeleton is a very complex structure. It is easier to understand its parts if they are addressed in natural sequence, from head to tail and from central core to outer extremities. In addition, biologists traditionally divide the vertebrate skeleton into two major parts whose responsibilities date back to the beginning of time. The axial skeleton consists of the skull and spinal column. The skull protects the brain and eyes, whereas the spinal column protects the central nervous system by creating a flexible tube to carry the spinal cord. The remaining bones make up the appendicular skeleton, which is responsible for locomotion. It is considered to be a later evolutionary development than the axial skeleton. As the name suggests, it includes the limb bones and their supporting girdles.

In birds, each of those primary roles has been adapted to meet the special demands of flight. In the axial skeleton, the skull has become a light shell and is no longer a thick helmet offering mechanical protection to the brain. In the spinal column, the neck is long and incredibly flexible. It enables the beak to reach all areas of the body for preening, and it assumes other duties normally reserved for the forelimbs in wingless animals. Central sections of the spine may be fused into rigid blocks that apparently help to cope with mechanical stresses generated by aerobatic flight, and a long section is fused to the hipbones. In the appendicular skeleton, the front limbs are completely modified for flight and the remnants of the fingers can no longer be used to grip food or carry out other functions. The hind limb retains its traditional function but it has been reorganized, with three long bones instead of two; in addition, the toes have become exceptionally long.

Before we can examine the various parts of the skeleton, we must understand something of the materials from which it is made and how those materials are used. The chemical components of bird and mammal bones are very similar, and both groups use the skeleton to maintain body shape and move the limbs. The architecture and mechanics of the two types of skeleton are very different, however, and even the skeleton of a highly specialized flying mammal, such as a bat, is very different from that of a bird. Thanks to medical research, mammalian skeletons are relatively well understood, but very few biologists have focused their attention on bird skeletons or on the properties of avian bone. In his recent book, J.D. Currey discusses some of the most important differences between bird and mammalian bones, but he is mostly interested in mammals and does not spend much time explaining how the avian skeleton works [39].

One of the few scientists to specialize in the study of avian bones was the late Paul Bühler [20]. In the 1980s, he wrote several papers about the mechanical function of the avian skeleton and the characteristic features of bird bone. He was particularly interested in the special adaptations for weight reduction and the ways in which avian skeletal architecture anticipated the development of light-weight hollow-core construction techniques that are now so important in modern buildings and aircraft. Although he is often quoted in ornithological texts, he tended to write for special publications that are difficult to find outside technical libraries.

What Is Bone?

Bone as a Material
Bone is a complex, compound substance found only in vertebrate animals. Mechanically and structurally, it is somewhat analogous to modern construction materials such as ferro-cement or fibreglass; like those products, it achieves great

strength by combining two materials that have very different properties. In the case of bone, however, the materials work together to form a living tissue with features unlike those of any artificial product. At the molecular level, the flexible component of bone is a complex web of collagen fibres. Collagen is a specialized protein whose fibres have a very high tensile strength and are able to resist stretching. The solid filler or matrix that contributes rigidity to bone consists of crystals of hydrated calcium phosphate. It is similar to the mineral hydroxyapatite[2] and, like many crystalline rocks, is very resistant to compression and can support heavy loads. Scattered throughout the matrix are the cells that make bone a dynamic living tissue [39].

If bone were a simple mineral product, it would be relatively easy to determine its mechanical capabilities. It is perhaps the most complex structural material in nature, however. In some ways, it acts like wood and has a distinct grain so that it is typically stronger or more elastic in one direction than another. Its behaviour depends on the ratio of matrix to filler. Bone with a high collagen content tends to be elastic and have a very high tensile strength. Bone with a high mineral content tends to have high compressive strength but is brittle. The long legs of a crane are high in collagen; the dense bones in the flipper of a penguin are high in minerals. Bone is also visco-elastic, meaning that it responds differently depending on how quickly a stress is placed on it. It may break under sudden stress or bend slightly if the same stress is applied slowly. Even when bone is elastic, it may exhibit a property that engineers call "hysteresis," returning to its original shape more slowly than it bent. It may also show "creep" and continue to deform or bend under a fixed stress even though the pressure on the bone decreases as it bends out of the way.

J.D. Currey points out two curious properties of apatite that make it more useful in bone than other candidate minerals, such as calcium carbonate or hydrates of silica [39]. Calcium carbonate is found widely in nature, especially in structures that support marine organisms. Such organisms are often sedentary and are not inconvenienced by being enclosed within a frame of large, brittle crystals. When calcium carbonate appears as a component of bone in terrestrial vertebrates, it comprises only 4%-6% of the bony tissue and the remaining mineral component is apatite. Unlike calcium carbonate, apatite resists the formation of large crystals, consisting instead of small crystals that enhance the flexibility of bone and help it resist breaks. Apatite also has the advantage of resisting dissolution in acids. The internal environment of active vertebrates often becomes extremely acidic and would quickly dissolve structures built from calcium carbonate.

Silica compounds are also widely distributed in nature and are resistant to acids. Like calcium carbonate, they tend to form large crystals or spicules that are most useful to single-celled organisms or primitive animals such as sponges. They have never been successfully adapted for use in the skeleton of a large mobile animal.

---

2 The chemical formula for hydroxyapatite is $Ca_{10}(PO_4)_6(OH)_2$.

A variety of other ingredients make bone far more sophisticated than any artificial construction material. Too often we know bones only as dry museum specimens. In the living animal, they are active tissues in a constant state of physiological flux. They are serviced by blood vessels and a variety of living cells, including osteons or osteocytes that reside within the bone, osteoblasts that build it up, and osteoclasts that break it down. Bones contain a varying amount of water that is an important determinant of their mechanical behaviour, affecting both their weight and their flexibility. Wet bone is more flexible than dry bone. Bone also includes various proteins and polysaccharides whose functions are poorly understood. They may affect the mechanical properties of bony tissue, but J.D. Currey believes that they may be more important in physiological processes such as the initiation and control of mineralization [39].

The significance of water and soft tissue in bird bones is easily seen in the generally poor condition of museum specimens, and may help explain the distorted and sometimes confusing shape of avian fossils. In living birds, the high protein and water content of flexible bones, such as the long sacral ribs, makes them very springy. As they dry, however, the proteins shrink and the bones become bent and twisted into unnatural shapes. They also become quite brittle and are easily snapped if not handled carefully. Sometimes during the preparation process, a larger bone will absorb water from surrounding tissues and twist out of shape if it later dries unevenly. The keel of the sternum seems particularly vulnerable to this form of distortion, and museum specimens often lack the smoothly rounded curve found in living birds. High organic content and a tendency to absorb water after death may be just as important as small size and light construction in accounting for a general scarcity of bird fossils.

Bony tissue performs a variety of functions in the vertebrate body that call for varying degrees of flexibility and strength. It always consists of the same basic compounds, but at the microscopic level, its components can be organized in a variety of ways to meet specific needs. The apatite crystals may be neatly lined up to form "woven" bone that is particularly highly mineralized and very dense. It is typically laid down during development of the embryo but also appears in the calluses around healing breaks. The fine structure of woven bone makes it relatively porous, and there is ample room for living tissue and blood vessels between the crystals. The crystals are even more neatly aligned in "lamellar" bone. As the name implies, it is laid down in thin sheets, like sedimentary rock, usually in alternating layers 0.01 mm and 0.05 mm thick. The layers tend to be better organized in mammals than in birds. In spite of its neat array of crystals, lamellar bone is actually less mineralized than woven bone. A third type, "parallel-fibred" bone, is also highly mineralized and has structural characteristics that lie between those of woven and lamellar bone.

Both mammals and birds tend to have bones that are richly supplied with blood vessels, and evidence from dinosaur fossils shows that their bones were also richly supplied with blood vessels. Other groups of vertebrates are not so well vascularized, especially "cold-blooded" types, whose growth is strongly seasonal. In large animals, including very large types of birds, parts of the original deposits of bone may dissolve and be replaced by specialized arrays of tubular structures called Haversian systems. These structures were once thought to indicate that the animal could control its body temperature and were put forward as proof that dinosaurs were warm-blooded. Haversian systems, however, also appear in crocodiles and other presumably cold-blooded animals such as the giant amphibians, known only from fossils. Modern amphibians are much smaller than their ancestors and their bones are able to develop without internal blood vessels. Their small bones have neither the space nor the need for complex structures.

The osteocytes within the bone are connected to each other by tiny canals. Variation in the organization of these canals has proven remarkably useful in classifying the basic groups of birds [91]. The paleognathous birds (ostriches and their relatives) have been recognized as a structurally discrete group since the beginning of the 19th century. All other kinds of birds are neognathous. Unfortunately, the characteristics of the palate, originally used to support the distinction between these two groups, are not particularly clear, and it has been difficult to decide whether the tinamous belong in one group or the other. Either tinamous are the only paleognaths that can fly or they are a slightly aberrant, chicken-like neognath. The canals offer evidence that may resolve the issue. In the long bones of neognaths, the canals form an almost random network between the osteons. In paleognaths, the canals are more orderly. They tend to run parallel to the long axis of the bone and are connected to each other by a few cross-members. Ostriches have a more regular layout than tinamous, but the pattern in the tinamous is not nearly as random as that in the neognaths. It seems most likely that tinamous are paleognaths and the only living members of that group capable of flight.

Histological examination of bone canals has proven useful in the study of fossil birds and may help in the investigation of animals proposed as remote ancestors of birds. For instance, the canals in the limb bones of theropods are quite similar to those found in paleognath birds, making it one of the few characteristic that justify the use of the prefix "paleo" in the name of those birds. The restriction of the more random arrangement to neognath birds suggests that it is a derived and advanced characteristic.

Peter Houde examined thin sections cut from fossils of *Hesperornis,* the giant, toothed, loonlike diving bird from the Cretaceous, to see whether it should be placed among the paleognaths or neognaths [91]. *Hesperornis* is so different from any modern bird that it has been difficult to classify ever since its description in 1880. At that time, its discoverer, Othniel C. Marsh, thought that its flat breastbone might place it among the paleognaths because ostriches and their relatives were the only types of birds known to lack a keel on their breastbone [120]. There are no other keel-less birds among living types, but several lineages of recently discovered fossil birds have a more or less smooth sternum. In 1973, P.D. Gingerich [72] supported Marsh's opinion when he described the palate of *Hesperornis* as distinctly paleognathous. Joel Cracraft, however, used a cladistic analysis to place *Hesperornis* close to loons and grebes among the neognathous birds [36]. Houde's study revealed that the bone canals of *Hesperornis* are clearly like those of a neognath [91]. He used the same approach to demonstrate that fossils of the Lithornithiformes belonged to an extinct group of paleognath birds, possibly related to tinamous. Like tinamous, the lithornithids could fly and had keeled breastbones [90].

Bone Structure

The external surface of a bone is often smooth and only its general shape offers clues to its role in the body. The type of tissue in tubular bones varies from one area to another. The walls of the sides are usually dense layers of lamellar tissue enclosing a hollow core that usually lacks much in the way of internal structures. The articulating surfaces at the ends are rounded masses of hard woven bone that contain areas of cancellous bone, a spongy bone that looks as though it were made from a foaming plastic. With air making up 85% or more of its volume, birds find cancellous bone particularly useful and use it in many different parts of the skeleton.

When cancellous bone appears within the bones of a joint, its role is to transfer stresses from the articulating surface to a location where they can be more easily absorbed. Beneath the articulating face of the humerus, cancellous bone leads stress away from the joint into walls of the tubular section, just as the arches beneath a bridge lead stress into its foundations (Figure 2.1). In the bizarrely enlarged casque of the hornbill's skull (Figure 2.2), the struts and cells of cancellous bone are oriented in a specific direction. They run from the rigid face of the casque, where blows land during head-butting jousts, to the base of the skull, where the impacts can be absorbed by connective tissue in the neck.

When cancellous bone is sandwiched between two thin plates, it can be used to construct large rigid units while maintaining a very low weight. Owls provide two of the best examples. The skull includes an internal core that cradles the eyes and brain (Figure 1.4). It also separates the ears and places the openings for the auditory canals at different levels. Owls depend on acute hearing when they are hunting in dim light. The widely spaced and asymmetrically placed sound detectors gives them "depth" of hearing just as widely spaced eyes gives us depth of vision. There are very few other examples of asymmetry in bird skeletons.

The owl's hips also play a role in the hunt. They look large and heavy but are greatly inflated masses of cancellous bone that are mostly air. They help to absorb the impact of the owl's attack on its prey, and their size also gives the owl a particularly wide stance. When striking at a small, running mammal in the dark, it helps to be able to spread the legs wide and maximize the area covered by the open talons.

Although most bird bones are filled with air, there are some important exceptions. One of the most unusual is medullary bone, a type unique to birds [43]. It fills the hollow core, or medulla, of certain long bones, hence the name "medullary."

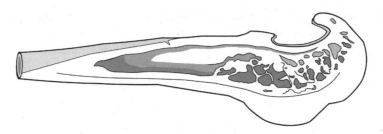

Figure 2.1 Sagittal section through the upper humerus of a Common Murre. The shaft of the bone is a hollow tube. Struts appear near the joint with the shoulder and quickly become a thick layer of cancellous bone.

Figure 2.2 Sagittal section through the skull of a Helmeted Hornbill, showing the specialized cancellous bone that transfers impact on the face of the casque to supporting structures in the neck (based on a photograph in W.P. Pycraft's *History of Birds* [168]).

It is unlikely that this bone has a mechanical function but it is extremely active physiologically. It is laid down as a loose and irregular meshwork of crystals that, under a microscope, looks like the more organized woven bone but it has less fibre and more mineral content. When the time comes for a bird to produce eggs, hormones stimulate osteoclasts to digest the medullary bone and release its calcium for transport to the shell gland. Producing eggshells can be a challenge for birds that live in calcium-poor environments, and medullary bone provides an opportunity for the bird to collect and store a supply before the onset of the breeding season. Needless to say, medullary bone is found only in female birds. Its peculiar structure may aid in identifying the sex of fossil birds or dinosaurs [187].

Significant reduction in the calcium content of human bones is immediate cause for concern. It is an important symptom of osteoporosis and deteriorating health. In birds, a form of reversible osteoporosis may just be part of the normal cycle of life. Biologists have noticed significant loss of calcium from bird bones during periods of feather moult and a gradual reconstitution of the bone during other times of the year. Feathers are mostly keratin and contain few minerals, so the purpose of calcium mobilization from bones during the feather moult is unclear. Perhaps the mobilized calcium acts as a catalyst for part of the synthetic process, or perhaps birds renew their skeletons cyclically, just as they renew their feathers [137].

Tubular Bones

If outright strength were the only consideration in the growth of bones, they would be solid masses of tissue, but all animals must balance the safety of heavier bones against the energetic costs of moving them. Most terrestrial animals require a certain degree of speed and manoeuvrability, resulting in the almost universal use of tubular bones. A tubular bone is not as strong as a solid bone of the same volume and it may break instead of bend, but a tube makes fewer demands on its owner because it is lighter and requires fewer resources for construction. Much of the success of tubular bones has been the result of a characteristic in the response of any object to a bending force. When an external force attempts to bend a bone, or any other object, the greatest tensile and compressive stresses are concentrated in the outer layers. There is actually very little stress on the central core. Consequently, wherever weight is a consideration, it is more efficient to have air or some kind of semi-fluid tissue in the central core of a bone rather than stronger but heavier bone tissue.

Most terrestrial animals have found that tubular bones are strong enough to absorb a lifetime of stress. The big difference between birds and other animals is that the walls of bird bones are spectacularly thin. The walls of the long bones in mammals and dinosaurs are typically about 25% of the bone's outside diameter. In large, lightly built birds, such as swans, the walls of the wing bones are only about 5% of the outside diameter. The walls in the wing bones of the huge pterosaurs, which flew among the dinosaurs, were even thinner, about 4.2% of the outside diameter. Paul Bühler calculated that these were near a theoretical minimum and that walls thinner than 3.5% could not withstand the pull of muscles [20].

In mammals, the bone tissue does all of the work of the skeleton. In birds, however, the bones are a compound structure and other tissues play important roles. Bird bones are bound, inside and out, by a tough layer of sclerotic tissue that is not much thicker than a layer of varnish but is unusual in that it is continuous with the thin outer layer of the skin. It has been suggested that the great tensile strength of this sclerotic tissue could help prevent a bone from becoming deformed under stress. Thin layers of paper on foam board work that way to create a stiff but very lightweight construction material. The sclerotic layer seems too thin to serve such a function, but it seems to be continuous with collagen fibres within the bone. Perhaps it works to prevent fractures by inhibiting the development of surface micro-fractures as the bone bends. The development of such micro-fractures is thought to be a precursor to an actual break. This can be a critical issue in a bird because any broken bone in the pectoral girdle is nearly always fatal in the wild.

Other injuries may be less critical, and birds that have survived broken legs and beaks are occasionally captured in banding operations.

The thin sclerotic layer may also protect the inner spaces of the bird from invasion by fungus. Like the space beneath the feathers, the hollow bones are warm and moist, creating ideal conditions for the growth of fungi or bacteria. Birds under stress are notoriously vulnerable to attack from the black bread mould, *Aspergillum.* The smooth layer of keratin may offer little in the way of nutrients while presenting too smooth a surface for spores to gain a toehold.

The avian humerus illustrates how a bone's thin walls can be combined with other structural elements to create a particularly strong functional unit (Figure 2.1). For much of its length, the humerus is a simple tube without internal struts or supports. In some ways, it is like a drinking straw that is able to maintain its shape under normal use but would suddenly collapse if subjected to too strong a bending force. Over millions of years, however, the forces of natural selection have given it sufficient stiffness for the stresses of a bird's day-to-day life. Only a collision or sudden blow is likely to break it.

As it nears the shoulder, the tube begins to widen and becomes a broad support for a very large articulating surface. This broadened area has a great deal of strength across its wide face but is not very strong in the other direction. The bone uses internal structures to overcome a potential risk of collapse. A few small struts appear where it begins to widen, and rapidly increase in number towards the articulating surface until they grade into cancellous bone. The relatively narrow bands of struts and cancellous bone demonstrate just how effectively such systems transfer the stresses on the articulating surface to the thin tubular walls in the body of the bone.

There has been some argument about whether the thickness of the walls varies according to the weight of the bird. Theoretically, larger birds might be able to function with proportionately thinner bones. In 1985, Currey and Alexander decided against such a relationship because the walls in the femur of a bustard were within the margin of error for a sample of 55 smaller species [40]. The bustard weighed 16.3 kg and the walls of its femur were 6.5% of the outside diameter. In the other 55 species, the walls of the femur varied from 6% to 9% of the outside diameter.

It seems likely that weight is not the only parameter that needs to be considered in this kind of study. Other aspects of a bird's lifestyle might come into play in ways that affect the thickness of the walls in bones. For instance, there are many different flight styles among the various orders of birds that have subjected the humerus to varying degrees of stress. Wing loading would seem to be a particularly important factor. Large-winged birds such as owls, eagles, cranes, storks, and pelicans that have low wing loading and use gliding or soaring flight should have thin-walled bones. In fact, the wall of their humerus varies from 6% to 9% of the outside diameter. Loons have very high wing loading and use very vigorous flapping flight. The wall of a loon humerus is about 21.3% of the outside diameter. Penguins subject their wings to unusual levels of stress by "flying" through a very dense medium. The wall of a penguin humerus is about 23% of the outside diameter, comparable to the walls in a leg bone of a mammal or dinosaur.

Why is the wall of a humerus in a flightless penguin only a little thicker than that in a flying loon? Paul Bühler [20] suggested that thick walls in the bones of loons facilitated diving by increasing overall body density; however, thicker walls create only a little extra weight and, in birds as large as a loon or penguin, the increase in density might not be significant. On the other hand, we would expect a bird that uses its wings for underwater propulsion to have a thicker-walled humerus because it requires a much stronger wing skeleton to withstand stress underwater. Nature offers us a natural experiment in the tube-nosed seabirds. Species of shearwaters, which use their wings for swimming underwater, have a humerus with thick walls (16%-30% of outside diameter). Similarly sized petrels, which use their wings only in the air, have a thin-walled humerus (about 12% of outside diameter).

Tendon

The major role of any skeleton is facilitation of locomotion. The bones cannot move themselves or maintain their owner's posture without muscles, and muscles are attached to bones by tendons. Tendon is famous as one of the toughest materials in nature. It transmits the force of the muscles as they contract. There seems to be no limit to the flexibility of tendon tissue, and it is also amazingly strong. A bone is more likely to break before a tendon snaps. Until recently, it was thought that a tendon could slide almost without friction through its sheath, but typists, pianists, and computer operators have made us all familiar with repetitive strain injury or tendonitis. Biomechanicists now realize that tendon deforms within limits and that it has the unusual characteristic of being able to stretch slowly but return to its original shape. In technical terms, it is "inhomogeneous, anisotropic, and hysteric."

Anatomists are able to tell much about the movement of bones in the skeleton from the attachment scars left by tendons, but, except for the plantar tendons of the foot, tendons have received short shrift. Traditionally, they have been seen as little more than the strings and wires that tie more important bits of the skeleton together. For centuries, museum curators carefully prepared bones by scraping tendons from specimens. Early paleontologists often polished fossil bones to make them more attractive for display, but further obscured points of attachment in the process.

Such cleaning rarely causes problems in the study of mammals, where tendons are usually short and simple; in birds, however, the route between a muscle and its eventual insertion on a bone can be long and complex and may not be obvious in a disarticulated skeleton. Most mammalian tendons pass across just one joint, but the peculiar centralization of muscles in birds has led to the development of several tendons that cross two joints to act on distant bones. Muscles in the upper leg flex the toes after passing over the ankle and digit joints. Mammalian tendons are usually short and connect to bones at right angles to the joint, but in birds the long tendons may be inserted at odd angles so that they can rotate or slide bones in the joint. There are several examples of complex movement in the hip and wing joints. In the hip, rotation of the femur is associated with the bird's posture that suspends its body weight from a flexed knee. In the wing, the hand bones rotate to perform the complex movements of flight control.

Sesamoid Bones

When the early bone collectors discarded everything except the major bones, they also lost or discarded tiny bones embedded in the connective tissue. Such bones are less frequently encountered in mammals and their significance in birds went unrecognized for many years. Ornithologists eventually realized that these nodules could be critical to the function of birds in spite of their small size. The small bones that form in tendons are usually called "sesamoids" because they were first described as small ossicles shaped like a sesame seed. Their description as an ossicle implies that they are somehow temporary or unimportant, but their occurrence in bird skeletons is regular and predictable. Although they are a special structure that is technically independent of the main skeleton, sesamoids begin their development much like regular bony tissue. Specialized bone-generating cells gather at the appropriate point and are nurtured by a locally enhanced blood supply. After initial mineralization, there is a period of remodelling so that the new bone can meet any special role it might play in the movement of neighbouring bones. Because all birds reach adult size soon after hatching, the sesamoid bones appear in both juveniles and subadults and cannot be used as indicators of maturity or old age.

Sesamoids that form in joints often act as skids that slide between the tendon and the bone of the joint. Technically, the kneecap (patella) is a sesamoid bone that serves such a purpose. Some birds have a comparable bone in the elbow, another in the shoulder, and the "os pisiforme" in the wrist. The latter appears to facilitate wing function by lengthening the route followed by a particular tendon and changing the angle with which it attaches to the metacarpus.

The lower leg also has sesamoids. In birds, the elongated tarsometatarsus has evolved to function as part of the leg and has left all the important responsibilities of the foot to the toes. As a result, the joint between the two has become very important. In many birds, its flexibility and adaptability have been increased by the addition of small wedge-shaped sesamoids to this new "ankle." Each is known as an "os cuneatum."

Sesamoids also appear in tendons some distance away from joints. Albatrosses have an interesting "sesamoid" in the wing [17]. It is a rather long bone that runs across the tendons within the sheet of skin (propatagium) that extends from the shoulder to the wrist. The exceptionally long wing of the albatross is designed for aerial efficiency. If the large propatagium flapped or luffed like a sail as the wing changed shape, it would cause localized turbulence and greatly increase the energetic costs of flight. The small spar of bone helps keep the surface of the patagium stiff and smooth so that flight remains efficient, just as similar structures help to smooth a boat's sail.

This development of the albatross's patagial sesamoid must be a recent evolutionary event because the bone takes different forms in other petrels. In the albatross, a long ligament attaches the simple ossicle to a point on the humerus, near the elbow. In a shearwater, a similar ossicle sits directly on a process of the humerus that arises at the same point. In true petrels, the ossicle has a wide, flat cartilage at its outer end. The whole structure is absent from fulmars and from members of the genus *Daption* even though they are similar in size and habits to other petrels. It is also absent from the tiny storm-petrels and the short-winged diving-petrels.

Another type of bony growth develops in the ligaments that lie along the backbone. These bones take the form of very long, flexible splints that lie parallel to the tendons. They help restrain the flexibility of the back without making it completely rigid. Similar supporting structures are widespread in the long tails and long necks of dinosaurs.

Later in life, many birds develop bony additions to their skeletons that increase their rigidity. Usually this new bony tissue does not have a predictable shape and may occur in unexpected places. It often appears as sheets within layers of connective tissue and may completely replace that tissue. When these growths extend from the edges of bones, they may obscure the shape of the original skeletal components or fuse joints that are free in younger specimens. For instance, in the adult Hoatzin, irregular bony growths develop along the edge of the ribs, changing them into wide plates, while others form a roof over the groove that runs along the inner side of the coracoid bone. In other large birds, such as cranes and pelicans, bony material may grow over the joints between thoracic vertebrae near the hips, effectively moving the joint between the fused mass of the synsacrum and the thorax towards the shoulders.

Flat Bones

The limb bones are long tubes, but they are typically anchored to larger flat bars and plates. Flat bars bend readily in one direction but are very stiff across their wide face. Such a design is very useful where the skeleton needs more strength in one direction than another. Ribs are perhaps the most familiar example. The skeleton may also make use of the flexibility of small plates of bone, but most plates are large and sculpted in ways that prevent folding. Often the broad surface of a plate will be interrupted by a flat bar set vertically into its face. In cross-section, these plates form a T, V, Y, or even an X. Bowl shapes are also inherently rigid and useful to protect important soft tissues. The skull, the sternum, and the hips are complex collections of fused plates.

In birds, some of the flat bones form broad sheets that are rigid but extremely thin for their width. They maintain their shape only because they are amazing examples of hollow-core construction on a microscopic scale. In a famous example described by Paul Bühler, the walls of the braincase of the European Nightjar were only about 1.0 mm thick. Most of this scant thickness was air bounded by an outer skin only 0.02 mm thick and an inner layer that varied in thickness from

0.025 mm to 0.050 mm. Even the articulating faces between the skull and the bones of the jaws (e.g., the basipterygoid processes) were only 0.075 mm thick. The two layers of the braincase were held in place by a few struts and patches of airy, cancellous bone. This type of hollow-core construction is exceptionally light but rigid in spite of its thinness. Instead of bending, it is likely to suffer catastrophic collapse if stressed. The degree of stiffness seems to be sufficient for the needs of the skull, however, and Bühler [20] found examples in which several layers of cancellous bone were arranged in well-organized, multi-storey galleries. The skull of an owl had four such layers between its inner and outer surfaces.

The Bony Parts of the Avian Skeleton

In the early embryo of birds, all the bones in the basic skeleton develop from discrete sources, or primordia, that can be traced back to homologous structures seen in the early chordates. The skeleton can be divided into two basic parts: the axial skeleton, which includes the skull, the vertebrae, and the ribs, and the appendicular skeleton, which consists of the limbs and the girdles of bones attached to them (Figure 1.1).

The Axial Skeleton

The Skeleton of the Head

The axial skeleton is supposed to be the oldest and most primitive part of the skeleton, with structures inherited from the earliest chordates. It begins with a brain capsule, jaws, and arches that support the gills. In the head of a human, the primitive structures appear as the large bony mass of the skull, a lower jaw, and three pairs of tiny ear bones. These components represent the fusion of many bones that were independent structures in the earliest vertebrates. You might expect the head of an adult bird to be just as simple, if only to maintain lightness. Exactly the opposite is true. In addition to the skull capsule that holds the brain, the skeleton of a bird's head may include as many as eight independent bones associated with the jaws, a pair of ear bones connected to the ear drum, and a collection of thin bones that form both the palate and the turbinate bones in the nostrils. Usually the turbinates, the paper-thin bones of the palate, and some other bones of the upper jaw are seamlessly fused at the edges so that any individuality is completely obscured. Birds also have an independent collection of hyoid bones that support the tongue and extend beneath the skull.

The structural differences between mammalian and avian jaws are responsible for fundamental differences in the way these two groups begin to process their food. Mammals have robust jaws, usually armed with teeth, and they begin the mechanical breakdown of food by chewing it before swallowing. They also begin chemical digestion by flooding the mouth with enzyme-rich saliva.

The jaws of birds tend to be delicate and flexible, and no living species has teeth. Birds feed on items that are either swallowed whole or broken into manageable chunks. Their saliva works as a lubricating fluid; both chemical and mechanical digestion are delayed until the food item reaches the muscular parts of the gut. In the absence of any hard grinding surfaces, many species swallow grit or small stones to aid the mechanical reduction. The delicateness of its individual components does not make the avian jaw weak or ineffective. A bird may not be able to chew, but its jaw can perform a variety of functions associated with both the capture of food or care of the body. In some cases, the jaws can be very strong. Geese, parrots, penguins, and many other large birds are capable of delivering a powerful bite. It is not so much a matter brute force exerted by large muscles as the sophisticated interaction of a number of components.

In a group as large and diverse as birds, it should not be too surprising that there are groups that have found ways to chew their food in spite of the apparent shortcomings of the typical avian mouthparts. Three groups – cuckoos and turacos from Africa and the Hoatzin from South America – have unique structures in their mouths and specialized tongues that enable complex processing of food in the mouth [109]. Although these birds lack teeth and do not have digestive enzymes in their saliva, the process is not much different from that in mammals, and the end product is a material that can be swallowed comfortably. Interestingly, these three groups have long caused trouble for taxonomists and have occasionally been linked to each other phylogenetically. Their specialized jaws and extraordinary food-handling techniques tend to encourage such a point of view.

The cuckoos are the least specialized of the three. They prey on large, slow-moving insects of the kind that depends for protection on bristles and spines more frequently than on cover and secrecy. When faced with a particularly well-defended prey, the cuckoo bites it vigorously and wipes it back and forth on some hard object. Once the remains can be moved safely to the back of the mouth, the cuckoo mashes it between special bony plates. These plates are unique to cuckoos and consist of wide spots on the lower jaw that match a similar structure on the jugal bar. Other parts of the palate have been strengthened to support these paired plates. After mechanical reduction, there is little risk to the cuckoo from the prey's protective spines and the item can be swallowed confidently.

The ability to process prey that is too well armoured for other small birds may explain the success of the cuckoo family in tropical and warm temperate regions. It is much more poorly represented in cooler regions with intense seasonality. In those regions, large terrestrial insects tend to be available for only brief periods. The cooler regions are populated by relatively small, secretive arthropods that hide in tunnels or under bark, where they are beyond the reach of a bird that takes only visible items. Cuckoos in the temperate zone use nest parasitism to exploit the foraging skills of other kinds of birds by conscripting them as unwitting parents. In the future, cuckoos may benefit from global warming if temperate regions warm up and begin to include more habitats for very large insects.

The turaco specializes in eating juicy fruit. It chooses types that fit into its mouths in one piece; it must then be able to slice through the skin to remove any pit too large to swallow. The beak of the turaco is large and its mouth is a high, domed cavity suitable for holding a large fruit. The beak has scalloped edges that work like scissors, while a crenellated ridge in the palate slices into the fruit as the upper jaw works back and forth. The action is quite vigorous, and special structures are needed to prevent the jaws from moving laterally. If the pit is too large to swallow, it is squeezed out of the mouth, near the base of the beak. Once the fruit is broken up, the tongue kneads it into a dough for easy swallowing.

The Hoatzin also produces a vegetable pulp in its mouth, but it eats foliage that is much more difficult to digest than fruit. It extracts nutrients by fermenting its food in a large chamber reminiscent of the rumen in a cow. Unlike the inflated beak of the turaco, the beak of the Hoatzin is small and pointed. Its shape is ideal for tearing off pieces of leaf,[3] and the edges enable the Hoatzin to process food in its mouth. The edge is doubled so that the outer can function as a kind of lip that keeps food in the mouth while the inner is scalloped to act as a cutting and grinding surface. As in the turaco, the tongue thrusts back and forth, doing most of the work. Even the toughest leaves are quickly reduced to a pulp that can be swallowed.

In most birds, the individual components of the jaw are difficult to distinguish because they are seamlessly fused to their neighbours. Extensive fusion does not increase rigidity because the bone in the joints is extremely thin and may be almost as flexible as a ligament. The apparent purpose of such extensive fusion is to reduce the overall weight of the head by getting rid of the relatively heavy connective tissue in ligaments.

Parrots are one of the few groups of birds in which none of the basic joints are fused; each joint is clearly marked by a pad of connective tissue (Figure 2.3). This retention of primitive joints may be related to the varied demands that parrots place on their exceptionally large jaws. Their jaws peel fruit, act as a third limb for climbing among branches, preen the feathers, participate in close communication with the mate, and serve as a handy tool for threatening and defending.

The system for moving the lower jaw in birds looks rather simple. A pair of joined bones is moved up and down by a set of muscles, just as in a mammal. Unlike in mammals, however, the capacity for sideways movement is greatly limited. In comparison, the arrangement for moving the bird's upper jaw looks amazingly complex, partially because we are more familiar with the mammal's upper jaw, which does not move at all. Even though the bird's upper jaw includes a surprising number of bones, its operation is straightforward (Figures 2.3). The force needed to move it is generated by muscles that stretch either from the sides of the braincase or from behind the eyes to the pair of quadrate bones on either side of the skull, near its base. When these muscles contract, they pull the quadrate forward

---

[3] The shape is much like that of a chicken's beak and contributed to the idea that Hoatzins might be related to the galliform birds. The musculature and tendons of the head are completely different, however, and galliforms lack the specialized internal structures found in the beak of the Hoatzin.

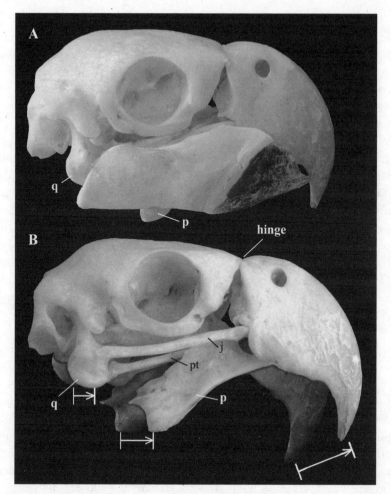

Figure 2.3 Parrots need to be able to move the upper jaw to handle difficult fruit. (A) The skull with the lower jaw in place. (B) In the nearer of these superimposed images, the muscles controlling the quadrate (q) have contracted, pulling it forward, raising the tip of the beak, and rotating the palatine bone (p). (C) The underside of the skull at rest. (D) The underside of the skull, in which the quadrate (q) has pushed on the jugal bone (j) and the pterygoid bone (pt) to exert pressure on the lower edge of the beak. The main role of the pterygoid may be to stabilize the movement of the beak along the centre line.

to raise the upper jaw; when the posterior muscles contract, they pull the quadrates back to lower it. When the quadrates move forward, they press on two sets of pushrods, the pterygoids and the jugal bars, that extend forward to the base of the beak. The pressure from the quadrates is transferred through the pterygoids to the lower edge of the palate, near its midline, while the jugal bars push on the outer edges of the base of the beak, just in front of the eyes. The only way for the beak to ease that pressure on its lower edges is to rotate at the hinge, or bending

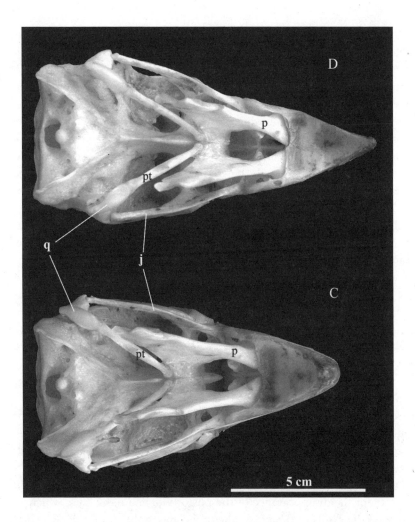

plate, on its upper edge, just ahead of the eyes. The quadrates can rotate only a short distance because their muscles are short. Even with the help of the pushrods, the movement of upper jaw increases the overall gape of the mouth by only an extra 5%-10%. This small advantage must be significant to birds, however, because they all retain this feature, regardless of the kind of food they eat.

Although fusion makes it difficult to identify individual bones or interpret their function in the jaws of birds, the structure of the avian jaw has been very important in taxonomy. In 1813, Blasius Merrem proposed a fundamental division of Class Aves according to the structures of the upper jaw [135]. The ratites and tinamous appeared to have a reptilian type of palate and were placed together in the paleognaths (old jaws), where they have been, off and on, ever since. All other birds were supposed to have a newer type of palate and were placed in the neognaths (new jaws).

At various times, there has been some discussion about the relationship of ratites and tinamous because the characters in the palate are not all that clear. In the paleognath palate, the pterygoid bones are supposed to meet the palatine bones at the midline in the roof of the mouth. In the neognath palate, the pterygoids do not meet. The character has not been a great help in the classification of the tinamous because they have a somewhat intermediate arrangement. Over the years, tinamous have moved from one group to the other and back again, until biomolecular analyses and the histology of their bone canals [91] placed them firmly among the paleognaths. Although the characteristics of the jaws are no longer the only method of distinguishing between the two basic lineages of birds, the names Paleognathae and Neognathae have precedence and remain in use.

Until very recently, it was believed that the relationships between the pterygoid bone and the palate affected the nature of the movement that occurred between the braincase and the upper jaw. Ratites (ostriches and the other flightless paleognaths with a keel-less breastbone) have a classic paleognath palate and exhibit a kind of "central kinesis," in which the upper jaw rotates at a narrow bending zone on the upper part of the beak. Tinamous have a broader bending zone that is only generally similar to the condition in the ratites. Neognaths exhibit "prokinesis," in which the hinge or bending zone is located further forward. In sandpipers, the hinge is so far forward that they can open the tip of the beak to pick up a small food item without moving the rest of the jaw. This extreme version of prokinesis is called "distal rhynchokinesis."

These structural differences may be more related to function than position in the family tree. Sander Gussekloo and his colleagues have studied the various types of jaw movement and concluded that variations in the pterygoid/palate relationship had more to do with reinforcing the skull than movement of the upper jaw [78]. The linkage of the various bones appears to create internal supports that compensate for the disappearance of the postorbital and nasal bars, which strengthened the skulls of the bird's ancestors among the dinosaurs. Recently, Gussekloo and his colleagues applied an X-ray technique called Roentgen stereo-photogrammetry to take extremely accurate three-dimensional photographs of the quadrate bone and the jugal bar as they moved the upper jaw [77]. Detailed measurements revealed that jaw movement was basically the same in both neognaths and paleognaths. Variations between the two groups may be related less to family history than to differences in the strength that the bird needs for other functions, such as handling prey.

The Neck

The avian neck is much more than a simple link between the head and the body. It is a complex structure that must perform many of the duties that other animals assign to the forelimbs. Therefore, it needs to be both flexible and strong at the same time. Mammals with long necks depend on massive shoulders to support a small number of very long vertebrae, stacked vertically on one another. The result

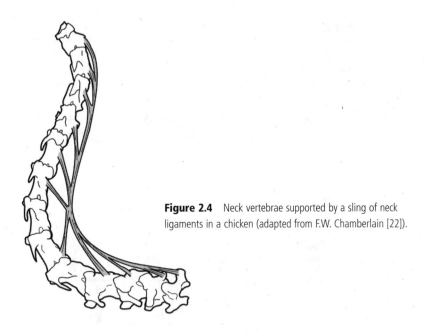

Figure 2.4 Neck vertebrae supported by a sling of neck ligaments in a chicken (adapted from F.W. Chamberlain [22]).

is a relatively stiff and inflexible structure that would be quite unsuitable in a bird. Among other shortcomings, it would not allow the bird's beak to reach all parts of its body for preening. The long neck of a bird consists of large numbers of small vertebrae[4] connected by a complex series of some 200 muscles and tendons, many of which pass across two or more vertebrae [212]. Of those muscles, only one, the *Musculus biventer cervicis*, extends all the way from the head to the body.

The neck vertebrae are not simply stacked one on top of the other to resist the force of gravity but hang from the shoulders in a gentle loop, supported by an elaborate sling of ligaments (Figure 2.4). In most birds, the neck appears short because it is completely hidden by the plumage. Many species rest with the sling fully contracted and the head tucked snugly against the body. Long-necked birds often fold the neck, so that the head can lie on the back, between the wings.

Angélique van der Leeuw has worked out three of the basic ways that the neck can move in birds (Figure 2.5) [212]. The simplest occurs in the chicken, but two more complex types of movement occur in very long-necked birds. One is found in the swan and the other in the rhea (a South American relative of the ostrich). The chicken does not have particularly long legs and often wants to reach out to an object on the ground. To move the head forward and down, each vertebra rotates

---

4 Most birds have more than 11 vertebrae, but swans have 25 and parakeets have only 9. The earliest known bird, *Archaeopteryx*, also had only 9 neck vertebrae, whereas the ball-shouldered birds (Enantiornithes) of the Cretaceous had 12 or 13.

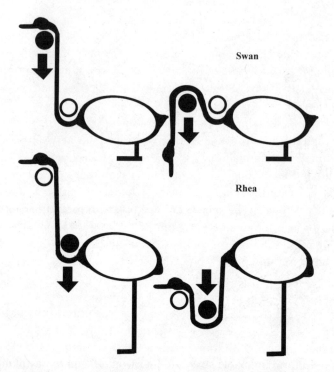

Figure 2.5   Two different methods of lowering the head in extremely long-necked birds. In the swan (above), neck bending starts near the head, as though the neck were a rope passing through a pulley (dark circle). In the rhea (below), bending begins near the body and the head moves up and down like the periscope of a submarine. Shorter-necked birds use a combination of the two methods to extend the neck forward [212].

a little as the neck straightens. No one section of the neck bends much more than another.

The long-necked birds move the head much further, and the folding and unfolding of the neck involves a more elaborate action. The swan has relatively short legs but often wants to reach straight down while it is swimming, to reach succulent roots directly beneath its body. Normally the head rests on top of an almost vertical neck. To reach food underwater, a wave of rotation begins at the top of the neck and passes back towards the body. As the wave passes, each vertebra is pulled through a rotation by its forward neighbour, until the whole structure points straight downward. The head seems to dive, slowly down the line of the neck, into the water. To raise the head, a wave travels from the body end of the neck forward; the tip of the beak is the often last part of the head to leave the water.

Like the swan, the rhea holds its neck vertically so that its huge eyes can detect approaching predators at a distance. Often the head is all that can be seen above the tall grass of its native habitat. The rhea also has very long legs and must collect food far below its body. To lower its head, a wave begins at the body and travels along the neck towards the head. The head drops sedately out of site, like the disappearing periscope of a submarine. It is just as likely to rise vertically into view a few seconds later. Cranes and some other long-necked birds use the same technique. Others combine it with the simple extension seen in the chicken, and there are many intermediate versions of these movements among the various groups of birds.

At its junction with the head, the thinness of the neck contrasts sharply with the width of the expanded braincase, but the attachment of the head to the neck is no more fragile than it is in most other vertebrates. The joint between head and neck is less complicated than in mammals. Except for a very special case discussed below, the avian skull has only a single point of contact with the neck, the occipital condyle. It articulates with the first of the neck vertebrae, the atlas. The real job of the occipital condyle is to keep the spinal cord lined up with the brain stem as it passes into the skull through the foramen magnum. The actual attachment of the skull to the neck is the responsibility of a complex set of muscles and connective tissue.

The avian atlas is a thin plate of bone that seems too small to restrain rotation of the skull. Connective tissue binds it tightly to the second neck vertebra (the axis), however, and it is locked in place by a unique bony process. The base of the atlas has a deep groove to receive an extension from the base of the axis called the "dens" process (Figure 2.6). The peglike dens fits snugly into the groove, locking the atlas in place and preventing any rotation between the two bones. The dens extends all the way through the atlas to meet the base of the skull, making birds the only animals in which two different vertebrae participate in the articulation of the head with the backbone.

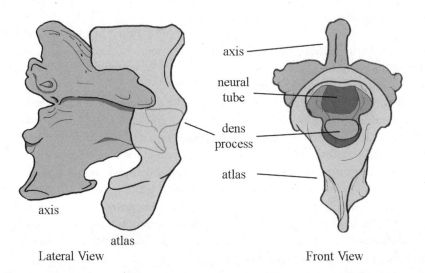

axis

neural
tube

dens
process

atlas

axis

atlas

Lateral View

Front View

Figure 2.6 The first two cervical vertebrae in a Brandt's Cormorant, showing how the dens process of the axis passes through the atlas.

In spite of the seemingly fragile link between the skull and the neck, there are birds that carry elaborate sexual advertisements on their heads. Usually these displays consist only of feathers, but hornbills grow an elaborate bony casque from the top of their bill. In male Helmeted Hornbills, the front of the casque includes a thick plate made of a material that looks like antler or ivory (Figure 2.2). It is an unusually heavy structure to find on the head of any bird, and it brings the weight of the hornbill's head close to 10% of its total body weight. The male hornbill is reported to use this plate in violent, aerial casque-butting jousts that often leave the front scarred and pitted. We have already mentioned how the elaborate array of bony struts within the head transfers the impact of these collisions to the back of the skull, where they can be absorbed by the neck. The internal design of the skull is not the whole story, however. The connection between the skull and the neck is also unusually strong, even in species of hornbills with small casques.

Violent movement of a heavy head during a head-butting contest inevitably generates a great deal of angular momentum and puts great stress on the neck joint. These birds have reduced the risk of injury by fusing the atlas to the axis and by creating a second point of articulation between the skull and the spinal column (Figure 1.7). This new articulating condyle sits above the foramen magnum, where it can prevent additional vertical movement and may help to restrain side-to-side movement. Hornbills are the only birds with heavy head casques, and the only birds with this special kind of neck articulation.

Although the bird's neck must be very flexible to perform its various functions, it must not be so flexible as to permit damage to the spinal cord. The conflicting demands of flexibility and protection in the spinal column seem to be a recurring challenge for vertebrates, which must solve problems caused by dependence on a string of individual bones. The solutions preferred by birds include dorsal and lateral extensions on individual vertebrae that limit bending at right angles to the spinal cord. Where additional rigidity has been needed, birds have added supports that lie parallel to the spinal column or fused series of vertebrae into solid masses that do away with flexibility completely.

None of these solutions has proven suitable in the neck, although many birds limit vertical bending with tall dorsal spines on some neck vertebrae. The neck vertebrae have a number of ridges and processes that make them much more elaborate and varied than vertebrae from other parts of the spinal column. Typically, the features on a neck vertebra are relatively small and compact, giving the impression of a highly sculpted bone carved from a small cylinder. As in dinosaurs, the structural complexity of some neck vertebrae is increased by a pair of ribs. These cervical ribs are usually on vertebrae near the junction between the neck and the thorax. They lack the two terminal articulations found on thoracic ribs and consist of a single piece of bone.

Perhaps the most important features of neck vertebrae are the two pairs of processes, one in front and one at the back, that articulate with adjacent vertebrae.

Together with the interlocking saddle shape of the main articulating face (hetero-coelous centrum), they prevent rotation on the articulating surfaces. Perhaps twist-ing is a more serious threat to the integrity of the spinal cord than lateral or vertical bending. Although the two small articulations above the spinal cord and the larger one below appear to form a rigid triangle, they permit a little vertical and lateral bending. A slight stretching of the ligaments at each of the articulating faces allows the neck to bend in a long, gentle curve.

Since the origin of the Neornithes, most of the neck vertebrae in birds have had a heterocoelic (saddle-shaped) surface on their centrum. If the surface were flat, one edge would pull away from the joint as the neck bent. The saddle allows adja-cent vertebrae to slide against each other while maintaining a constant area of con-tact within a limited range of movement. A ball-and-socket joint would also offer greater mobility as well as a constant area of contact in the joint, but directionality would be more difficult to control. On the forward face of the vertebra, the "saddle" of the centrum lies on its side; on the rear face, it stands on end so that opposing faces lock into each other at right angles. The alternating arrangement helps con-strain movement to the vertical and horizontal axes. When birds twist their necks, the resulting bend is the sum of many small rotations between vertebrae.

Although the neck functions as an extendable arm in birds, very few species depend on it for more than a moderate amount of strength. Boobies, cormorants, and pelicans are an exception. These birds drive their heads underwater, sometimes while plunging out of the air, and attempt to scoop up fish in an extendable throat pouch. Pelicans carry the technique to an extreme, pushing a huge pouch through the water and lifting it into the air full of water and struggling fish. The neck in these birds is designed like a derrick. The front half is the arm of the derrick, and its vertebrae look relatively normal although they are large and highly pneumatized. The rear half is a massive supporting column. Its vertebrae are exceptionally large and have adopted a very unusual shape. Their forward end has become an enlarged dome, modified for the attachment of numerous ligaments and tendons attached to muscles in the shoulders. When these neck bones are assembled in sequence, they resemble a stack of inverted cups in which each supports the next in line.

The Thorax

The thoracic skeleton surrounds and protects all of the bird's vital organs except the brain. It also carries the muscles that power flight and supports the pectoral girdle and the wings. This is a very different function from the neck but much of the work is accomplished by a series of vertebrae that look very similar to those in the neck. To produce a stiffer segment of backbone in the thorax, the processes that extend from the main bones tend to be more robust and the centra are wider and flatter. Nonetheless, the neck vertebrae blend almost seamlessly into the tho-racic series. Each individual vertebra is unique, but it is also very similar to its im-mediate neighbours and the peculiarities of one can be followed through the string

of vertebrae as a series of gradual changes. Typical thoracic vertebrae are independent bones, tightly bound to their neighbours by connective tissue and supported
by thin bony splints that lie parallel to the spine. Often some are fused to their
neighbours, but fusion is not always consistently distributed within a family.

Living birds have about a dozen thoracic vertebrae. Opinions about the actual
number vary from writer to writer. Hans Gadow one of the foremost authorities
on avian anatomy in the late 19th and early 20th centuries, lumped the thoracic
and cervical vertebrae together in his counts (Table 2.1) [63]. Estimates of the
number of vertebrae between the head and the hips depend on criteria for distinguishing the thorax from other parts of the body. Most modern writers use the
first vertebra with a two-headed rib to mark the end of the neck and the beginning
of the thorax. The two-headed thoracic ribs are distinguishable from any ribs on
neck vertebrae by having distinct dorsal and ventral sections that meet at a flexible
connection near the rib's midpoint. The outer end of the ventral portion of a thoracic rib is usually attached to the edge of the sternum.

The thoracic region of the bird's skeleton includes the sternum (breastbone) as
well as a variable number of ribs and vertebrae. At their distal end, the first five or

Table 2.1

Hans Gadow's allocation of vertebrae to zones in the avian spinal column

| | | Allocation within the spinal column | | | |
| | | | | Caudal | |
Species	Number of vertebrae	Cervical and thoracic	Sacral	Fused to the synsacrum	Free	Fused in the pygostyle
Archaeopteryx	50-51	20-21	2	8	20-21	0
Hesperornis	47-48	26	2	7	10-12	0[a]
Ostrich	56	34	3	7	12	0[a]
Rhea	47	33	2	6	6-7	0[a]
Cassowary	58	37	1	8	12	0
Kiwi	50	34	2	4	5	5?
Tinamou	48	34	1	6	7-9	0
Chicken	46-47	29-30	2	4	5	6
Turkey	46-47	28-29	2	4	4	8
Swan	64	41	1	9	6	7
Raven	44-45	26	2	>3	>6	6-7
Dove	44-45	26	1	5	6	6-7
Grebe	57	30	2	9	5	11
Cormorant	52	30	2	7	5	6

Note: Flightless species are shown in boldface type.
[a] Gadow did not consider these birds to have a true pygostyle.
Source: [63]

six thoracic ribs are attached to the sternum and help suspend it in place. In most birds, any sense of zonation disappears as you approach the synsacrum, because some of the thoracic vertebrae have been absorbed by the hips and fused to the synsacrum. The series of vertebrae immediately in front of the hips may lack ribs, but other species have ribs that stick out from beneath the hipbones. Unlike anterior ribs, these do not link back to the sternum. Close examination of their point of attachment usually reveals a suggestion of the double articulation that is typical of a true thoracic rib.

Whatever the real total, the number of thoracic vertebrae in modern birds is much smaller than in *Archaeopteryx* or any of the dinosaurs. The evolutionary trend in birds seems to have been towards a shorter length of back with greater stiffness. Historically, birds appear to have begun shortening their back about the time that their caudal vertebrae were fused into a pygostyle. It seems likely that the changes in both areas are related to the development of more powerful and more controlled forms of flight.

Features of the thoracic skeleton are frequently misrepresented in textbooks because the domestic chicken has been widely used as a teaching aid for over a century. Chickens are important birds in our society and make a convenient example. Portions of the thorax in galliform birds are often fused, possibly to absorb the shock of the explosive escape flight that is typical of the group. Fusion also appears among raptorial falconiform birds that have exceptionally vigorous flight styles requiring both speed and power. The amount of fusion is not consistent across the whole group, and may be related to the impact with which a raptor strikes its prey. Falcons strike very hard and have more extensive spinal fusion than hawks or eagles, which typically overpower prey on the ground. Many families of birds do not have fused segments in their thorax. For instance, owls also attack prey on the ground, but all their thoracic vertebrae are independent.

Sometimes the exact purpose of spinal fusion is not clear. Fusion of thoracic vertebrae in large grebes may be related to the stresses generated in underwater swimming, but it is uncommon in smaller grebes and generally absent from other kinds of diving birds such as loons, auks, and cormorants. Perhaps it is related to the deep wing strokes that large grebes need during take-off, but comparable strokes appear in birds without fused backbones.

Many birds have a peculiar structure on some of their thoracic vertebrae called a hypapophysis[5] (Figure 2.7). These are processes of bone that extend into the abdominal cavity and may create an extensive internal skeleton. In 1933, Hans Gadow commented that hypapophyses "serve for the thoracic origin of the *musculus longus colli anticus*" and therefore assisted forward movement of the head, especially in species that darted at prey [63]. Like his comments about spinal fusion in birds

---

[5] This Greek name is the only word for this structure. If I offered some descriptive term in English, the reader would not be able to find references to it in other books.

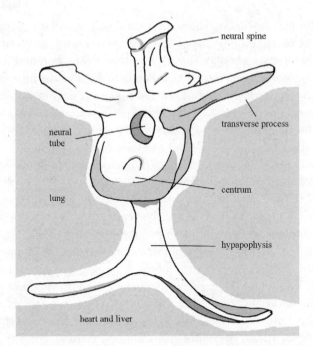

Figure 2.7   Thoracic vertebra from a Common Loon showing the unusually large and flared hypapophyses below the centrum that provide internal support for the lungs, heart, and other vital organs.

in the same book, this statement has been widely repeated without careful consideration. The muscle is attached as he described, but such structures also appear in birds that do not dart their head at their prey. It seems likely that such large and elaborate structures have other important functions.

Hypapophyses occur regularly among types of birds whose internal organs are subjected to a great deal of stress from vigorous flight, sudden stops, or rapid changes in pressure. For instance, auks that dive underwater and jaegers that assault other birds have them, but other charadriiform birds do not. They appear in most wing-propelled and foot-propelled diving birds, including loons, diving-petrels, and penguins, but they are absent from ducks. Owls, swifts, and humming-birds have small three-headed examples. Falcons have a simple set, as part of a string of fused thoracic vertebrae, but eagles do not. Their presence in kingfishers may be related to the shock of diving into water. There are unusually long hypapophyses in relatively inactive colies or mousebirds, but these birds roost communally and often hang upside down. Perhaps the long hypapophyses help reduce the risk of crushing or suffocation when an unusual posture redistributes the weight of the internal organs.

Hypapophyses appear to be either absent from or extremely small in paleognath and galliform birds as well as other kinds of birds that spend most of their time walking. They have not been reported in *Archaeopteryx* or any of the dinosaurs proposed as ancestors of the birds, but they appear in some fossils of modern birds.

O.C. Marsh found a large hypapophysis on vertebra number 18 of the loonlike *Hesperornis,* and another on one of the thoracic vertebrae of the more ternlike *Ichthyornis* [120]. If the development of these structures is connected to vigorous flight or wing-propelled underwater locomotion, their absence from the thoracic vertebrae of the entirely terrestrial dinosaurs is not too surprising.

Thoracic hypapophyses safely separate the pair of immobile lungs in a bird. In a mammal, with dynamic lungs, such bones might abrade the tissues as the lungs inflate and deflate. We can only guess at the respiratory system of dinosaurs, and the appearance of well-developed thoracic hypapophyses would support the notion of something birdlike. None have been reported, however.

The Ribs

Bird ribs are quite distinctive and have features not found in mammals. The bird's rib is not a single bone but has separate dorsal and ventral components joined in the middle. The ventral component is always a simple bar that makes a connection between the sternum and the dorsal component. The dorsal component is more complex. It has the double-headed articulation that is found only in birds and dinosaurs. It also has an associated bone, called the uncinate process, that is attached near the middle of the upper unit and usually extends over the next rib in line. The uncinate process may be weakly attached to its rib. As a result, it is occasionally lost during specimen preparation or during the more uncertain processes of fossilization.

The ribs of most terrestrial birds are rather simple bars, whereas those of diving birds are highly specialized. You might expect them to be heavy and rigid to resist changes of pressure underwater, but loons, grebes, auks, diving-petrels, and even penguins have exceptionally long and thin ribs (Figure 2.8). In smaller species, they are little more than long bristles. The thoracic ribs arch back from the spine, past the hip joint, before they bend back to join the sternum. They are accompanied by a series of sacral ribs that extend from beneath the hipbones almost to the vent. These sacral ribs have only a dorsal section; the ventral section is missing. Sheets of connective tissue bind them firmly to the abdominal wall so that they act like the whalebone ribs sewn into antique corsets. They are flexible but do not stretch, and the abdominal wall cannot be distorted or become distended by internal gases when the bird dives.

A "Lumbar" Joint

Many structures in the bird's backbone are critical for survival, none more so than the junction between the front and back halves of the body (Figure 1.1). The front half must remain inert when the bird walks on its hind legs, while the back half is pulled along passively by the thrust of the wings. No other animal is so thoroughly divided into discrete locomotory modules. Not surprisingly, the point of contact between the two halves requires exceptional architecture. Surprisingly, its anatomical location is not fixed. It may lie between the last thoracic vertebra and the

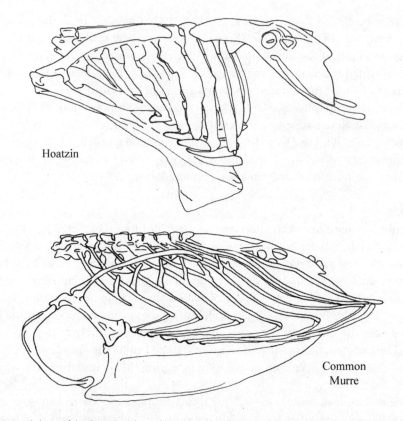

Figure 2.8 Skeleton of the thoracic and sacral regions in the Hoatzin and Common Murre. Large sections of the skeleton of the forest-dwelling Hoatzin are rigidly fused, and the ribs are massive and almost turtle-like. The skeleton of the deep-diving murre is more flexible, consisting of many independent bones, even along the thoracic spine; the ribs are greatly elongated and extremely slender.

anterior sacral vertebra, but the prevalence of fusion in birds often shifts its location forward into the thorax.

Because birds have poorly defined boundaries between the regions of their body, they have no lumbar vertebrae in the sense of a group of bones with distinctive character. Nonetheless, "lumbar" is a useful term for describing the contact between the last free thoracic vertebra and those fused to the synsacrum. The problem for avian anatomists is that the number of vertebrae fused to the hips varies, almost from species to species. Often the first few retain their two-headed thoracic ribs and, except for their rigid connection to the hip section, are typical thoracic vertebrae. Frequently, a series of thoracic vertebrae will lose their ribs and become completely fused to the hips, and the only indications of their origin are the paired passages for nerves going to the hind limb. Because the number of thoracic vertebrae fused to the synsacrum varies from group to group, the anatomical position of the "lumbar" joint also varies. In some birds, it is far forward of the hips; in others, it is only a few vertebrae in front of the hip joint, or acetabulum.

Although birds lack a clearly defined lumbar area, the interface between the thoracic vertebrae and the synsacrum marks a basic division that is reflected in one of the most ancient and fundamental structures of the body. In primitive vertebrates, pairs of short epaxial muscles are attached to each vertebra, forming a continuous series along the whole length of the spinal column. In birds, the series is unbroken along the vertebrae of the neck and thorax, but is suddenly interrupted at the lumbar interface by the fused bones of the hips. The series resumes behind the hips, and continues along the free caudal vertebrae into the tail.

This discontinuity in the epaxial muscles marks a major division in skeletal responsibilities. In the front half, between the lumbar interface and the neck, the body is dominated by activities of the forelimbs and a series of flexible structures associated with the pectoral girdle. Although the rib cage protects many vital organs, such as the heart, lungs, and liver, it is the massive flight muscles attached to the sternum that contribute most of the weight to this area. In the rear half of the body, between the lumbar interface and the tail, the skeleton and musculature are largely devoted to movement of the hind limbs. Although the thigh muscles are very large, most of the weight in the area comes from the great mass of the digestive and reproductive systems. From the outside, the tail may look like an important structure, but it is only a lightweight fan of feathers attached to a small, fleshy nub.

There is some evidence that the interface between the thoracic and sacral vertebrae may be the Achilles heel of avian design. Most forest birds spend their entire lives in areas protected from strong winds, and the transition between the anterior and posterior regions of the spine is gradual and unspecialized. Flocks of songbirds, however, are often attracted to roadside salt and are killed when air turbulence from passing semi-trailers whips them into the air. Violent movement twists the body, concentrating stress at the relatively small section of the backbone between the two locomotory modules until it gives way. The casualties typically show massive internal bleeding near the lumbar joint, where the dislocated backbone has ruptured the blood vessels near the kidneys. Birds that live in open areas are often exposed to strong winds and may be very vigorous fliers. Their lumbar areas often have some form of structural reinforcement.

Because the front half of a bird's body may be moving in one direction while the rear half tries to move in another, the lumbar interface must be able to cope with angular forces that try to bend or twist the spine. Most skeletons resist these forces by using leverage. Long arms of bone interact with each other far from the point of articulation. The vertebrae on either side of the lumbar interface are usually the largest in the body. Tall processes on top, or long hypapophyses below, restrain dorsoventral bending, and connections between adjacent processes limit rotation of the joint. Long lateral spines prevent excessive sideways bending. In many birds, the hipbones, particularly the anterior plates of the ilium, provide additional support by extending forward towards the edge of the rib cage.

Pelicans and some other birds have taken a different approach to their lumbar architecture. They have moved the lumbar interface far forward by fusing four or five additional thoracic vertebrae to the synsacrum. This greatly shortens the flex-ible region of the back and moves the joint forward to the point where it gets addi-tional support from the ribs and the ends of the shoulder blades.

Diving underwater exposes a bird's spine to strong bending and twisting forces because water is very much denser than air. Loons, large grebes, and auks all have elaborate hypapophyses on their thoracic vertebrae. These birds are noted for their speed and manoeuvrability when they chase fish, so it is not surprising that many have developed specialized structures to prevent bending between the thorax and sacrum. One of the most elaborate adaptations appears in the Cassin's Auklet. The saddle-shaped face between the last thoracic vertebra and the synsacrum has be-come a deep tongue-in-groove structure. A vertical ridge on the posterior face of the last thoracic vertebra fits into a matching groove on the anterior face of the first sacral vertebra. The saddle shape may allow a little horizontal flexion but it resists twisting completely. In addition, a long ligament runs along the tips of the thoracic hypapophyses, anchoring them to a prominence on the lower surface of the synsacrum. This string of connective tissue may serve other functions in the abdominal cavity, but it also prevents vertical flexion of the spine at the "lumbar" joint (Figure 2.9).

The two basic groups of diving birds have solved the problem of twisting in two very different ways. The foot-propelled divers, such as cormorants, are typically long, thin, and flexible, and their body weight is distributed in a roughly equally manner between the thorax and the abdomen. Vertebrae near the lumbar interface

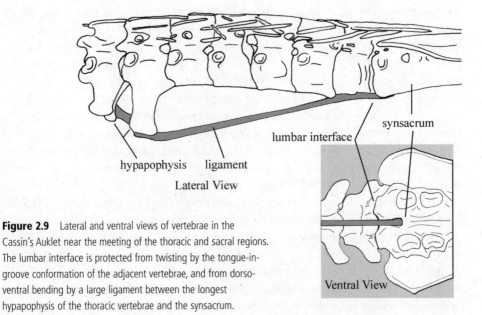

Figure 2.9   Lateral and ventral views of vertebrae in the Cassin's Auklet near the meeting of the thoracic and sacral regions. The lumbar interface is protected from twisting by the tongue-in-groove conformation of the adjacent vertebrae, and from dorso-ventral bending by a large ligament between the longest hypapophysis of the thoracic vertebrae and the synsacrum.

are extremely tall and narrow. The lateral spines are shorter, suggesting tolerance for sideways flexibility but strong resistance to arching up or down (Figure 2.10). The hipbones are exceptionally long and narrow, pushing the legs far to the rear but offering little support to the sides or belly. Forces generated by swimming movements of the feet are distributed forward, throughout the body. Wing-propelled divers, such as auks, have short, round bodies and are much more compact. Swimming forces are generated near the centre of the body, but the birds' nearly spherical shape reduces the risk of twisting because the extremities are not exposed. The head is kept tucked close to the chest and the hindquarters are small. Even the legs and feet are pulled beneath the abdominal plumage, except when they are needed for steering.

Like wing-propelled divers, vigorous aerial fliers seem to have chosen to combine flexibility and rigidity in their design strategy. They often retain a long thorax but its structure and interface with the hips vary from group to group. In aggressive hunters, such as falcons and accipiter hawks, some of the central thoracic vertebrae are fused together to create an exceptionally strong central thorax. Their close relatives, soaring hawks and eagles, have a more flexible string of independent vertebrae. Falcons, which strike their prey in the air, have an articulating face on the first sacral vertebra that seems specialized to resist horizontal bending or twisting. Its saddle-shaped face is deeply carved and quite a bit wider than tall. The wide face on the centrum of this vertebra may also be related to the stresses imposed by the falcon's vigorous flight style. A similar widened face appears in

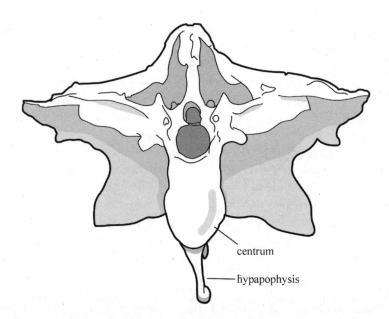

centrum

hypapophysis

Figure 2.10 The lumbar interface in a Double-crested Cormorant. The tall, slender shape prevents bending dorsoventrally, while ligaments attached to the hypapophyses and the lateral spines reduce twisting.

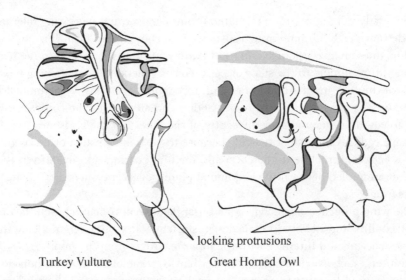

Turkey Vulture   Great Horned Owl

Figure 2.11   The lumbar interface in a Turkey Vulture (left) and a Great Horned Owl (right). Small processes on the lower part of the centrum help prevent twisting. The dumbbell-shaped structures are the points of attachment for double-headed ribs, indicating that the synsacrum of both birds has captured thoracic vertebrae and moved the lumbar interface forward.

jaegers, albatrosses, and shearwaters, which also need to make rapid movements with large wings. In the soaring hawks and eagles, the lumbar joint is large but not exceptionally so, and it has no obvious structural specializations. Large dorsal spines and long lateral processes on the vertebrae may help to support it in a general way.

The lumbar joint of owls is also supported by lateral extensions of the vertebrae, but these birds have another specialization that helps them withstand the shock of striking prey on the ground. They hunt in the dark and may twist their hips violently when they hit the prey with only one foot or are forced to wrestle with a struggling animal. Later vertebrae in the thoracic region have pairs of short protuberances on their lower edges. They reach forward and fit into shallow sockets on the edge of the previous vertebra (Figure 2.11). This simple device locks the adjacent faces of the vertebrae in position and prevents them from twisting against each other. A similar structure appears at the same point in the skeleton of the New World vultures, such as the Turkey Vulture. Although these birds do not use their feet for foraging, they have unusually long legs, the weight of which could contribute to inertia as the bird takes flight or as it attempts to land at low speeds.

Sacral Region
The sacral portion of the axial skeleton is the centre of pedestrian locomotion in birds and is an important part of their general body plan. It consists of a short

series of vertebrae attached to the three primitive hipbones: ilium, ischium, and pubis (Figure 1.1). The component bones are so thoroughly fused and modified that the individuality of the component parts is almost completely obscured. The resulting structure has been given the technical name "synsacrum."

The shape of the synsacrum varies from group to group according to the way the bird uses of its legs. It is long and narrow in both running birds such as the ostrich and emu, which cannot fly; fast swimmers such as grebes, loons, and cormorants; or those that merely walk, such as the kiwi. Perhaps the mechanics of vigorous foot-propelled swimming have much in common with running. The cuckoos are another group consisting almost entirely of walkers and runners, but their synsacrum is wide and boxlike, even in the speedy Roadrunner. In hawks and owls, which spend a lot of time sitting in wait for prey, the synsacrum is arched downward. The shape may help these birds perch vertically, and it also places a thick layer of airy cancellous bone over the spinal cord (Figure 2.12). Many of these birds live in cooler parts of the temperate zone, and the insulating effects of the dead air trapped in the hollow hipbones could be important in reducing the cost of keeping the spinal cord warm during long periods of inactivity.

One of the unique features of the sacral region is the glycogen body, which appears to be part of the spinal cord. It sits in an enlarged portion of the neural tube behind the acetabulum, or hip joint. Its function has long been a puzzle, but current research suggests that it is an organ of balance [138] (see Chapter 5). Cormorants and storks have another peculiar structure between the glycogen body and the lumbar joint. At that face, the neural canal opens up into a second tube, which lies parallel to and above the neural tube. This second tube travels through three or four vertebrae and ends abruptly. Before its end, there are two pairs of large openings that lead into the space above the transverse processes of the backbone.

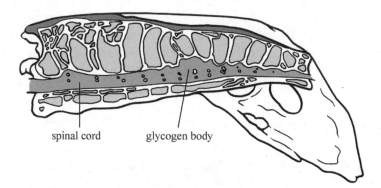

spinal cord   glycogen body

Figure 2.12 Cross-section through the hipbones of a Bald Eagle (*Haliaeetus leucocephalus*), showing the series of spaces and exits for paired nerves that mark the original vertebrae and the enlargement of the neural canal that houses the glycogen body.

This strange tube is actually part of the circulatory system (R. Necker, personal communication). In most birds, this "internal vertebral venous sinus" lies immediately above the spinal cord, but in cormorants and a few other birds, it has been isolated in a separate tube that passes through part of the synsacrum. Such tubes appear in a variety of unrelated groups, including giant hornbills, boobies, tropicbirds, frigatebirds, and some storks. There are other poorly understood structures associated with the spinal cord. Paired pockets or depressions occur in the walls of the neural tube in some groups of birds, usually in particularly large species.

Caudal Region
Many of the original tail vertebrae have been caught up in the synsacrum. You can count them by looking at the ventral surface of the hipbones. The bulges and lumps representing the remnants of their centra are difficult to interpret, but the struts that support the ischium are the remnants of their ribs. Many birds have increased the width of their hips by lengthening these ribs.

The free caudal vertebrae give flexibility to the tail. They are often quite small, but in woodpeckers, tropicbirds, and some other groups, the tail skeleton is exceptionally large and contributes significantly to the total weight of the skeleton. Woodpeckers use their large tail as a prop to brace themselves against tree trunks, but the purpose of a large tail skeleton in other groups is less clear. Songbirds do not have a particularly large tail skeleton but it participates in their respiratory process. The tail vertebrae are drawn down when the bird compresses the abdominal air sac. This pumping action of the tail becomes exaggerated when birds pump a lot of air through their system in order to sing. Many birds use flicks of the tail to add a visual component to their display.

The most distinctive feature of the caudal skeleton in most birds is a peculiar plough-shaped bone at the tip of the spinal column called the pygostyle. It represents the fused remains of 6-11 caudal vertebrae and is the entire skeletal foundation for the feathered tail. The feathered tail consists of 10-12 pairs of large feathers (remiges) mounted in a fleshy bulb that is all that remains of the long reptilian tail seen in *Archaeopteryx*. The two halves of the bulb sit on either side of the pygostyle. Muscles attached to the caudal vertebrae and hipbones control the movement of the remiges, and can open them like a fan by distorting the fleshy bulb.

The Appendicular Skeleton

Pectoral Girdle
The pectoral girdle is a collection of fairly large and important bones attached to the forelimbs. In mammals, it is firmly attached to the backbone. In birds, the connection is achieved through muscles and ligaments. When the bird is at rest, the pectoral girdle appears to hang from the vertebral column at the end of the ribs. In flight, the position is reversed and the weight of the body rests on the springy ribs.

The pectoral girdle's peculiar structure reflects this division of roles. The large sternum, or breastbone, is the foundation for support during flight. Its broad inner surface supports and protects the vital organs while its leading edge supports two girders that carry the shoulders. These girders are the coracoids, a pair of very large, stiff bones that lift the shoulder up and forward. The tendons from the massive flight muscles pass over the tips of these bones on their way to the wing skeleton. Frigatebirds and other species that use unusually long, deep strokes of the wings in flight have very elongated coracoids. Swifts and hummingbirds, which beat their wings very quickly, have relatively short and stubby coracoids.

The coracoids meet two other paired bones in the shoulder, both of which appear to be part of the flexible suspension system that helps to redistribute the stresses of flight. The furcula is familiar as the wishbone, but its function in the bird is poorly understood. In fact, it is the fused product of the paired clavicles, or collarbones, and a number of otherwise typical birds have retained those independent bones. When present, the furcula comes in a wide variety of shapes and sizes, suggesting that its function may vary from family to family according to changes in flight style. Typically, it is a U-shaped structure that hangs down in front of the breast; its lower end is often attached directly to the sternum. It may even be fused to the sternum. More often, it is suspended in the breast muscles and held in place by a ligament.

The furcula may be absent or reduced to a pair of clavicles in some piciform birds, some parrots, the mesites of Madagascar, some owls, and some songbirds, such as Australian bushlarks and scrubbirds. In owls, the furcula may be little more than a narrow ribbon that marks the connection, or phare, between different parts of the pectoral muscles. In accipiter hawks and falcons, which expend a great deal of energy in high-speed pursuit, it is a very large, rigid bone whose hollow arms sweep forward and then back. The broad curve carries the pectoral muscles forward for greater power in flight. The anterior air sac extends into the hollow arms of the furcula, where it can collect excess heat radiated by the pumping flight muscles and expel it as the bird exhales. Some soaring birds also have massive hollow arms on the furcula, suggesting that they might also benefit from an ability to get rid of excess heat during flight. In the pelican, the hollow arms of the furcula are rigidly and thoroughly fused to the front of the sternum.

The furcula gained a reputation as a spring that aids respiratory activity because it was seen to flex in and out in X-ray cinematography of a starling. Such an important role seems unlikely for such a variable structure, however, even among songbirds. It seems too weak to function as a spring that exerts significant force on surrounding tissue, and although its movement is suggestive, it may be more passive than active [5].

Recently, Clifford A. Hui analyzed the shape of the furcula in relation to flight style [96]. Puffins and other wing-propelled diving birds that balance the demands of aerial and submarine "flight" have a furcula that is more V-shaped than other birds. It is also strongly curved forward. The sweep of the forward curve greatly

increases the area of attachment for a branch of the pectoral muscle. The arrangement increases the ability of that muscle to pull the wing forward while it generates thrust during the downstroke. The forward movement is called "protraction."
It helps the bird counteract the effects of profile drag.

Profile drag is generated by friction over the wing's surface. It is strongest during glides or at the end of a wing beat, when thrust is greatly reduced. In the air,
puffins flap their wings very rapidly, and there is only a brief pause in the thrust as
the wings change direction. Underwater, puffins propel themselves with deep
strokes of partially folded wings. The swimming action is not as rapid or continuous as flight in the air. Auks often pause briefly with their wings closed tightly
against their body. The pause may help them conserve energy by resting the muscles briefly, but water is such a dense medium that profile drag slows them very
quickly and the "glide" period is extremely brief.

Other birds also make use of a strongly curved furcula. A falcon protracts its
wings to dramatically increase their stroke as it accelerates for an attack. Deep,
rapid wing strokes quickly increase its air speed. Albatrosses and petrels, noted for
their energy-efficient soaring flight, also have a strongly curved furcula. The narrow profile of their wings minimizes profile drag and facilitates soaring, but their
flight is also dependent on fine control of wing movement. A curved furcula may
be related to the power required for easy manipulation of wings with such very
large surface areas. When winds are calm, petrels also need strong muscles to make
deep, powerful strokes for take-off.

Although the short, oval wings of small forest birds leave them almost entirely
dependent on deep wing strokes and flapping flight, they often have a rather flat
furcula. A significant number of species, including some owls, make do with independent clavicles and no bony furcula at all. Such birds tend to fly fairly slowly
and never achieve speeds at which profile drag becomes a significant cost. Many of
them fly only short distances, while others are noted for bounding flight in which
bursts of wing beating alternate with pauses. The pauses allow the flight muscles
to rest briefly. Significantly, only a few of these birds attempt to glide. Their wings
are just the wrong shape and the bird quickly loses air speed. The long-distance
migrants among the forest birds and pursuit hunters such as swallows have longer,
pointed wings and fly at higher speeds than their relatives. They always have a
well-developed furcula. Sometimes it has a forward curve, suggesting that profile
drag could be a significant factor in their energy budgets.

The tip of the coracoid also meets the scapula, or shoulder blade, in the shoulder joint. The scapulas are a pair of flat, springy bones that seem to have little to
do with flight. They extend far back, across the upper part of the ribs, and are
genuinely blade-shaped. They look like curved sabres attached by muscles and
ligaments to the vertebrae. They may help to hold the pectoral girdle in position
relative to the backbone and increase the overall strength of the shoulder. The
scapulas' long tips almost reach the forward extension of the hipbones near the

"lumbar" joint and must offer some support to the vertebral column. Their long reach suggests that they might be useful in flight by redistributing stress across the tops of the ribs. The remaining gap between their ends and the leading edge of the hips emphasizes the isolation of the front and back regions of the body.

In modern birds, the shoulder end of the scapula ends in a ball-shaped boss that fits into a socket created by the coracoid and the furcula. It is a structure that clearly distinguishes modern birds from the Enantiornithes, which became extinct with the dinosaurs at the end of the Cretaceous. It is one of the characters that makes enantiornithean birds "opposite," as their scapula ended in a cup that held a ball-shaped boss on the coracoid – hence the term "ball-shouldered birds" that I have used throughout this book.

The trailing edge of the sternum reflects its role as a structure for carrying internal organs. In tinamous, grouse, and other walking birds, the trailing edge is deeply cut away and the support for the abdomen is reduced to four long splints of bone in a sheet of connective tissue. Auks are active underwater and may need extra abdominal protection for unusually large eggs. Just as their ribs extend far back towards their vent, the trailing edge of the sternum is extended as a broad, continuous plate covering much of the lower abdomen.

In its role as part of the pectoral girdle, the sternum is the point of origin for the flight muscles. Its keel forms a point of attachment for the pectoral muscle, which is always the largest of the flight muscles. It pulls on the humerus to depress the wing. The supracoracoideus muscle lies against the sternum, beneath the pectoral. To reach the humerus, its tendon passes through the triosseal gap, formed where the coracoid, scapula, and furcula meet in the shoulder. When the supracoracoideus muscle contracts, the gap acts as a pulley that reverses the direction of the muscle's action and raises the wing. No other animal depends on an apparatus as complicated as this pulley system for its major form of locomotion.

The remaining bones of the pectoral girdle make up the skeleton of the wing. Of these, the humerus is the heaviest and most variable in shape. Other than the furcula, it is the only bone whose shape varies greatly according to flight style. It is never a straight and simple bone but curves subtly to transfer stresses towards the joints. The head that articulates with the shoulder is distinctly curled over like a cresting wave so that most of the wing stroke can occur below horizontal. The physics of flight suggests that the lower part of the wing stroke is the most important because its power is vectored inward and downward against the other wing, instead of dispersing outward. The inward curl of the head of the humerus is a relatively recent development in modern birds that separates them from their primitive ancestors. It is probably associated with overall improvements in flight capability.

The humerus of a modern bird comes in one of three basic shapes. By far the most familiar is the relatively simple shape found in the domestic chicken. It begins as a large flat head at the shoulder that blends smoothly into a slightly curved cylindrical shaft. It ends in an unexceptional way at the elbow joint. A second form, closely related to the first, is the stubby rod found only in swifts and humming-birds. Its distinctive nature supports the proposed taxonomic relationship between these otherwise dissimilar groups. The third style is a long, flat arch that is found in all wing-propelled diving birds, including auks, diving-petrels, penguins, and some fossil forms.

The stubby humerus of swifts and hummingbirds seems to be an example of dwarfism or some genetic defect in which only the length of the bone has been lost. Its other features have become exaggerated to accommodate unusually thick and short muscles. The short, thick bone is appropriately shaped to absorb the stresses generated by the unusually rapid twitches of these birds' flight muscles, stresses that would shatter a longer bone.

The humerus found in auks and other wing-propelled diving birds is relatively long – that is, significantly longer than the radius or ulna. It is also arched and has a distinctive flat cross-section. Unfortunately, wing-propelled underwater locomo-tion has not attracted a lot of attention from biomechanical analysts and this un-usual bone is still poorly understood. A flattened bone might contribute to decreased drag by presenting a thinner cross-section as the wing passes through the water, but there is only a marginal difference between the thickness of an auk's humerus and the thickness of a typical round humerus of similar weight. It seems more likely that the bone is designed to cope with large stresses generated by wing movement through a dense medium like water. Shorter bones are generally stiffer than long bones of the same diameter, but the arch of the diving bird may transfer some of the stresses towards the joints, where they can be absorbed by elastic connective tissue. The walls along the bone's edges are much thicker than the broader surfaces and may provide additional strength to the overall structure. The bone's flatness may also enable it to take up strain by twisting slightly without breaking.

Automated Action of the Wing
In the 19th century, ideas about the flight of birds were simple and full of mis-understandings that helped to delay the invention of mechanical flight until 1903. We can see one example of the shortcomings of Victorian observational methods in attempts to understand the automatic operation of the wing skeleton. In 1839, Carl Bergmann wrote that the action of the wrist extended the wing into the flight position automatically as the elbow was straightened. The idea was attractive and should sound familiar. Students are still presented with this phenomenon as an example of specialized mechanical design in avian skeletons. Unfortunately, it is wrong, or at least not the whole story. Bergmann and later writers assumed that the radius and ulna acted like the linked arms of a parallel ruler and exerted force on the elbow to shift the radius in a way that also extended the manus (hand).

Bergmann's version of automated wing extension became widely used as an example of how birds reduced weight to conserve energy. In 1871, Elliott Coues explained how "muscular power is correlated and economized" by the function of

the "parallel ruler" model of wing extension [34]. As late as 1990, U.M. Norberg
stated that it helped reduce weight and inertia because muscle power was not
needed to extend the hand [139]. The parallel ruler model, however, does not explain the existence of some two dozen muscles that move the wing, or why some
of them act across both the wrist and the elbow to move the hand.

The flaw in the earlier mechanical model of wing operation was identified in
1936, when Maxheinz Sy noticed that the functions proposed for some of the wing's
articulations were wrong [206]. The structure of the wing was not thoroughly re-
examined until 1994, however, when R.J. Vasquez determined that movement of
the radius was not a simple matter related to the shape of the articulating surfaces
(dorsal and ventral condyles) of the humerus [215]. He did not reject the automa-
tion of the wing skeleton completely, but described a much more sophisticated
process that integrated the activity all the components of the wing skeleton. He
proposed functions for small protuberances in the joints and drew special attention
to the tendons that cross two, connecting the shoulder to the wrist through the
patagium.

Unfortunately, Vasquez's model of wing mechanics is very complex and lacks
the simplistic visual appeal of the parallel ruler model. It requires a detailed understanding of avian anatomy and a technical terminology that is far beyond the
scope of this book. For that same reason, it tends to be ignored in most textbooks.
The true nature of wing function is not just an unimportant technical detail, however. Vasquez's explanation separates the different stages of wing movement in
flight, and suggests the presence of some automatic function in the circular motion
of the wingtip in flight. His model makes it easier to understand why birds need
some two dozen muscles, an array of specialized carpals and sesamoid bones in the
wing joints, and the multiple articulating facets on the major wing bones. By relating the specialized shapes and structures of the wing keleton to the mechanics of
flight in modern birds, he offers a method of interpreting the flight capabilities
of fossil forms.

Pelvic Girdle

Unlike the suspended or floating pectoral girdle, the main components of the
avian pelvic girdle are thoroughly fused to the backbone. Its specialized structures
are unique, and its development as a point of attachment for the muscles that
move the hind limbs is one of the fundamental steps in the early evolutionary
story of birds. It will be examined more fully in Chapter 4 and in many of the subsequent chapters. The ancient change in the hips affected the size and shape of
the femur and contributed to its being both the largest and the most conservative
bone in the bird skeleton. It is easy to identify a femur as that of a bird, but extremely difficult to assign an isolated example to a species, or even a family, with
confidence.

The bones of the hind limb were introduced in Chapter 1. The tibiotarsus, the
fibula, the tarsometatarsus, the tarsals, and the digits of the foot vary from group

to group according to reasonably well-understood rules based on the locomotory style of the bird in question. Many land birds have very similar legs with unspecialized bones. Predators and diving birds use their legs for purposes other than walking and often have bones with unusual shapes, especially in the lower leg (see, for example, Figure 1.8).

Perhaps the most eccentric shape among the leg bones is found in grebes and loons. As mentioned in Chapter 1, the tarsometatarsus of these diving birds is a flattened box reminiscent of the flattened humerus in wing-propelled divers. Drag may well be the primary issue in its bladelike design. The similarity between these bones in grebes and loons is surprising because the two groups appear to use entirely different styles of swimming and you would think that the stresses on the bones would be very different. Loons swim by pumping their feet in a synchronous backstroke, whereas grebes whip their feet around in tight circles so that the peculiar lobes on their toes can act as hydrofoils [103]. The foot action propels the grebe forward at high speed. The lobes on the toes need to be firmly anchored, and characteristic grooves can be seen in the bones of the digits. We do not know whether fossil waterbirds, such as *Hesperornis,* had lobed toes or webbed feet. The absence of grooves from the digits suggests, however, that their swimming action was not that of a grebe but was probably more like that of a cormorant or loon.

Automated Function of the Foot

The bird's leg ends in a highly variable number of toes arranged in several different formats (Figure 1.9) according to the bird's lifestyle. Just as early ornithologists were misled about automatic features in a bird's wing by their manipulation of dead specimens, so they misinterpreted the mechanisms that enabled a bird's foot to grip a branch automatically. Misinterpretation of the operation of a bird's foot is not as significant a problem as misinterpretation of the wing action, but it appears in just as many textbooks as another example of avian structural efficiency. Earlier investigators had based their observations of the actions of feet on birds that had stiffened in rigor mortis or been preserved in chemicals that hardened the tissues. A simple tug at the tendon appeared to close the toes. For this reason, fresh chicken feet were a popular toy for schoolboys in the days before shrink-wrapped chicken bits. The action of the toes suggested that the main tendon had to travel further as the leg bent.

It was not until 1965 that W.J. Bock [11] conducted a series of simple experiments on the foot of a living bird to work out the real mechanism. Like the action of the wing, the foot's operation depends on the integrated operation of several distant muscles. Bock never published the full details of this study, but in a 1974 review, he listed a number of other peculiar misconceptions about skeletal function in birds [12]. Most of them are technical details, of interest only to specialists in the field and not as dramatic as a reinterpretation of automated wing or foot function.

In 1990, Thomas H. Quinn and J.J. Baumel found another automatic feature of the bird's foot [169]. There are structures in the toes that lock them into position and take the strain off the leg muscles during long periods at the roost. Basically, a section of the toe tendon has transverse ribs or ridges, while the complementary section of the tendon sleeve has matching transverse grooves. When the weight of the bird presses the toe against a branch, the ribs fit into the grooves and the tendon is effectively locked in place. As the bird takes off, the weight is lifted from the toe, the ribs leave the grooves, the tendon is free to move, and the toe releases its grip. The mechanism is surprisingly like the simple tongue-and-groove device that locks plastic sandwich bags.

This mechanism occurs in a wide variety of birds, including types that use their toes for swimming, wading, grasping prey, clinging, hanging, and tree climbing as well as perching. Quinn and Baumel even found an analogous structure in the toes of bats. The device is also widespread in web-footed birds, where it may help cup the toes for a more effective swimming stroke. Many birds with simple tendons that lack ribs or grooves are those that spend a great deal of time walking, such as ostriches and other ratites, some shorebirds (plovers, oystercatchers, stilts), and penguins. The structure also appears to be missing in swifts, nightjars, flamingos, and grebes. Grebes are also members of the recently described Metaves, a subclass within Aves that may represent an early radiation of groups within modern birds. The significance of this new group is discussed further in Chapter 6 [59].

The tendons of the toes play an important role in the design of the hypotarsus, a structure characteristic of birds. The tarsometatarsus is derived from bones of the foot, so the joint at the end of this bone that lies nearest to the body is equivalent to the heel in mammals. The hypotarsus is a box-shaped structure at the top of the tarsometatarsus, located behind the joint. Anatomically, it is on the bottom surface of the foot – hence "hypo" (beneath). The tendons that connect the muscles on the upper leg to the toes are guided through passages in this structure (Figure 2.13). The two largest passages serve the tendons that go to the longest toes (number 3) and the hallux (number 1). These toes are probably the most important in perching. Up to four smaller passages serve the other toes. Sometimes these passages are not completely covered and appear as grooves on the surface of the hypotarsus.

The number and arrangement of the passages varies from group to group and can be useful in classification. The Passeriformes (songbirds) are divided into two major groups, the oscines and suboscines, which seem to have very ancient roots. The only living songbird that does not belong to either group is the New Zealand Wren, which appears to be a survivor of a more primitive stock. Each of these types has a distinctive hypotarsus. Fossil songbirds, found in early Miocene rocks in Europe, have a hypotarsus that is somewhat similar to that of the New Zealand Wren and very different from those of either the oscines or suboscines. The presence of songbirds related to the New Zealand Wren in Europe suggests that such fossils represent an early radiation of primitive Passeriformes that were eventually replaced by more modern forms. They left no living relatives in the Northern Hemisphere [132].

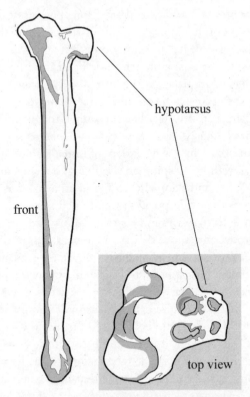

Figure 2.13 In many birds, the large tendons that manipulate the toes pass through a boxlike structure, the hypotarsus, on the back of the tarsometatarsus. Often the canals form a distinctive pattern that can be used to identify the bird.

Conclusion

Working within the constraints of the basic avian skeleton, birds have exploited a huge number of niches in forested habitats and have also taken advantage of other opportunities. They have evolved into a variety of long-legged and long-winged species that can cope with the open spaces of estuaries and grasslands. They have also evolved into a group of powerful divers and fliers that have invaded the oceans. Surprisingly, they are all readily identifiable as birds. Ancient sailors might have thought penguins were exotic geese, but they never called them fishes.

Further Reading

In this chapter, I have relied heavily on the technical literature, particularly one of the most informative books on bones of both mammals and birds:

Currey, J.D. 2002. Bones: Structure and mechanics. Princeton University Press, Princeton, NJ.

I also strongly recommend finding a copy of Paul Bühler's essay on thin bones. His illustrations are spectacular.

Bühler, P. 1988. Light bones in birds. Pages 385-93 *in* K.E. Campbell (ed.). Papers in avian paleontology honoring Pierce Brodkorb. No. 36, Science Series. Natural History Museum of Los Angeles County, CA.

There are several university-level textbooks that introduce the relationship between structure and function in animals. Although it gives short shrift to birds except when it discusses flight, one of the most useful and readily available standards is:

Hildebrand, M., and G.E. Goslow, Jr. 2001. Analysis of vertebrate structure, 5th ed. John Wiley and Sons, New York.

A recent compilation by V.I. Bels and his colleagues offers an introduction to the more technical aspects of biomechanics. A chapter on the beak of parrots and another about the beak of finches illustrate how complex anatomical function can be, and how important it is in understanding an animal. Homberger uses a knowledge beak operation and fruit characteristics to create an evolutionary story for parrots, and Bout looks at how the ability to move the upper jaw may limit the ability of birds to adapt to foods that require a particularly strong bite:

Homberger, D.G. 2003. The comparative biomechanics of a prey-predator relationship: The adaptive morphologies of the feeding apparatus of Australian Black Cockatoos and their foods as a basis for the reconstruction of the evolutionary history of the Psittaciformes. Pages 203-28 *in* V.L. Bels, J.-P. Gasc, and A. Casinos (eds.). Vertebrate biomechanics and evolution. BIOS Scientific Publishers, Oxford, UK.

Bout, R.G. 2003. Biomechanics of the avian skull. Pages 229-42 *in* V.L. Bels, J.-P. Gasc, and A. Casinos (eds.). Vertebrate biomechanics and evolution. BIOS Scientific Publishers, Oxford, UK.

In the same volume, there is a discussion of the evolution of wing structure that will play a role in later chapters of this book:

Rayner, J.M.V., and G.J. Dyke. 2003. Origins and evolution of diversity in the avian wing. Pages 297-317 *in* V.L. Bels, J.-P. Gasc, and A. Casinos (eds.). Vertebrate biomechanics and evolution. BIOS Scientific Publishers, Oxford, UK.

A Bird Is Like a Dinosaur

3

*Features shared
by birds and their reptilian ancestors*

A quick look at the animals in the modern world will tell you that they can be divided into two big groups: those that crawl on their belly and those that carry their bodies on their legs. All members of the latter group are either birds or mammals, but until 65 million years ago, you could have added dinosaurs to that list. Whether or not you believe that birds have evolved from dinosaurs, it is obvious that birds and dinosaurs share a great deal of their basic body structure. That does not mean that it is easy to contemplate a relationship between the tiny hummingbird in your garden and one of the gigantic tyrannosaurids of the Mesozoic. Nonetheless, a broad similarity in anatomical structure and architectural strategy suggests a strong link between these two groups, while the many obvious differences may be the product of evolution during the unimaginably long time since they shared a common ancestor.

Basic Architecture

No features appear to link dinosaurs to birds more closely than their posture and their method of terrestrial locomotion. At some point, very deep in the history of animals, a creature got off its belly and propped itself up on its legs. This first step into the third dimension was a primal event in vertebrate evolution. Only two groups of vertebrates can boast of this simple act – the diapsid archosaurs, which includes dinosaurs and birds, and the synapsid reptiles, which gave rise to mammals. Walking while propped up on four legs is a great advance over sliding on the belly squamosally, like a lizard or salamander. In squamosal animals, the legs operate like the oars of a boat, swinging back and forth, dragging the body forward. The limbs are sprawled out to the sides and lie parallel to the sagittal plane that divides animals into a top and a bottom. In the parasagittal stance of birds and mammals, the legs move back and forth under the body, and "stand" at right angles to the sagittal plane (Figure 3.1).

In humans, the sagittal plane passes through both ears, both hands, and both feet. It appears to divide our bodies into a front and a back, but the anatomists who mapped the body and named its parts paid respect to our very ancient squamosal

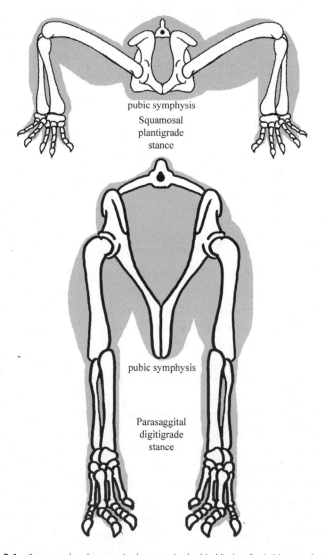

pubic symphysis

Squamosal
plantigrade
stance

pubic symphysis

Parasaggital
digitigrade
stance

Figure 3.1 Squamosal and parasagittal stances in the hind limbs of primitive vertebrates.

ancestors. They described humans and all other animals as though they were lying flat on their belly – face forward and palms flat on the ground. In such a position, the thumb becomes digit 1. Humans, as far as I know, cannot lie on their belly with the soles of their feet flat on the ground. The best we can do is splay our feet sideways with the biggest toe (digit 1) on the ground and little toe (digit 5) up in the air. From the squamosal point of view, the directional information widely used in anatomical texts makes topographical sense. The head is in front (anterior), the back is on top (dorsal), and the belly is on the ground (ventral). If we had a tail it would be posterior, or towards the tail (caudal) (Figure 3.2).

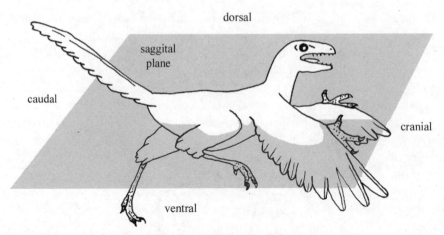

Figure 3.2   The sagittal plane and directional orientation in a bird.

The importance of getting into the third dimension cannot be overstated. Even in the oceans, it separates advanced life forms from the bottom-dwellers, and has enabled some fish, crustaceans, and squids to become active hunters in the water column. On land, it is a feature of birds and mammals as well as the most successful invertebrates, such as flying insects or ballooning spiders. It is often related to increased coordination and greater intelligence. Free-living and active mollusks, such as squid or octopus, are better at problem solving than the sedentary clams. Among mammals, dolphins, which live three-dimensionally in the oceans, and primates, which swing through the three-dimensional forest canopy, are considered to be brighter than the cows and other grazers on the two-dimensional plains.

Successful application of the parasagittal stance is not without costs. Lifting the body off the ground requires a sense of balance, and higher vertebrates have had to adapt some muscle groups to new responsibilities and strengthen shoulder and hip joints. Elevation also entails some risks. Lizards and other squamosal animals do not need to worry about falling over and breaking a leg; for animals whose limbs carry their body weight, however, a broken limb bone can be fatal. Evidence of leg and hip injuries in the fossils of large predatory dinosaurs such as *Albertosaurus* and *Tyrannosaurus rex* suggests that they could be knocked over and disabled. *Diplodocus* and other giant herbivorous dinosaurs may have taken advantage of this vulnerability and defended themselves by using their long, whiplike tails to bowl attackers over.

In general, the advantages of elevation have outweighed any risks. The parasagittal stance lifts the animal's head so that more of the surrounding habitat is visible. Taller animals are more likely to spot distant risks and spend less of their time searching for food. Tall herbivores can reach distant branches; tall predators can watch for the best moment to catch a meal.

The height advantage may also have helped many predators reduce the risk of hunting by allowing them to specialize in smaller, less offensive prey. Small prey species often occur in large numbers and can be taken with much less risk than a struggle with a single large animal would entail. Predators that specialize in over-powering relatively large prey are not known for their longevity, but storks and other tall birds may live for the better part of a century by subduing nothing larger than a hapless frog. Height not only offers a clear view of the numerous small crea-tures that scurry about on the ground but may also place the predator beyond the detection distance of its prey. A small fish or mouse may remain completely una-ware of the sharp tip of a heron's beak, poised a metre overhead, until it is far too late. Throughout history, so many predators have used long legs and long necks to get their toothy mouths well above prey species that edible animals from *Stegosaurus* to porcupines have armoured their upper surfaces with a variety of spikes and plates. Many others have developed elaborate camouflage to reduce their visibility from above.

Another practical advantage of the parasagittal stance is the overall reduction in body width. The legs are pulled beneath the body instead of sticking out sideways, where they can be snagged by passing branches. Squamosal animals are very good at scooting across open surfaces such as mudflats or rocky deserts. Such open ter-rain must have characterized much of the world before vascular plants began to grow vertically and gave the environment a degree of three-dimensional complex-ity. Narrow parasagittal animals would have been much more nimble in the new kind of habitat, slipping between upright stems more easily than the wider-tracked squamosal types. Squamosal animals have also attempted to adapt to complex ground cover. Many make do with very small legs, while snakes and some other animals have no limbs at all.

Once up on four legs, it was only a matter of time before some creatures learned to balance on just two legs and became bipedal. Some modern lizards make a clumsy attempt at bipedalism, but their short gallops on hind legs are only slightly better than a controlled forward fall. Coordinated, stable locomotion using the hind legs alone is a very rare locomotory strategy among vertebrates. It has evolved in only a few mammals, such as humans, kangaroos, and some small rodents, and in dinosaurs and birds. Dinosaurs achieved a bipedal stance very early in their his-tory and it became one of their most noteworthy characteristics even though some very large species reverted to a four-footed stance. In mammals, bipedalism is a very recent trick, but it is only one of several unusual forms of locomotion that have recently developed in that group. Like birds, many mammals have learned to fly and swim, whereas dinosaurs remained successful pedestrians during the whole span of their existence. As far as we know, none of the dinosaurs became special-ized swimmers, but a few of the smaller, feathered species appear to have become arboreal and learned to glide among the trees. Perhaps it is only a matter of time before we discover one that was capable of flapping flight.

Among living animals, only humans and birds are specifically designed to walk on two legs. The type of bipedalism currently used by birds, and formerly by dinosaurs, is very different from our own, however. We are perched on top of our legs. Our body is poured into the bowl of our hipbones and propped up by our backbone. In dinosaurs and birds, the body hangs down from a spinal column that is balanced across the hips at the top of the hind legs, somewhat like a child's teetertotter. Bipedalism had a major effect on the evolutionary history of birds and mammals because it has left the forelimbs free for other duties. Humans developed hands that use tools; birds evolved wings that power their flight. Surprisingly, most other bipedal animals did not exploit this opportunity to develop specialized forelimbs. Some dinosaurs used their forelimbs to manipulate prey, but others allowed them to degenerate into vestigial nubs. Similarly, bipedal kangaroos and rodents have retained fairly small, generalized forelimbs.

Wing-powered flight is a highly specialized form of locomotion that has evolved in several different lineages of vertebrates. Birds are the only bipedal animals that managed to achieve it, but a version evolved separately in quadrupedal relatives of the dinosaurs to produce pterosaurs, and in small insectivorous mammals to produce bats. Both pterosaurs and bats depend on membranes of skin to create a wing. In pterosaurs, the membrane stretched from a greatly elongated outer finger back to the body. In bats, all four of the outer fingers are unusually long and the membrane is stretched between them and from the last digit to the outer edge of the leg. Only birds and their relatives among the dinosaurs have developed such a specialized flying surface as the feather.

In both reptiles and mammals, there are many examples of groups that have stopped short of powered flight but are capable of gliding. They are suggestive of the kinds of intermediate steps necessary for a quadruped to evolve from a purely terrestrial form into a flier, but it has never been certain that birds had such gliders among their ancestors. Only recently have we found feathered dinosaurs, suggesting such intermediate forms; so far, however, they are all more recent than the early birds (Chapters 7 and 8). The oldest accepted representative of the avian lineage is *Archaeopteryx.* It is clearly a bird in spite of its many dinosaurlike features, and appears to have been capable of flapping flight. Nothing is known about its ancestors. The best-accepted theory is that the ancestors of birds were small, predatory dinosaurs, not too different from the *Compsognathus,* a tiny theropod found in rocks near the first *Archaeopteryx* fossil.

Other aspects of the locomotory apparatus are similar in birds and dinosaurs. For example, both are digitigrade: they support themselves on their toes rather than on the flat of their foot (plantigrade). Bears and humans are plantigrade, as are many squamosal animals, but there is no clear indication as to whether the ancestors of dinosaurs became bipedal before they became digitigrade or vice versa. *Euparkeria,* one of the best-preserved fossils of an ancient reptile, or archosaur, appears to have had a parasagittal stance while remaining more or less quadrupedal and plantigrade. Even early dinosauromorphs, such as *Lagerpeton* and

Lagosuchus, appear to have been bipedal and digitigrade, like all their descendants. Walking on their toes may have given these early predators the advantage of speed. Certainly many of the more recently evolved, fast-running mammals are also digitigrade. On the other hand, so many of the primitive mammals and mammal-like reptiles are plantigrade that foot structure may represent one of the basic architectural differences between them and the dinosaurs. Perhaps the digitigrade stance originated with some simple evolutionary advantage, such as the greater stability offered to a bipedal animal by a broad fan of toes, while true quadrupeds were supported by a larger number of legs and did not need oversized feet.

Appropriately enough for a large group of walking animals, a second ancient and fundamental difference in foot structure splits the primitive archosaurs into two distinct lineages. The pseudosuchian lineage produced crocodiles, while its sister lineage, the ornithosuchians, led to both dinosaurs and birds. The basic design of the skeleton of the hind limb in the ancestral tetrapod included a femur connected to the hip and a lower leg consisting of a pair of parallel lower leg bones, the tibia and the fibula. The presence of two bones in the lower limb facilitates rotation of the foot and is a universal characteristic of primitive vertebrates. Each of the lower leg bones has a matching tarsal or ankle bone. The tibia articulates with the astragalus and the fibula with the calcaneum. In pseudosuchians, a hinge line passes between the astragalus and the calcaneum so that the latter functions as part of the foot while the former moves with the leg. In ornithosuchians, the hinge line lies beyond both the astragalus and the calcaneum, so that they both function as part of the leg. In birds, the location of that hinge was likely an important factor in the evolution of the fused tarsal bones and the development of the tarsometatarsus as a third major unit of the hind limb.

The exact position of the foot hinge may seem a minor technical detail, but it is important to the theory that birds belong within the lineage of theropod dinosaurs. When the astragalus and the calcaneum fused to the tibia, they became the tarsal components of the tibiotarsus, the central long bone in the bird's unique three-unit hind limb. Various levels of fusion can be seen in the legs of advanced theropods, suggesting the evolutionary steps that might have occurred. The calcaneum remained a small bone in the ankle, but part of the astragalus became a much larger and important piece of bone known as the ascending process. It extended up the face of the tibia in a way that probably helped the dinosaur reduce problems with torque generated in the ankle. By resting against the tibia, the extended astragalus would have stabilized the active face of the joint. In birds, the growth of the ascending process and its fusion to the face of the tibia can still be traced during the development of the embryo in the egg.

Bipedal locomotion requires a great deal of balance and control, especially during rapid movement. Modern reconstructions portray *T. rex* and other giant bipedal predators as nimble animals, but their long tails and relatively heavy heads may have reduced their agility by generating large amounts of rotational inertia when they changed direction. David Carrier has developed a series of laboratory exercises with a special backpack that imitates the distribution of weight in a large theropod [21]. It greatly reduces the performance of human athletes and leads to a loss of speed during turns. Because the backpack lacks the flexibility of the theropod's tail, it cannot counteract shifts in the body's centre of gravity effectively, but it helps to illustrate potential limitations in the design of bipedal theropods. They may well have had difficulty negotiating rugged terrain or maintaining their balance during the twists and turns of a chase. The vertical posture of humans, and their lack of a heavy tail, helps them avoid much of the problem.

In spite of this apparent limitation to their agility, theropods were able to hold on to their position as top predators for millions of years. Late in the Cretaceous, some lineages began to experiment with modifications to improve agility by centralizing their body mass, becoming much more birdlike in the process. When the feathered *Caudipteryx* was discovered, it had such a short tail and birdlike legs that, for a while, its relationship to other dinosaurs was in doubt. *Caudipteryx* and the similar oviraptors suggest that part of the evolutionary impetus for inner-bird architecture was related to pedestrian locomotion and not to flight. Those centralized dinosaurs did not anticipate the design of modern birds, however. They already had two subclasses of modern-looking birds among their contemporaries. Today, even the largest birds are so highly centralized that they have no problems with rotational inertia.

In the long evolutionary story of dinosaurs, various lineages have experimented with alternative forms of locomotion. Some larger dinosaurs abandoned bipedalism to return to a type of four-legged stance, just as some birds have abandoned flight to become obligate pedestrians. All birds have remained bipedal, however, and some have completely lost their forelimbs. Only penguins have come close to being quadrupedal, when they scoot along on their bellies by pushing with their feet and flippers.

It has been suggested that birds cannot return to quadrupedalism because they have lost the capacity for independent control of the forelimbs and are obliged to use their wings synchronously. It is uncommon for most adult birds to move their wings independently but there are some examples. Most waterfowl stretch one wing at a time, and adult shearwaters use all four limbs, more or less independently, while swimming underwater [224]. Independent arm movement is widespread in nestlings, and the young of most species move about the nest using their limbs alternately, just like any other crawling vertebrate. Young African turacos and fledgling South American Hoatzins are able to clamber from branch to branch, hand over hand, like small monkeys. They even have tiny claws on their fingers to help them maintain their grip [155].

Bipedalism is linked to a whole suite of physiological and behavioural changes that represent fundamental events in the evolutionary history of vertebrates. Unfortunately, such events are not clear in the fossil record; often, all we can do is infer the existence of a new kind of behaviour or an important physiological advance from indirect evidence. At some unknown point in their long history, both birds and mammals escaped from climatic constraints by becoming warm-blooded (homeothermic); they developed the complex neurological linkages required for coordinated running, swimming, and flying; and they acquired the sophisticated respiratory systems needed for extended periods of vigorous activity. These major adaptations have received a lot of attention because of their role in the debates about the evolution of birds from dinosaurs, but others are less well known. As we shall see, much of the success of birds has depended on the development of a reproductive system capable of producing very large eggs; birds have also been able to modify their digestive and excretory systems to meet special demands.

Being upright on two (or four) legs enables you to run, but running demands a great deal of oxygen. Mammals meet the demand for oxygen with continuous-action lungs, whereas birds have a unique flow-through respiratory system that includes lungs and air sacs. Primitive respiratory systems force other terrestrial vertebrates to be much more sedentary. Crocodiles pump their liver back and forth to inflate their lungs and must hold their breath while moving. When they are in a hurry, the best they can do is make a series of short rushes between deep gulps of air. Although modern illustrations make dinosaurs look too active for low rates of oxygen exchange, we must be careful not to be overly influenced by clever reconstructions. The nature of dinosaur lungs is still a guess. If they used a primitive hepatic pump, predatory dinosaurs might have led a crocodile-like existence and just lurked in the trees until food passed by. If they were able to run prey to exhaustion, in the way that wolves chase moose or birds chase fish underwater, they must have possessed some advanced type of respiratory system that allowed for the continuous consumption of oxygen. If dinosaurs had such an advanced respiratory system, perhaps it was a feature they inherited from a common ancestor that they shared with birds.

So far, evidence for a specific type of respiratory system in dinosaurs has been very sketchy. In 1998, a beautifully preserved fossil of a small theropod, *Scipionyx*, was found in Italy. Even some of the internal organs seem to have been preserved. A team led by John A. Ruben thought that there was enough structure present to prove that *Scipionyx* had a hepatic pump respiratory system [181, 182]. A hepatic pump in an advanced theropod like *Scipionyx* would be a strong argument against an evolutionary relationship between theropods and birds. Other researchers disagreed with Ruben and concluded that the structures were mere artifacts of preservation [156].

Dinosaur respiration is still an area of active investigation. Papers appear on the subject regularly [162], but until better fossil material is available, the precise nature of theropod respiration will remain in doubt. The discovery of a theropod fossil with hypapophyses on its thoracic vertebrae might help to rule out a respiratory system based on inflatable lungs. So far, such structures have been reported only in birds, the only animal with immobile lungs.

The avian respiratory system is built around a system of impermeable sacs that hold the air before and after it is pushed through a set of spongelike lungs. The lungs do not move at all during respiration. In most birds, the air sacs pass through special openings into the interior spaces of the larger bones, which are said to be "pneumatized." There is no direct fossil evidence that dinosaurs had a system of air sacs, but the vertebrae of theropods were deeply sculpted and very complex, like those of birds, and they appear to have been pneumatized [147]. Other bones suggest that pneumatization was limited in dinosaurs. The long leg bones of theropods were hollow, but there is no sign of passages that would lead the branches of air sacs into their central spaces.

Even if dinosaurs lacked a pneumatized skeleton, they might still have had a respiratory system that included air sacs. Pneumatization is common in birds but its extent and location vary. In general, small birds exhibit less pneumatization than large ones because many bones are reduced to mere bristles. Large aerial predators such as falcons or frigatebirds, and soaring birds such as cranes or pelicans, are exceptionally highly pneumatized, with air sacs extending throughout their whole skeleton. In some soaring birds, the bones are little more than inflated shells with paper-thin walls. Although pneumatization is widespread among birds, it is not absolutely essential for success. The charadriiform birds (sandpipers, gulls, terns, and auks) lack pneumatized bones but have successfully adapted to a wide variety of lifestyles and make energetically demanding flights [38].

In vertebrates, energetic lifestyles and specialized respiratory systems are linked to the ability to control the body's temperature. External layers of fur, fat, or feathers help them conserve body heat and maintain their temperature, whereas endothermic physiological processes generate heat. For many years, it was argued that dinosaurs could not be the ancestors of birds because they were cold-blooded reptiles that lacked an endothermic physiology and could not control their body temperature. A cold-blooded animal did not seem capable of absorbing enough oxygen to support a continuous activity, such as flapping flight. It is still unclear when, why, or how animals developed the ability to generate and control their own body temperature. It is not the sort of feature that shows up directly in the fossil record, although it may be linked to the development of some kinds of bone tissue, rapid growth in young dinosaurs, or some behaviours implied by the specialized architecture for running seen in advanced theropods such as *Velociraptor*. Flight in the pterosaurs may also have required an advanced respiratory system and homeothermic physiology.

Recent studies of large and small crocodiles suggest that we may have learned the wrong lessons from reptiles by looking at smaller types. Homeothermy may not have been critical for giants like dinosaurs. By looking at a range of crocodiles, Frank Seebacher found that very large ones operated at a higher body temperature than smaller ones, and that the large animals were able to maintain that temperature within a narrow range [188]. The data suggest that very large dinosaurs maintained a body temperature over 30°C and within a range of ±2°C. When he compared his model of their physiology to reconstructions of ancient climates, he found good matches to the distribution of some 700 species. They may have been able to withstand moderate winters, and their rapidly growing young could have reached critical size during the warmer seasons.

Smaller dinosaurs would not have been able to maintain high body temperatures in the climates typical of mid-latitudes. They would have needed an endothermic physiology to generate body heat and insulation to retain it. Perhaps the two evolved together, which would help explain the existence of small, feathered dinosaurs in the company of larger, unfeathered species. Seebacher suggests that endothermy evolved among small, birdlike Coelurosauria but not among the sauropods, stegosaurs, or other giants.

Although the presence of feathers among small, slender theropod dinosaurs, such as *Eoraptor,* suggests that these tiny animals were warmer than their surroundings and made use of insulation, they may still have been cold-blooded and used the feathers only to delay the cooling process in the evening. Insulation is most useful, however, if you are generating your own heat for at least part of the day. Some modern birds, such as hummingbirds, are able to turn their internal temperature controls on and off. They become inactive at night and on cold, cloudy days, when it is particularly expensive to keep warm. They quickly become active with a little heat from the sun. Perhaps full homeothermy developed from some intermittent form; it is going to be difficult to get firm evidence one way or the other.

There is fossil evidence that several lineages of dinosaurs developed birdlike feathers (Chapter 7), but the widespread use of these by dinosaurs presents a bit of a paradox in the fossil record. In the Tertiary, when birds became abundant and numerous, loose feathers became a fairly common form of fossil. This is not surprising, considering that feathers are made from a rather tough material and every bird replaces its supply twice a year. In some places, dropped feathers collect in great windrows during the moulting seasons. Among Cretaceous fossils, feathers are relatively rare. A Russian-Mongolian expedition has reported a large accumulation in a sample of Lower Cretaceous deposits, but most are small contour feathers typical of birds. Where are all the fossils of dinosaur feathers? They should be much larger than bird feathers and are just as likely to have collected on the edge of water bodies during the moult or become embedded in seeping tree resins. The peculiar fossil *Paeornis sharovi* is one of the largest candidates [110]. It seems to be a 15 cm section from the tip of a primary feather from a bird about the size of a raven. It is not perfectly preserved, however, and has also been interpreted as a plant fossil. There are no other examples of solitary large plumes that might be attributed to dinosaurs.

The fact that both birds and dinosaurs share bipedalism and a general similarity of form is not sufficient evidence to suggest that birds and dinosaurs are related. It is entirely possible that solving the biomechanical problems posed by bipedal locomotion could have led to broad evolutionary convergence in the two groups. It is the numerous minor skeletal similarities that birds share with small predatory dinosaurs that suggest a common ancestry.

Shared Characteristics of the Head and Neck

Sensory Receptors

Many of the most suggestive similarities between birds and dinosaurs are associated with the head and neck. Many dinosaurs had a long birdlike neck, and some, especially the bipedal predators, shared the bird's dependence on jaws to carry out functions that other animals accomplish with their hands. They also shared some of the solutions to difficult biomechanical problems created by the attachment of the head, with its sensitive brain and specialized sensory apparatus, to the end of a long, flexible neck assembled from many small pieces.

All vertebrates combine feeding, respiratory, sensory, and cognitive functions in the head. Many animals have additional structures for display and communication that increase the head's overall size and weight. If an animal makes rapid movements with a heavy head on the end of a long neck, the build-up of momentum increases the risk of dislocation or injury to the cervical vertebrae. On the other hand, a small, lightweight head may impose design limitations on the eyes, ears, and olfactory receptors. Smallness might also impair the ability of the animal to capture prey and ingest food.

Both birds and dinosaurs depend on specialized skeletal architecture to reduce the weight of the skull without decreasing its overall size. The heads of rodents and other mammals that weigh between 6 g and 300 g appear to be similar in size to those of birds but account for about 25% of the total body weight. A bird's head contributes only about 14% to the body weight. The brains are similar in size, so much of the difference in weight can be attributed to the bird's lack of teeth and heavy muscles to operate the jaw.

It is difficult to estimate the weight of a dinosaur's head. The brain was proverbially tiny, and the head was either very small, as in the giant herbivorous apatosaurs or stegosaurs, or included a great deal of empty space between bony struts, as in the predatory theropods. Ceratopsians and other "big-headed" dinosaurs that carried horns and heavy bony frills were squat animals with rather short necks and large, muscular shoulders. The hadrosaurs had moderately large heads atop long necks; many of them sported elaborate bony resonators on their skulls. These animals may not have been as active or as fast-moving as other species, but even low-density bony structures would have added significantly to the head's weight and could have been adaptive only if they served some important function in the survival of their owner.

Animals keep most of their sensory systems close to the brain to take advantage of the brief response time offered by short nerve connections. Quick reflexes are important survival tools. In most dinosaurs, the small heads appear to have been accompanied by relatively small sensory organs. Only a few specialized predators had particularly large eyes, and there is no evidence of either large ears or large noses. In birds, the ears and olfactory organs are also small, as we saw in Chapter 1, but the eye is always large.

In contrast to the visual system, the olfactory system has often been the loser in evolutionary trade-offs in birds, and it has almost disappeared in some groups. It has been a simple matter of evolutionary competition for space between the two systems. Both need a forward position on the skull to be effective, and both need specialized receptors that may take up a lot of space. The absence of elaborate ol-factory structures on the heads of most birds suggests that a sense of smell is usu-ally not very important. The fossil evidence suggests that dinosaurs did not face such a drastic choice between two sensory systems.

In the few groups of birds for which olfaction is important, there is ample evi-dence in the form of specialized structures on the surface of the beak or in the space beneath it. For instance, petrels are known to depend on their sense of smell to track down food on otherwise featureless expanses of ocean. The edges of their nostrils form a characteristic funnel that gives the group its vernacular name of "tube-nosed seabirds." Because it points forward, the funnel gives directionality to incoming olfactory signals. In spite of its simplicity, it seems to have a wide variety of other functions. Many petrels use them as effective nozzles for injecting stom-ach oils into their nestlings. Prions and other petrels have specially modified acces-sory structures that help them retain moisture and humidify inhaled air. The funnels of albatrosses may contain receptors that help them exploit subtle changes in air pressure as they fly through the tricky air currents above oceanic waves [161].

The external nostrils of a bird lead to the turbinate bones, a complex internal structure in the nasal chamber below the beak. These folded sheets of bone are covered with a thin layer of soft, moist tissue and play an important role in both moisture retention and olfaction. Complex turbinate structures appear in many vertebrates. They are particularly large and elaborate in desert-dwelling mammals, such as camels, which depend on their respiratory system to keep cool but cannot risk the loss of moisture. Such structures are absent from or poorly developed in amphibians and reptiles, which do not have physiological control of their body temperature.

The apparent absence of comparable bones from the fossils of dinosaurs was once considered proof that dinosaurs could not have been warm-blooded and were therefore an unlikely choice as the ancestors of birds. It is not certain that turbinate bones were actually absent from dinosaurs, however. They are extremely fragile structures and could have been supported by weakly mineralized cartilage, making them unlikely candidates for fossilization. Modern reconstructions of dinosaur heads suggest that there was ample room for such structures in the nasal area.

During the 170 years or so since the discovery of dinosaurs, there have been many attempts to reconstruct the living animal. In the process, they have been given a wide variety of noses. Some of the early drawings showed fiercely flared, bearlike nostrils. Then there was a period when dinosaurs were seen as sluggish reptiles, and the noses became more crocodile-like. Recently, dinosaurs have been reinterpreted as birdlike animals, and the nose and its nostrils have changed shape yet again.

The shape of a dinosaur's nose may seem like a profoundly unimportant topic, but it is one of those minor issues that may lead us to a better understanding of some general principles behind animal structure.

Because small bony structures are rarely preserved in dinosaur fossils, our understanding of these animals' olfactory system is generally weak. Often, we have only the major bones from the skull. They provide ample evidence of the importance of weight reduction in the head. A dinosaur skull often looks more like a collection of large empty spaces than a specialized container for important organs. The opening for the nasal passages is often the largest space. It may take up most of the face and be half as long as the whole skull. Without other evidence, reconstructing the position of the fleshy nostril within such a large space has been as much a matter of artistic taste as biology. Some artists have increased the birdlike appearance of a dinosaur by placing its nostril high on the face, near the top of the internal opening of the nasal passage. This may be consistent with current ideas about a dinosaur's lifestyle, but there is fossil evidence that such nostrils significantly misrepresent the animal's appearance in life.

In birds, the external nostril appears to lie high on the face because the lower part of the face is unusually long. The premaxilla, which forms the tip of the beak, has extended very far forward and, in a sense, given the bird a very long "upper lip." In other animals, this bone occupies the narrow gap between the nose and the upper teeth. In all animals, the rear edge of this bone forms the forward edge of the nasal opening in the skull, and the external nostril lies just above that edge.

Birds make use of this long forward extension of the face to feed and preen. Its length does not interfere with olfaction because birds rarely depend on a sense of smell to locate their food. Even when food is close to the mouth, they make do with an array of shorter-range chemoreceptors along the tip of the beak. There are two notable exceptions. The petrels, mentioned above, use their tubular nostrils to locate distant food, but there is still quite a wide gap between the funnel-like openings and the tip of the bill. The other noteworthy exception is the flightless kiwi of New Zealand. It depends on olfaction to find earthworms in loose soil, and its nostrils are at the very tip of its bill. Its design helps illustrate the fundamental similarity of birds to other kinds of animals and may be useful in understanding the construction of dinosaurs.

In most animals, the nostrils lie closer to the mouth than to the eyes for straightforward physiological reasons. If the nostrils were located near the top of the nasal opening, the air would pass directly into the upper respiratory system, bypassing any olfactory structures in the nasal cavity and losing the opportunity to yield its chemical information. Lawrence Witmer examined facial structures in 26 families of birds, 4 families of lizards, a family of crocodilians, and some turtles

[231]. In every case, the external fleshy nostril lay at the front edge of the bony nostril or was located in some fleshy extension that carried it even further forward. The human nose is just such a structure. Among living animals, there are no examples of a fleshy nostril lying near the back of the opening to the bony nostril. Even in whales, the fleshy nostril is in a typical location in relation to the bony nostril, even though the latter has migrated to a position high on the skull.

Witmer also found clues to the correct position of the fleshy nostril in fossil animals. The olfactory responsibilities of the nostrils are assisted by a mass of fleshy sheets and whorls called cavernous tissue. The activity of these tissues requires a substantial blood supply, which leaves marks on the adjacent bone that can be detected in fossils. In some cases, vessels actually pass through the bone; in others, they merely leave grooves on the surface. These traces indicate the position of the fleshy nostril in relation to other facial features. In all the living taxa that Witmer examined, the traces indicated that the nostril lay near the front of the nasal opening in the skull.

When he examined dinosaur skulls, he found the remains of a complex of blood vessels near the nasal opening. In life, this "nasal vestibular vascular plexus" would have supplied a rich flow of blood to the area and supported important physiological structures behind the nostrils. Its size implies that dinosaurs, like most other vertebrates, had a well-developed olfactory system; its location suggests that there was no reason to portray dinosaurs differently from living animals. Reconstructions should show a fleshy nostril near the front of their head, immediately above the edge of the nasal opening in the skull, where it is in living animals.

The auditory system is the third of the long-range sensory systems mounted on the skulls of vertebrates. Unlike the olfactory and visual systems, the actual receptors are quite compact and are typically packed into small otic capsules at the base of the skull. In mammals, the performance of the receptors is enhanced by external ears, large fleshy reflectors that direct the incoming signal to the small entrance of the auditory canal. Birds lack fleshy external ears, and there is no evidence in the fossil record of such structures on dinosaurs.

In some ways, it is surprising that birds lack external ears. Most make extensive use of songs or calls as part of their communication system, but only a few have any special adaptations to enhance hearing. In most birds, the opening of the ear is a rather small hole near the base of the skull; there are no external structures of any sort. Even the Oilbirds and cave-swiftlets, which use a basic form of echolocation to navigate in the dark, lack the elaborate external ears and other facial structures that are so characteristic of sonar-using bats. The only comparable avian hearing "aids" are constructed from feathers. The specialized facial plumage of owls reflects high-frequency sound into the ears. Beneath those feathers, some owls also have asymmetrically arranged ear openings that may help them locate prey very precisely in total darkness. Humans have symmetrically placed ears and can easily identify sounds coming from the left or right. In owls, one ear sits higher on the head than the other and may enable them to "hear in three dimensions" by improving their ability to separate sounds that are close from those that are a little more distant.

Although dinosaurs roar and scream on film and television, there is little evidence that hearing was critically important to them. On the other hand, momentum generated by the head was an important design consideration; like those of birds, dinosaur skulls show only small otic capsules. For most species, external ears would probably have been unnecessary additions to the weight of the head. In the absence of fossil evidence one way or the other, paleontologists have taken a conservative approach and modelled their dinosaurs on earless lizards and birds. *T. rex* and *Velociraptor* already had relatively large heads with big eyes and heavy teeth, however. The extra weight of external ears would have been negligible, while acute directional hearing could have been very useful for locating unseen prey. Perhaps external ears will be discovered on some exceptional fossil in the future.

Feeding Apparatus

The greatest trade-offs between weight and function in the skulls of both birds and dinosaurs have involved the feeding apparatus. Predatory mammals have massive, thick skulls to carry killing, cutting, grinding, or crushing teeth as well as the heavy-duty muscles needed to operate them. A long, fleshy tongue manipulates their food towards those teeth, and digestive processes begin with a flood of saliva from large glands in the mouth. Some species can even store excess food in expandable cheeks or beneath heavy lips. Both birds and dinosaurs are designed to handle food in an entirely different way.

Birds lack both large muscles on their head and heavy grinding surfaces on their jaws, so most cannot use their mouths to begin the process of mechanical reduction of food. Only cuckoos, turacos, and the Hoatzin come close to chewing their food; other birds lack suitably specialized structures in their beaks [109]. The last of the toothed birds appears to have died out 65 million years ago. Their simple peglike teeth were useful for gripping prey but not for grinding it up. Birds typically eat items that can be swallowed whole or are easily broken into suitable mouthfuls, but vultures, hawks, and even gulls tackle tough items that are far too large to swallow whole. They depend on the sharp edge on their beak to tear off convenient chunks of food. If saliva is secreted into the mouth, it is only for lubrication; chemical digestion does not begin until the food reaches the foregut. Mechanical reduction begins in the muscular stomach or crop, where a store of gravel often helps to break up the food.

Birds get some help in handling large food items from the hinge between their brain case and the beak. The hinge allows the upper jaw to rotate against the skull (prokinesis), and the mouth can increase its gape another 5% or so. Dinosaurs also got some help ingesting large items from a hinge in the skull. Their bending zone lay directly above the eyes, near the middle of the skull (mesokinesis). Dinosaurs had robust lower jaws, but the bird's lower jaw is relatively weak and its simple construction leaves it useful only for holding food in place. The avian tongue is often not very flexible but may be armed with arrays of spines that pull food into the mouth and down the throat.

Dinosaurs appear to have eaten in much the same way as birds. Large predatory species appear to have swallowed prey whole or in large pieces, while specialist herbivores had teeth designed to cut and chop food or grind it to a pulp. All the teeth in a dinosaur jaw were usually very similar in shape and never showed the variety of specialized designs found in mammals. Dinosaurs also lacked the massive cranial muscles that mammals use for crushing bone, and no dinosaur teeth were robust enough to have taken such abuse. Large herbivorous species swallowed sizable rocks to help the mechanical reduction of their food, and these highly polished "gastroliths" are a very common form of trace fossil. Interestingly, some dinosaurs abandoned teeth and developed a horny, birdlike beak.

The Neck

Perhaps the feature that makes birds most like dinosaurs is the long, flexible neck. In both birds and dinosaurs, a longer neck means more vertebrae and usually more flexibility. In mammals, long necks are stiff and constructed from a small number of elongated bones. All birds depend on the flexibility of the neck to enable the beak to meet its wide range of responsibilities. The neck may have had a similar but less critical role in dinosaurs. The small meat-eating theropods had a flexible neck even though they had long-fingered hands and an opposable thumb for handling prey, and the giant herbivores had a heavily muscled neck that balanced the weight of their massive tail.

We can only guess at the normal posture of the long neck in dinosaurs. A hundred years ago, the brontosaurus was frequently shown with a vertical neck, but modern reconstructions of large herbivores suggest that the neck was held horizontally, suspended by tendons from the massive bones of the shoulder. Swans, ostriches, and the few other birds that habitually carry their necks in an extended posture hold it vertically, reducing the need for heavy musculature. A vertical posture also keeps the neck close to the centre of gravity, near the midpoint of the body, and eliminates the need for a counterbalancing tail. Most other birds pull their head close to the centre of gravity by slinging their neck in a loop, tucking the head close to the body. A thick layer of breast feathers hides the neck, making it appear much shorter than it actually is.

Shared Features of the Body

Forelimbs

Paleontologists believe that the ancestors of birds lie among the maniraptoran theropods because this group included many small, active predators. They had unusually well-developed forelimbs, and recently discovered species, such as *Buitreraptor,* had elaborate plumage [118]. It is difficult to understand how fingers encumbered by long feathers functioned successfully as grasping hands, but the maniraptoran dinosaurs also had the characteristically birdlike way of folding the wrist. In most mammals, the wrist folds downward so that the fingertips can approach the elbow, depending on the flexibility of the joint. In birds and theropod dinosaurs, the hand rotates outward and back when the wrist bends. If you place your hands flat on your chest, you cannot keep them flat and rotate them so that all your fingers point at your navel. A bird could, if it had a navel.

The unique action of the wrist in theropod dinosaurs is dependent on the presence of a semi-lunate carpal. In 1969, John Ostrom's understanding of the importance of the semi-lunate carpal made his interpretation of *Deinonychus* one of the most important events – perhaps the most important – in dinosaur paleontology during the 20th century [154]. His claim that this bone was homologous to similar structures in the wrist of birds reignited the debate over the origin of birds among dinosaurs and led to general acceptance for their origin among theropods. Now, the semi-lunate carpal is seen as only one of many structural characters shared by the two groups.

When a long-fingered theropod ran after its prey, it probably spread its fingers in a fan, much as an owl or hawk opens its talons as wide as possible to increase the likelihood of a catch. If the prey tried to escape into thick shrubbery, the fingers of the theropod could simply fold back and lie loosely along the side of the body. In this position, they would not snag on passing branches and might offer some protection from sharp sticks and thorns, especially if they were covered with robust feathers. In the bird's wing, the fingers automatically rotate forward, to spread the feathers at the tip, as the arm is extended. A similar mechanism could have automatically flicked the fingers of the theropod open as it reached for its prey. The unusual wrist joint also allows the folded hand, and later the folded wing, to rest in the appropriate position for incubation of eggs or the brooding of nestlings. One of the most famous fossils of *Oviraptor* appears to have died while incubating its nest, just like a bird.

The origin of the characteristic forelimb downstroke in birds is much more difficult to comprehend. No one has proposed a credible use for a powerful, synchronized downward pulse of the forelimb in a terrestrial theropod. Perhaps the early ancestors of birds used a coordinated two-handed snatch to seize their prey.

The Furcula

Many other similarities between birds and dinosaurs are rather general and could be the result of evolutionary convergence responding to an architectural strategy based on bipedal locomotion and a body hung from the backbone. Both groups have two very specialized structures not found in other vertebrates, however: the furcula (wishbone) and the neural plexus.

For many years, no dinosaur fossil was found with a furcula, and its absence was important in the argument that birds could not have evolved from dinosaurs. Furculae have now been found in so many species of bipedal carnivores that the bone is considered to be a synapomorphy, or shared ancestral character, of all the tetanuran (stiff-tailed) dinosaurs, including *T. rex*. The function served by the furcula in the theropods is unclear, however, and it only vaguely resembles the wishbone of a flying bird. To be important evolutionarily, it must be derived from the same bones and tissues in the embryo. Perhaps we will be able to gain some insight by examining the remains of embryos extracted from fossil eggs that are being found in increasing numbers.

The theropod furcula is little more than a simple bar with a slight bend in the middle that may have helped to spread the shoulders and keep them in place. In modern flying birds, the typical furcula has a deep U-shape. The responsibility for spreading the shoulders lies with the large coracoid bones resting on a wide sternum. Significantly, the furcula found in theropods looks much like the furcula found in the oldest known bird, *Archaeopteryx,* and in the later Confucius birds. The adoption of the U-shaped furcula is an important part of the transition from early dinosaurlike birds to more modern forms, and is closely linked to the development of advanced flight capabilities. We will examine it in detail when we look at fossil birds in Chapter 8.

The Neural Plexus and the Glycogen Body

The other special structure that links birds and dinosaurs is the neural plexus that lies in the hips. Early anatomists thought that the brain in the head of a dinosaur was far too small to control the body of such a giant. They decided that a widening of the neural tube, near the hips, provided space for the extra brain that a primitive and slow-witted animal must have needed to control the after part of its body. If this organ had been an actual brain, it would have been about 20 times the size of those found in primitive mammals. The presence of a very similar structure in tiny birds was much more puzzling and its explanation continues to challenge biologists [47]. Some modern theories about its function come surprisingly close to the original brain-in-the-tail theory.

We cannot know much about the internal structure or metabolism of the organ that occupied the neural plexus in dinosaurs. None of the soft tissue survives but the structural details on the lining of the bony capsule that housed the organ suggest that it was a glycogen body, just like those found at the same location in modern birds [71]. As the name implies, the glycogen body is a mass of glycogen-rich tissue lying alongside the spinal cord. Lobes, called accessory bodies, extend from it into the space adjacent to the spinal column.

The activities of the glycogen body and its accessory lobes continue to puzzle avian physiologists even though it has been extensively studied, in part because of its possible effect on profit in the poultry industry. The fact that this body stores a lot of energy in the form of glycogen suggests that it represents a major commitment of resources by the embryo. Its growth could affect the development rate or eventual size of the hatchling and could therefore be economically significant in the production of chickens. In spite of careful examination, the glycogen body has not given up its secrets easily. It does not appear to supply nutrients to the body or shrink during periods of stress or starvation. Biochemical examination has revealed that it is not part of the hexose (six-carbon sugar) metabolism cycle that controls much of the body. It lacks appropriate enzymes and stops its metabolism at the pentose (five-carbon sugar) stage.

This unusual metabolic characteristic has led to other theories about its role. The presence of glycogen cells along many parts of the spinal cord suggests links between the glycogen body and the autonomic nervous system and the possibility of a neurological function. For a while, the glycogen body was believed to help in the maintenance of the myelin sheath. Myelin is fatty material that surrounds nerve fibres and insulates them so that they can carry an electrical signal without interfering with adjacent tissues. The glycogen body appeared to be relatively inactive, however, and no neurological function was ever demonstrated in a conclusive way.

Reinhold Necker has been leading an investigation into the physical structure and neural connections of the glycogen body and the pairs of accessory lobes attached to it. His findings echo some of the earlier ideas about its function as a spare brain [138]. It has long been known that the glycogen body is not a solid mass but includes a complex system of fluid-filled canals. The smaller accessory lobes also have fluid-filled spaces; electron microscopy has revealed that fine hair-like structures extend into the fluid in much the same way that tiny fibres extend into the canals of our inner ear. Necker and his colleagues are certain that the tiny fibres are sensory and imply that the glycogen body is an organ of balance. In a bipedal animal such as a bird, it makes a great deal of sense to have a major organ of balance close to the hips. It is also midway between the wings and the tail, so it could play an equally important role in front-to-back balance and orientation of the body during flight. The neural plexus in a living bird may not be an actual "brain in the tail," but it may still be responsible for control of movement in the after section of the body.

The Shell of the Egg

In the past few years, paleontologists have uncovered a large number of dinosaur eggs, some even containing recognizable embryos. The eggshells have also proven useful and suggest that theropods were a more primitive type of animal than a bird. The shell structure of theropod eggs is very similar to that of bird eggs but not exactly the same. Theropod shells have only two layers, while birds seem to be more advanced, with a three-layered shell. The fact that the eggs were frequently laid in pairs suggests that the theropods had two functioning ovaries. The reproductive system in birds evolved from a two-ovary structure, but birds lay eggs one at a time and have only a single ovary [75].

It would be interesting to compare these dinosaur eggs with those of early birds to see at what point birds evolved a more elaborate eggshell. Unfortunately, there are very few fossils of eggs from the earlier birds and they may not be suitable for analysis.

Seeing Dinosaurs as the Ancestors of Birds

Dinosaurs as Lizards

Dinosaurs have not always been seen as birdlike animals. In the 1840s, Richard Owen invented Class Dinosauria and reconstructed its known members as squat, bearlike animals (Figure 3.3) that walked on the flats of their feet. Even though Thomas Huxley and others recognized dinosaurlike features in *Archaeopteryx*, dinosaurs remained lumbering "thunder lizards" for another hundred years. A genetic relationship between such animals and the backyard sparrow flew in the face of common sense and was contrary to Gerhard Heilmann's more convincing theory of a relationship between birds and ancient thecodonts [82]. Reptile-like reconstructions of dinosaurs remained popular, and there was little new support for a link between dinosaurs and birds until the end of the 20th century [6].

Ironically, fossil evidence for giant birdlike animals in the Jurassic and Triassic was older than the word "dinosaur" and much older than knowledge of *Archaeopteryx*. The Reverend Edward Hitchcock described the Ornithicnites as a new class of animals in 1836. He extrapolated his description of the group from birdlike footprints that he found in Jurassic and Triassic rocks in Connecticut. From the shape of the track, he deduced many of the features that we now associate with the advanced theropod dinosaurs – the same group that we now think may have given rise to birds.

Although the footprints came in a variety of sizes – from the crow-sized *Grallator* to the two-tonne *Grandipus,* they all shared certain birdlike features. The tracks were narrow, like a bird's, and were left by a fan of toes that met at a narrow central joint. The fan shape implies that the "ankle" connected to a thin bone similar to a bird's tarsometatarsus. There were only occasional impressions of a hand, and little evidence of a heel. Impressions of the hallux suggested that it was

Figure 3.3 A Japanese carver visiting the 1855 Festival of Britain captured Richard Owen's reconstruction of an iguanodon in this 19th-century netsuke.

mounted higher on the leg than other toes. By examining the impressions left by fleshy pads under the feet, Hitchcock worked out the toe formula (number of phalanges in each digit) for his giant birds. Working outward from digit 1 to digit 5, he decided that the formula was 2, 3, 4, 5, 0, not 2, 3, 3, 3, 3 as found in mammals or attributed to dinosaurs in Victorian reconstructions. In addition, he found occasional marks that were likely left by a long reptilian tail. The possibility that a bird could have such a tail or that an avian hand could have independent digits would not arise until the discovery of *Archaeopteryx* 25 years later. Now that the birdlike features of dinosaurs are better understood, the footprints in Connecticut provide important insight into species diversity in early communities of dinosaurs [233].

Conclusion

For many years, biologists rejected the idea of a relationship between birds and dinosaurs simply on the basis of size. The known dinosaurs were large, ponderous animals, whereas even flightless birds were smaller and more lightly built. Only recently have we found truly birdlike dinosaurs. One of the most important is *Eoraptor.* This fossil got off to a bad start as the back end of a fraudulently constructed chimera that fooled many paleontologists and was detected only by the application of sophisticated forensic tests [180]. It has now been legitimized and redescribed as an exceptionally small dinosaur, one that could have lived in the trees and may have had its own form of flight. It was also covered in feathers, and reconstructions suggest that it could have behaved something like a modern flying squirrel, competing with Mesozoic birds for resources in the forest canopy. *Eoraptor* is important to the story of birds and dinosaurs because it shows that there were no lower limits to the size of dinosaurs. It was small without being a bird.

   Eoraptor and a few slightly larger dinosaurs, such as *Caudipteryx,* could mimic the lifestyles of birds without losing their identity as dinosaurs. The other side of the coin has many more examples. Birds have frequently taken advantage of special circumstances (usually the absence of mammals) to become large, flightless bipeds that mimicked dinosaurs but never lost their identity as birds. After the dinosaurs became extinct at the end of the Cretaceous, some of these flightless birds became ferocious predators that competed with early mammals in a head-to-head struggle for biological domination of the globe. The flightless giant birds failed and became extinct, but it is not clear whether this failure was due to their being too much like birds or too much like dinosaurs.

   Although some of the flightless birds bear a superficial resemblance to dinosaurs, the similarity ends with their appearance as bipedal animals. They have not abandoned the highly centralized structure that characterizes the inner-bird strategy. They never redeveloped teeth or gripping fingers to handle prey, and they did not return to a more dinosaurlike leg. Certainly none of them has redeveloped the long reptilian tail. They have remained birds and have retained all the differences between birds and dinosaurs that we will examine in the next chapter.

Further Reading

Since 1995, there has been a steady stream of books about dinosaurs and birds that are both technical and readable. In the following list, the most purely technical are marked with an asterisk:

Benton, M.J., M.A. Shishkin, D.M. Unwin, and E.N. Kurochkin (eds.). 2000. The age of dinosaurs in Russia and Mongolia. Cambridge University Press, Cambridge, UK.

Chatterjee, S. 1997. The rise of birds: 225 million years of evolution. Johns Hopkins University Press, Baltimore, MD.

Chiappe, L.M., and L.M. Witmer (eds.). 2002. Mesozoic birds: Above the heads of dinosaurs. University of California Press, Berkeley, CA.*

Currie, P.J., E.B. Koppelhus, M.A. Shugar, and J.L. Wright (eds.). 2004. Feathered dragons: Studies in the transition from dinosaurs to birds. Indiana University Press, Bloomington.*

Dingus, L., and T. Rowe. 1998. The mistaken extinction: Dinosaur evolution and the origin of birds. W.H. Freeman, New York.

Feduccia, A. 1996. The origin and evolution of birds. Yale University Press, New Haven, CT.

Gauthier, J., and L.F. Gall (eds.). 2001. New perspectives on the origin and early evolution of birds: Proceedings of the international symposium in honor of John H. Ostrom. Peabody Museum of Natural History, Yale University, New Haven, CT.*

Paul, G.S. 2002. Dinosaurs of the air. Johns Hopkins University Press, Baltimore and London.

Shipman, P. 1998. Taking wing. Simon and Schuster, New York.

A Bird Is Not So Like a Dinosaur

4

*Features that distinguish
birds from their reptilian ancestors*

While it is clear that birds are somewhat like dinosaurs and share such fundamentals of structure as the bipedal stance and the diapsid skull, it is also clear that they are very different. *Archaeopteryx* and other early types of birds were much more like dinosaurs and differences could be determined only by close examination. At some point in their long history, however, birds abandoned the dinosaur's reptilian profile and evolved into the modern bird, with its highly centralized body. Mysterious events at the end of the Cretaceous period swept away the dinosaurs and the remaining dinosaurlike birds, but modern birds, with their "inner-bird" body plan, survived to become some of the most successful vertebrates in the history of the Earth. In this chapter, we will examine the structural changes that contributed to the success of birds and that now distinguish them from their ancestors.

The Ancestral Condition: Hips, Legs, and Tails

It seems obvious that flight should dominate any story of avian evolution but it does not, at least not in the opening chapters. The long legs and sleek lines of *Archaeopteryx* make it clear that a quick escape across the ground was just as important for survival as getting into the air. In many ways, *Archaeopteryx* was just another small, scurrying ground dweller. The bones in its body are slender and offer little space for the attachment of the large muscles that we associate with the flight of modern birds. There is nothing to suggest that it could generate sufficient muscle power for a quick take-off. In fact, only its feathers suggest some flight capability, and there is considerable doubt about its skills in the air. It may have been only an inefficient glider.

In some ways, the differences between *Archaeopteryx* and the inner bird are exactly what would be expected between an earliest known form and its modern descendant. Long before the ancestors of *Archaeopteryx* could fly, they must have successfully eluded terrestrial predators by running away. It is the only way for them to have passed on their characteristics to modern descendants with more significant flight capabilities. The hind limbs remain an important survival tool that still helps birds to walk, run, hop, swim, and climb away from danger. Not surprisingly, the history of the evolutionary changes to the hips, legs, and tail is a major part of the history of the whole bird.

Fossil footprints sparked an early interest in the ability of dinosaurs to run and walk, and played an important role in shifting opinion from dinosaurs as sluggish reptiles to dinosaurs as active animals. Examination of the implications of bipedal locomotion is a more recent event. In 1996, Stephen Gatesy and Kenneth Dial outlined a general theory of pedestrian locomotion in dinosaurs and birds based on the organization of the body into discrete functional units that they called "locomotory modules" [67]. The crawling action of a primitive tetrapod illustrates the use of the body as a single locomotory module. The forelimbs and the hind limbs are linked to each other neurologically but they move alternately while the head and tail swing from side to side to compensate for shifts in the centre of balance of the body. The bodies of bipedal dinosaurs were far more sophisticated but also functioned as a single locomotory module, in part because their forelimbs abandoned a role in locomotion for new responsibilities in the capture of prey while their bodies still moved as an integrated unit. When such a dinosaur walked, it needed to shift its weight onto the supporting leg and would have rocked from side to side as it moved forward. To keep its balance, it needed to make compensating movements with the head and tail. We can still see echoes of the dinosaur's rolling gait in walking birds. Most small modern birds hop about on the ground but crows and other larger species often walk, swinging their tail back and forth. The tip almost, but not quite, touches the ground, as though the weight of the feathers could counterbalance the lightweight bird just as a massive, fleshy tail would have counterbalanced *Tyrannosaurus rex*.

Although birds have much in common with the bipedal dinosaurs, their earliest known ancestor, *Archaeopteryx*, already possessed two independent locomotory modules. One was based on the wings attached to the forelimbs; the second took in the whole rear segment of the body. In modern birds, the tail functions as a third, partially independent locomotory module, but there is no fossil evidence that the action of the tail in *Archaeopteryx* was coordinated with the movement of its wings. On the ground, the movement of the long frondlike tail would have been dinosaurlike and coordinated with the movements of the legs. In the air, its shape might have provided additional lift but it would not have been a particularly effective rudder. Consequently, *Archaeopteryx* is not likely to have been very manoeuvrable in the air and may have flown in long straight lines, much like a modern pheasant. The replacement of the long reptilian tail with the manoeuvrable tail fan of modern birds is one of the most significant events in avian evolution.

In modern birds, the fleshy tail has become a mere nub, but its size is out of proportion with its importance as a locomotory module. Unlike the ancestral frond, the modern tail fan has the ability to change shape and perform a variety of unrelated functions. It does not operate independently but it can open or close to match the action of the wings in the air or coordinate its actions with the motion of the legs on the ground. While they were developing their highly specialized tail, birds also undertook a complete redesign of the entire abdominal area, including the skeleton of the legs and hips and the massive muscles responsible for moving the hind legs.

The changes that accompanied the separation of locomotory modules in birds must have provided important evolutionary advantages because we see the beginning of comparable trends, particularly a shortening of the tail, among the advanced theropods of the Upper Cretaceous. Some of these changes had been anticipated in *Archaeopteryx* some 10-25 million years earlier. These parallels with developments in avian architecture are so widespread and the structural similarities so comprehensive that Gregory Paul has suggested that they represent a genetic legacy. He proposes that at least some of the advanced theropods were flightless descendants of early birds [157].

Becoming a Bird

The Avian Tail

The profile of the folded fan of tail feathers may bear a superficial resemblance to the long tail of dinosaurs but any similarity ends there. The modern avian tail has a broad range of functions that can be achieved only in a structure that can change shape and be moved rapidly by a small array of muscles. Some of these functions may require the tail to be long and strong, but none of them requires it to be heavy.

Although the bird's tail has been reduced to a fleshy nub that is often derided as the "parson's nose," it still plays an important role in the locomotory system. The fleshy part is far too small to counterbalance the weight of the body's front end but it anchors 10 or 11 pairs of large tail feathers (rectrices). These feathers can be folded or fanned during flight, and birds have learned to use them as a control surface. By changing the tail's size, they can manipulate air resistance to steer or change the amount of lift, just as they adjust the wingtips for finer control.

The skeletal support for the tail is called the pygostyle, a bone that offers little hint of its descent from the massive tails of dinosaurs. During embryonic development, the pygostyle forms when a string of vertebrae fuse together. The result is usually a simple vertical plate housing the end of the spinal cord. Lateral blades near the bottom edge often give the pygostyle a shape that is somewhat reminiscent of a plough or the tail of an aircraft. *Archaeopteryx*'s dinosaurlike tail had no terminal structure and its 12 pairs of rectrices looked more like the frond of a fern than a fan. There was one pair of rectrices for each caudal vertebra, only one more pair of feathers in total than the tail of some modern birds.[1]

---

[1] There is another similarity between the tail of *Archaeopteryx* and that of the modern bird. In *Archaeopteryx*, the five tail vertebrae closest to the hips have transverse processes that are absent from the later vertebrae. Similar transverse process are present on the free tail vertebrae of modern birds but are only hinted at in the pygostyle.

Archaeopteryx had about 26 caudal and pelvic vertebrae, and it is still possible to account for about two dozen bones between the hip joint and the tip of the tail in modern birds (Figure 4.1). At the anterior end of the tail series, some of the original tail vertebrae have become anonymous components within the synsacrum, and only passages for the paired spinal nerves indicate their onetime independence. At the end of the spine, the vertebrae are so completely fused into the pygostyle that even passages for the lateral nerves have disappeared. Between the two fused sections, 5 to 8 caudal vertebrae survive as independent bones to give the tail the mobility it needs to function as a rudder in aerial flight (Table 2.1).

The appearance of a pygostyle in all living birds does not necessarily mean that the genetic information for the construction of the long reptilian tail has been completely erased. It may be so deeply buried, however, and the development of the modern tail so intimately linked to other developmental events that the ancient genes are very rarely expressed. In spite of the billions of chickens produced every year, there is only one plausible record of a reptilian tail appearing in a bird. Ulisse Aldrovandi (1522-1605) described a rather intimidating rooster with a long, fleshy tail that was kept by Duke Francesco de' Medici of Tuscany and he included a reasonably convincing illustration of the bird in a chapter on freak chickens in his *Ornithologia* [115].

The pygostyle is often very tiny and can be difficult to see, especially in birds that have given up flight or fly very rarely. It can be very difficult to find one in any of the paleognathous birds. It is often missing from prepared skeletons of the ostrich; if the kiwi has one, it is not very different from the other rather shapeless vertebrae that end the tails of ratite birds. Hans Gadow states that both the ostrich and the kiwi have a pygostyle [63]. There is little sign of a pygostyle in the tinamous even though they can fly. In 1898, Frank Beddard wrote: "There are at most faint traces of a ploughshare bone" [8].

The whole flight apparatus of the tinamous is more primitive than that of the galliform birds and they rarely put it to use [197]. They cannot take off as explosively as galliform birds, and when they use sudden flight to escape from predators, tinamous travel in a more or less straight line. Perhaps they fly so rarely and for such short distances that they can get along without much ability to steer. The tinamous appear to have had relatives in the early Tertiary that were stronger fliers. These lithornithids also had small, almost insignificant pygostyles and may not have been very manoeuvrable in the air.

Although they are often strong fliers, waterbirds often have poorly developed pygostyles. Auks fly at high speeds but are not noted for their aerial manoeuvrability. They have very short tails and their pygostyle may be so small that it is difficult to see. Grebes also fly well but have virtually no tail fan. Their pygostyle is as undeveloped as that of a kiwi. When short-tailed waterbirds need to steer or produce extra lift at low air speeds, they fan out their webbed toes instead of their tail. Many long-legged wading birds also have small pygostyles and short tail feathers, and they also use their feet during flight. When herons or cranes fly, they trail their long legs behind so that the drag of air on the feet can stabilize the bird's posture, just as a dragging tail stabilizes a kite.

A typical bird with some degree of aerial manoeuvrability changes the shape of its tail fan by deforming the pair of fleshy bulbs that act as a base for the tail feathers. Abdominal muscles squeeze the bulbs against the central ridge of the pygostyle, causing the tail feathers to fan out horizontally. Predictably, the pygostyle is large and most complex in birds such as the frigatebird, which makes use of its dramatically elongated tail feathers in highly aerobatic flight.

Small forest birds often take advantage of their long tails as a display. When the bird is standing on the ground or resting on a perch, the tail can be fanned out vertically as a signal to potential mates or a threat to competitors. The tail often participates in song by twitching up and down, driven by the remnant of the caudo-femoral muscle, which once powered the hind legs of dinosaurs. The flicking of the tail may help to pump blasts of air through the syrinx (voice box), but its movement also catches the eye and augments its value as a visual signal. When performing these auxiliary functions, the tail behaves as an independent unit, but when it is involved in locomotion, the tail always acts in concert with one of the two main locomotory modules. The abdominal muscles that open the tail fan, swing it sideways, or twist it to steer act in coordination with either the wings or the legs.

The small size of the caudal muscles suggests that the tail can be held in a certain position but otherwise it is merely a passive surface, acted upon by moving air. Even though its muscles cannot generate a great deal of power, the tail may assist in take-off in very small birds. Small forest birds have tails that are so large in proportion to body weight that a sudden downward flick might well provide extra lift when the bird springs off its toes, into the air.

Even the shortest tail may have a useful function in the air. Ducks, loons, auks, and other fast-flying birds, and highly efficient fliers such as the albatross, seem to make little use of their small tail for steering but it may play an important role as a splitter plate [81]. A splitter plate is a thin trailing edge that extends behind a body moving through a fluid medium. It helps to extend the zone of laminar flow in the streams of air passing over and under the body, and delays the point at which the two flows meet and cause turbulent flow. Some otherwise short-tailed birds, such as the Long-tailed Duck have a long, flexible central plume. This streamer may lead vortices of turbulence even further back, to a point far behind the bird where their departure no longer generates drag.

Feathers are fragile, non-living structures that must be replaced when exposure to the elements causes their deterioration. In spite of potential limitations in the use of feathers, birds have adapted the tail feathers for some unusual purposes and even exploited their supposed weaknesses. When the tip of a woodpecker's tail feather wears away, it exposes the sharp end of an exceptionally stiff central rachis. Woodpeckers use these spikes like a mountaineer's crampons when they climb tree trunks. Sharp claws on the feet grip the surface while the tips of the tail feathers brace the bird against the trunk, keeping the body stable while the bird excavates insects from under the bark. Surprisingly, no bird uses its tail as a weapon. Perhaps it is too light and fragile a structure, but it seems surprising that no modern birds have porcupine-like bristles to ward off attackers. Only a few species, such as the lyrebird, have a tail with feathers large enough and long enough to use as a noise-maker or rattle.

There could be a connection between feathers and defensive structures in the ancestors of birds. It was once thought that bird feathers evolved from simple reptilian scales, but feathers are so specialized and have such a unique chemistry that this seems unlikely. In some ways, feathers are much more like claws. Both are tubular structures excreted by the skin under the control of certain hormones and genes. Perhaps in some ancient ancestor, elongated, erectile "claws" grew in rows along the side of the body for protection, somewhat like the peculiar scales that appear in the modern-day scaly anteater or pangolin. If rows of such claws were the precursor of modern feathers, they might explain the peculiar arrangement of body feathers in discrete tracts (pterylae), and the occasional appearance of claws instead of feathers on the digits of the wing in modern birds.

The fleshy body of the bird's tail has one other duty for which there is no evidence in the fossils of dinosaurs. Its upper midline is the location of the uropygial gland, which produces quantities of concentrated sebum. A preening bird can collect sebum from the gland on its bill and apply it to any feather tips that have not received enough directly from the skin. This gland and a pair of smaller and less well-known cloacal glands (i.e., not mentioned in most textbooks) are the only sebaceous glands in a bird's skin. Except for feather follicles, they are the only external gaps in the sclera that covers a bird's body inside and out.

Developing a New Walk

In squamosal animals, large lateral muscles generate the sinusoidal movement that enables them to wriggle across a surface. Some of those muscles are attached to the legs and pull them back to push the body forward. When the ancestors of dinosaurs adopted the parasagittal stance, those muscles retained their locomotory duties, and one of the major responsibilities of the tail was to house the powerful caudo-femoral muscle, which drove the hind leg. The story of how the caudo-femoral exchanged a major role in locomotion for a minor role in respiration is a major part of avian evolution but it separated modern birds from ancestors that had already become distinct from dinosaurs.

When the ischium extended back to surround the base of the bird's tail, the muscles attached to the tail either disappeared or were assigned new duties. The elongation of the hipbone simply took up the space needed by the large caudo-femoral muscle and other muscular linkages between the tail and the hind limbs. At the same time, a forward extension of the ilium prevented the development of new muscular connections between the hind limb and anterior parts of the abdomen or thorax. This expansion of the two major hipbones created a broad surface with ample space for the attachment of a new set of muscles. The subsequent use of large thigh and hamstring muscles to move the hind leg offers one of the clearest differences between modern birds and their immediate ancestors.

When a dinosaur wanted to take a step with its left foot, the massive caudo-femoral muscle on the right-hand side of the tail pulled the right leg backward and swung the tail out to the right, shifting the weight off the left foot and freeing it to take a step forward. A mirror image of the process would then move the right foot. Repetition would have generated large swings of the tail and a swaying from side to side that we believe was characteristic of the dinosaur gait. A series of epaxial muscles that was continuous along the chain of vertebrae in dinosaurs generated a characteristic sinuous movement in their spine, still seen in lizards and crocodiles. On a smaller scale, the same series of events occurred when *Archaeopteryx* walked or ran across the ground. When fusion of the sacral vertebrae stiffened the bird's back, it interrupted the series of epaxial muscles and birds were forced to replace the sinuous walking movement of the dinosaur with a more rolling gait, in which the whole body swings from side to side as a unit. Usually this movement is most noticeable in short-legged birds such as ducks, where it reaches comic proportions.

Gatesy and Dial teamed up with Kevin Middleton to sort out the various walking gaits of dinosaurs and birds [67, 68]. They found that the two groups had fundamentally different suspension systems. Dinosaurs hung from hips perched on top of pillar-like legs that stood vertically to absorb the compressive forces of gravity. When a dinosaur walked, the femur was exposed to relatively simple bending forces by the contraction of the caudo-femoral muscles, and the leg rotated in the hip socket. This process is known as hip flexion. In contrast, a modern bird hangs from flexed knees that are always under tension. The femur rests in an almost horizontal position, and walking is achieved with knee flexion. The thigh muscles wrap part-way around the femur, and their contraction attempts to twist the bone when the bird begins to walk. Such torsional forces generated by the thigh muscles must be a persistent and constant phenomenon in the locomotion of birds because avian femurs look very much alike, despite great variation in the size and shape of birds.

Although the flexed knee helps the bird compensate for a shift in the centre of gravity away from the hip joint, the bird must still keep its point of balance over the toes. Consequently, walking is only slightly less complicated than flight and the musculature of the hip is as complex as that of the shoulder. In the very pedestrian chicken, 22 muscles operate the leg and 14 more operate the foot, while 29 muscles move the arm and another 23 shorter muscles operate the hand and wingtip [20].[2]

---

2   The number of muscles in these complex structures varies from species to species and from source to source, depending on the number of small slips counted as independent muscles.

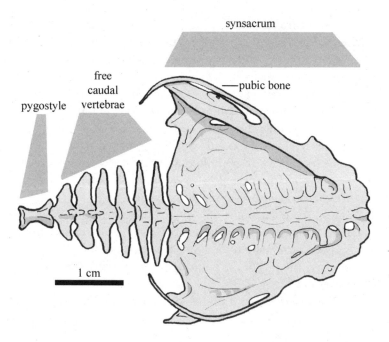

Figure 4.1 Expansion of the hipbones in a Belted Kingfisher to create an "abdominal vault" over the intestines and reproductive system.

Because the bird's hipbones have been fused into a solid mass called the synsacrum, the entire hip and tail region is much shorter than it was in ancient archosaurs. Many of its component parts appear as independent elements only briefly in the developing embryo and appear to be absent in the adult bird. Usually, they have not been lost so much as converted into specialized sections of larger, compound bones.

The rearward extension of the hipbones reduced the role of muscles attached to the tail in terms of walking and moving the hind limb, but it also created new responsibilities. Because the trailing edge of the hip lay close to the sides of the tail vertebrae, muscles between the two could be short and attached to the tail at a steep angle. When they contracted, the tail could be flicked quickly from side to side or up and down. This arrangement gave the fan of tail feathers its effectiveness as a rudder and is another feature that may distinguish modern birds from all their earlier relatives.

Redesigning the Hips: The Abdominal Vault
The evolutionary history of the hips in birds and dinosaurs has attracted a great deal of attention, and any book that compares dinosaurs with birds will describe the gradual rotation of the pubis in great detail. The pair of long pubic bones pointed forward in early dinosaurs and often ended in a large fused structure known as the

pubic apron or, more correctly, a pubic symphysis. In *T. rex* and other theropods, this structure ended in a pad that the dinosaurs may have propped themselves up on when they wanted to rest. In later species, the pubic bones pointed more directly downward, and eventually backward in some advanced theropods.

The hip region of *Archaeopteryx* was constructed much like that of an advanced theropod and included rather small hips before the long tail. Its pubic bones pointed backward, as in other dinosaurs, and their tips were fused together. The fused area, or pubic symphysis, formed a small cup-shaped structure that is well preserved in some specimens. In both dinosaurs and *Archaeopteryx,* the oviduct passed over the pubic symphysis and between the pubic bones. Consequently, the eggs had to be small enough to fit through that space.

In modern birds, the pubic bones are simple riblike structures whose ends no longer meet. Each lies parallel to the lower edge of the ischium and often ends without any sign of a terminal foot or cup, but sometimes the tip is flared out as a simple plate. Because the tips of the pubic bone are unfused, they no longer constrain the diameter of the egg.

Reducing the history of the bird's hip and tail to a simple story of reduction and fusion leaves out the creation of some very important skeletal structures. One of these is the large, undivided abdominal vault that has come to play a major role in the bird's life history. The creation of this space begins early in the development of the avian embryo. The growth of the tail vertebrae slows or ceases, while growth of the hipbones increases both in front of and behind the hip joint. The vertebrae near the hip joint remain as simple flattened discs while the hipbones grow alongside, eventually overtaking them and capturing their tissues. Although the caudal vertebrae behind the hip joint cease to grow in length, they continue to grow in width and they extend their lateral processes outward. The lateral processes push the hipbones away from the midline of the body and create a wide space, roofed by the vertebrae and walled on both sides by the broad plates of the ischium. The rafters of this roof are the remnants of the caudal ribs, which become strong, arched struts fused to their vertebrae. Their gently curved arches may not be as regular as those in a Gothic cathedral but their organic shape has its own elegance and they are the rafters that support the very large open space beneath the hipbones (Figure 4.1).

This large space is critical to the life history of the modern bird. Because the pubic bones are no longer fused, there are no skeletal elements limiting the expansion of the internal organs into the area. There is ample room to hold the enlarged and complex intestine needed by herbivores or a reproductive system that expands hugely during the breeding season. In many species, the fully developed egg fills this entire space for a day or so before it is laid (Figure 4.2). Surprisingly, no bird has been able to take advantage of this space either to hatch the egg internally (ovoviviparity) or to produce free-living young (viviparity).

The changing skeletal architecture that allowed the abdominal cavity to extend far to the rear of the hip joint carried the bird's cloaca or vent to a unique, almost

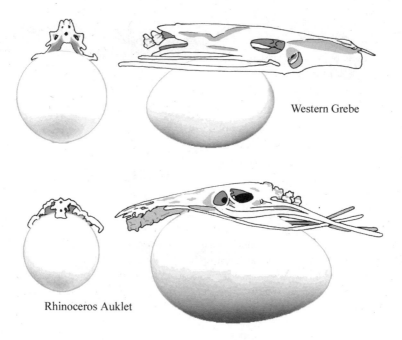

Figure 4.2 Relative size of the hipbones and an egg in the Western Grebe (top) and the Rhinoceros Auklet (bottom).

terminal, position. In every other vertebrate, the vent lies in front of the tail, just behind the hip joint. In birds, the cloaca lies below the last of the free caudal vertebrae, just in front of the pygostyle.

Although dinosaurs had the tail to maintain their balance, the gut and eggs had to be carried in front of the hip joint (Figure 4.3), and the size of the eggs was limited by the gap between the paired pubic bones. Consequently, each egg represented only a minor investment by the parent and the tiny hatchling required a great deal of time and food to reach adult size. Such small hatchlings may also have eaten very different foods from the adults and may have needed to live in several different habitats as they matured. Because birds reach adult size very quickly, they can be born into the adult environment and use its resources. Mammals derive a similar advantage from being suckled when they are helpless. Once they are weaned, they may live their entire lives on the same set of resources as their parents. This strategy may explain why mammals have developed such a variety of specialized teeth. Dinosaurs may have been forced to keep primitive types of teeth because their basic shape served well enough for the wide variety of foods encountered as they grew [159, 165].

Either dinosaurs invested in long periods of care for their young or they produced lots of young that were independent upon hatching and counted on abundance and luck for survival. Fossil nests, with large clutches of small eggs, suggest

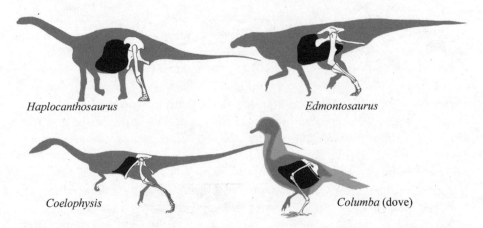

Haplocanthosaurus

Edmontosaurus

Coelophysis

Columba (dove)

Figure 4.3   The relationship between the distribution of the heavy digestive organs and the angle of the pubic bone in four members of the archosaur lineage. *Haplocanthosaurus* was a quadrupedal vegetarian saurischian with a very large digestive system. *Coelophysis* was a bipedal predatory saurischian with a small digestive system. *Edmontosaurus* was a partially bipedal ornithischian with a reverted pelvis. *Columba* is a modern pigeon with a fully reverted pelvis.

that at least some dinosaurs followed the latter strategy. It is not a popular strategy among birds.

Most birds have a reproductive strategy based on a period of intense parental care for a few young. Only a few species, such as Australian megapodes or bush turkeys, incubate large clutches with the systems favoured by crocodilians or other reptiles. Megapode eggs are warmed by heat generated in a mound of fermenting vegetation. It is effective enough to incubate a large clutch of young that are independent from the moment they leave the nest. Ducks incubate their large clutches with body heat but are very haphazard about later maternal care. They may lay a dozen eggs but only one or two of their brood survive to fledge. Birds that are more attentive to the needs of their young tend to lay fewer eggs. Clutch size typically varies from four to six, but some species lay only one or two very large eggs per season.

The broad roof of the abdominal vault gives birds room to produce huge eggs that are many times larger than those of any other vertebrate of comparable size. Some birds invest a whole season's reproductive effort in a single gigantic egg. Several species lay an egg that is more than 15% of the adult's weight; a storm-petrel egg may reach 28% of adult weight. These large eggs fill the abdominal cavity completely and stretch its walls (Figure 4.2). The benefit is that they contain such large amounts of nutrient that the hatchlings can easily reach adult size within a single season, and often within a few weeks.

Egg size has profound effects on life history strategies. Storm-petrels are tiny birds that raise their young in the comparative safety of small oceanic islands. They have a limited flight range and must schedule their breeding activity to fit within brief periods of local food abundance. The Cassin's Auklet and other auks face a similar problem. The large albatrosses also nest on offshore islands but their ability to forage over huge distances frees them from the restrictions of local or seasonal variations in food abundance. They can have a greatly extended breeding period and may care for their young for more than a year. Their eggs are unusually large. Nestlings reach adult weight fairly quickly but must get much heavier than the adult before they can begin to mature and develop flight feathers. These and many other kinds of seabirds depend on the absence of mammalian predators to keep their young relatively safe.

The ancestors of birds were not the only dinosaurs to look for advantages in wider hips; several lineages experimented with designs that were quite different from that of birds [156, Appendix A]. Dinosaurs appear to have avoided any changes that affected the function of the caudo-femoral muscle and continued to depend on hip flexion in the hind limb. Often, the increases in hip width were restricted to the area in front of the hip joint, but brachiosaurs, hadrosaurs, and heterodontosaurs widened the entire pelvic structure by supporting it with massive development of the lateral processes on the vertebrae. The strangely birdlike therezinodont, *Nanshlungosaurus,* made some of the most extreme changes. Its unique construction retained the narrow hip but widened the abdominal area with greatly elongated pelvic bones that flared out both before and behind the hip joint. The peculiar hourglass shape enlarged the abdominal cavity but would have left ample space for the unimpeded operation of a large caudo-femoral muscle. The eggs would still have passed through the narrow gap between the fused pair of pubic bones, and remained relatively small.

In the forward part of a bird's body, fragile organs such as the heart, liver, and lungs are thoroughly shielded by the flat expanse of the sternum and the basket of ribs. In the rear section, the intestines and reproductive system are far more exposed. They get some protection from the relatively immobile femur; the massive musculature of the thighs and the widened hipbones protect the upper surface but do not extend far down the sides. There are no bony elements at all on the lower surface, and most modern birds have no more protection for their lower abdomen than that offered by the skin of the abdominal wall. They have completely abandoned the use of gastralia (floating ribs), which were very common in dinosaurs and primitive birds.

Surprisingly few birds have experimented with specialized architecture to protect their lower abdomen. Penguins, loons, and auks subject their internal organs to rapid changes in pressure, and there is some danger that gases trapped in the gut might shift during a dive and distend the body as the bird rises to the surface. In these birds, the rear edge of the sternum is unusually long and extends towards the hip joint. It creates a bony plate that protects the upper portions of the intestine. The loon has increased the lateral support for its abdominal vault by binding the femur directly to the body wall. It is the unexpectedly long ribs, however, that provide the most protection to the abdomen in deep-diving birds. These birds have exploited the avian rib's characteristic hinge to extend both the vertebral and sternal halves almost as far as the cloaca. Connective tissue binds the ribs to each other and to the abdominal wall, so that the wall cannot be stretched out of shape by gases moving about as the bird dives. Because the ribs lie more or less parallel to the long axis of the bird, the abdomen can expand like a bellows during egg production.

Many diving birds also have sets of sacral ribs ranged along the edge of the hips. It is not clear whether these ribs are wholly new structures or something recalled from deep within the genetic memory. They look like genuine ribs and are associated with the fused remains of hip vertebrae, but they are simple structures and always lack the ventral section that connects to the sternum in thoracic ribs.

The Femur and the Tarsometatarsus

The modifications to the hip region were accompanied by changes to the bird's locomotory gait and the shape of its leg bones. Primitive species, such as *Archaeopteryx*, had unmodified hips and dinosaurlike legs with a long, thin femur and a short metatarsal region. Like dinosaurs, their hind legs rotated at the hips and they were said to have hip flexion. The abdominal vault widened the body and appears to have reduced the efficiency of long, straight legs. The solution found by the evolving modern birds involved the shortening of the femur and the lengthening of the tarsometatarsus. In effect, the femur lost much of its mobility and became a forward extension of the hipbones. Walking was achieved by knee flexion or rotation of the limb at the knee. Increasing the length of the tarsometatarsus creates a new long bone in the leg and increases the length of the bird's stride.

Knee flexion exerts a great deal of angular stress on the femur, and a short bone is less likely to break than a long one of the same thickness. The modern avian femur has also developed an array of special features that are reflected in complementary structures on neighbouring bones. For example, the femur has several trochanters. These are props that angle up out of a bone to create a firm foundation for the attachment of muscle. Sometimes they may butt against adjacent bones to limit movement in a joint. The head of the femur has a prominent crest at the top, the greater trochanter, which limits rotation of the femur. The greater trochanter protects the integrity of the hip joint by interacting with a large structure above the acetabulum called the antitrochanter. Because variations in the position and shape of the trochanters can be easily seen, they are often used by both ornithologists and paleontologists as characteristics in classification.

The trochanters are particularly important for the clear distinctions they may offer between birds and dinosaurs. The crest of the greater trochanter sits above the pneumatic fossa, a large opening on the rear surface that allows the passage of an air sac into the femur. The pneumatic fossa occupies the space once taken up by the lesser trochanter, a significant structure found on the femurs of dromaeosaurs. The absence of a pneumatic opening in the dinosaurs is further evidence that they had a less complex respiratory system than birds.

Where dromaeosaurs had a trochanter on the back of the femur for attachment of the caudo-femoral muscle, birds have a long, low structure called the obturator ridge. There is a related obturator process on the pubis and an obturator foramen on the edge of the ischium. The small obturator muscle arises on the edge of the obturator foramen, passes by the obturator process, and attaches to the obturator ridge on the femur. The only action of this muscle is the rotation of the thigh outward, but it is needed to help the bird walk with a flexed knee. The absence of an obturator ridge in dinosaurs is another feature that has attracted the attention of paleontologists and gained special significance in evolutionary debates.

The meeting of the femur with the hip is perhaps the most important joint in the body. Very few animals can survive an injury or deformation in this area. In birds, the head of the femur is a large, robust ball that fits into the socket of the acetabulum in the hip. It is one of the densest pieces of bone in the avian body. Consequently, femurs are rather common fossils, and the small differences in structure can be helpful in assigning a fragmentary specimen to the correct group of animals.

A peculiar pit on the face of the ball that forms the femur's articulating surface with the hip joint accommodates a ligament that binds the femur to the sacrum. The pit is called the capital ligamental fossa, and it holds the short, thick "round ligament." Every movement of the femur deforms it, but its stubby shape limits rotation and allows a little movement towards or away from the body. In the Enantiornithes, or ball-shouldered birds (see Chapter 8), an intriguing group that became extinct at the end of the Cretaceous, the head of the femur is more cylindrical than spherical and there is a distinctive elongated groove across its face. The groove suggests that the ligament in ball-shouldered birds was a short, flat strap of connective tissue. Perhaps it could be less robust in this group because the cylindrical head of the femur already limited the bone's ability to bend away from the body. Fossils of dromaeosaurs and other small theropods have a round head on the femur but lack a pit for a ligament.

At the distal end of the femur, birds have a deep "rotular" groove for the passage of tendons to the lower leg. The groove passes between the two large condyles that fit into matching cotyles on the tibiotarsus to form the main articulating structure in the knee. The large size of the condyles indicates the importance of the knee in birds. The entire weight of the body is suspended from this joint whenever the bird is on its feet. In dinosaurs, such as the dromaeosaur, the resting position of the leg was more vertical and there were significant muscles in the lower parts of the limb. Consequently, the dromaeosaur knee looks much less robust and its tendons travel through shallower rotular grooves.

Birdlike Dinosaurs

Robert Storer hypothesized that the wide hips of a bird were an adaptation for climbing trees; indeed, many groups, especially the parrots, make use of the breadth of the hips to reach out and grab branches [201]. The corollary is that the narrow hips of dinosaurs are an indication that they did not climb trees; tree-climbing lizards, however, tend to be slender animals with unspecialized hips, and overall size seems to be a more important factor in tree climbing. Unfortunately, small size and arboreal habits work against fossilization. The recent discovery of *Eoraptor* demonstrates that even narrow-hipped theropods could become arboreal. *Eoraptor* was hailed as the four-winged bird that filled *the* gap in the fossil record. It was a spectacular find because it had such well-developed plumage on its hind legs and looked very much as though it were an efficient forest glider, not unlike a flying squirrel. Its arboreal specializations appear to be limited to its plumage, however, and its skeleton is similar to those of other small theropods. Perhaps other small theropods lived in a similar way but their arboreal habits remain unknown because their fossils lack evidence of feathers.

In spite of their lengthy history, dinosaurs exhibit surprisingly little variation in the design of their legs compared with birds. Gatesy and Middleton thought that dinosaurs might have been so dependent on their legs for basic locomotion that they had few opportunities for experimentation [68]. Basically, all dinosaur legs are designed for walking or somewhat for running. Some of the fastest runners among them had legs that were very similar to those of the modern ostrich and have been given the appropriate family name, Struthiomimidae (ostrich mimics). Many predatory theropods used lean versions of the typical dinosaur leg. Such conservatism might reflect the fundamental robustness of an ancestral design that supported theropods in their long reign as top predators.

We can only speculate about the theropod lifestyle. Their legs and slender build suggest that they could achieve considerable speed. They were clearly not designed to overpower larger animals but this does not mean that they habitually ran prey to exhaustion. We can only guess at their endurance. They may have been a forest animal that hunted like a cougar rather than a cheetah – spending a great deal of time locating a meal but very little time chasing it down. To this day, the flightless birds of the forests – such as cassowaries and kiwis – have relatively unspecialized legs, while the grassland and desert birds – such as emus, rheas, and ostriches – are more strongly modified for running.

One of the peculiarities of birds is the narrowness of their walking gait. In spite of wide hips, walking birds tend to place their feet one after the other in a line. The longer the legs, the narrower the line appears to be. It may be the most logical adaptation for walking through tall vegetation. Where one foot has gone easily, another can follow. To accomplish this feat, the tibiotarsus is angled inward so that the bird's ankles are kept close to the midline of the body. When walking through grass, cranes lift the stepping leg and push it forward with the toes hanging limply so that they do not snag on obstructions. This action reduces the muscular effort required, and prey are not startled by moving vegetation. The tracks of predatory dinosaurs suggest that the use of a narrow gait has a very ancient history.

Another peculiar habit of both long- and short-legged birds is the tendency to stand for long periods on one foot. The long, splayed toes provide a broad base of support, but it seems an unlikely posture given the multi-part limb. The body's centre of mass must be in a stable position, perhaps cantilevered by the relative immobility of the head of the femur in the hip socket. It would be much more difficult to achieve this posture with a heavy reptilian tail, so it seems unlikely that this is a habit inherited from short-toed dinosaurs.

Adaptations for Flight

Avian Toes and Fingers

The evolution of specialized limbs in birds has regularly been the focus of discussions about the origins of birds among dinosaurs. The origin of the bird's remaining fingers and toes has been particularly important. Paleontologists have been able to reconstruct the history of dinosaurs from a relatively rich supply of fossil material, but the fossil history of birds is very poor. Much of the evolutionary story for birds has been inferred from the subtle events that take place during the development of the avian embryo. These events involve tiny, poorly formed structures that exist for very short periods and whose interpretation varies from observer to observer.

Both limbs of birds differ from those of dinosaurs but the reduction series that shaped the bird's wing has attracted the most attention. For many years, the hands of birds and dinosaurs appeared to tell different evolutionary stories because three-fingered dinosaurs and three-fingered birds seemed to have lost different digits. Both groups lost digit 5, the outermost member, but dinosaurs went on to lose digit 4 while birds apparently lost digit 1. In dinosaurs, digit 3 was always the largest but digit 1 had a special role as an opposable thumb (pollex) in their hand. Unlike the human thumb, which is offset at the wrist, this dinosaur thumb was offset at the first knuckle. In birds, embryonic evidence appeared to show that digit 1 was completely lost and that the small digit that makes up the alula on the leading edge of the bird's wing is actually digit 2. Digit 3, which was so large in dinosaurs, is also the main component of the bird's wingtip and carries all the large primary feathers. The feathers give this digit responsibility for attitude control, especially during low-speed flight, when there is a risk of stalling. In the bird wing, digit 4 is a minor supporting element fused to digit 3.

Richard Owen first proposed the "2-3-4 formula" for the structure of the bird hand in 1836, and it remained a strong argument against a close relationship between birds and dinosaurs throughout the 20th century. Other interpretations of the osteology were possible, however, and W.K. Parker suggested a 1-2-3 structure for the bird hand as early as 1891 [155]. He had had the opportunity to study the embryo of the Hoatzin, whose young have independent, clawed fingers on their wings. Embryologists argued back and forth until 1999, when G.P. Wagner and Jacques Gauthier proposed an entirely new theory. They suggested that an event called a "homeotic shift" in the embryonic hand explains how birds could have a 2-3-4 hand and still be related to dinosaurs [220]. The homeotic shift is true sleight of hand. Embryonic birds lose digit 1 but retain carpal 1, and digit 2 slides over to sit on it. Digit 3 slides over to sit on carpal 2, and digit 4 onto carpal 3. In this way, digit 2 is in position 1 and is available to become the alula, with the independent muscles that enable it to contribute to flight control. Digit 3 becomes the main support for the wingtip. Digit 4 becomes a small bone that supports digit 3 but has no independent function.

The process sounds extremely complicated, but there appear to be other examples that occur during the embryonic development of birds and the use of homeotic shifts may be an important evolutionary strategy for structural modification. For most biologists, the complexity seems counter-intuitive and the idea has not been widely accepted. It shows, however, that there could be alternatives to the more conventional interpretations.

In 2004, Alexander Vargas and John Fallon found a way to bring biomolecular techniques to bear on this issue [214]. It turns out that the embryonic development of digits in the hand is controlled by Hoxgenes d12 and d13.[3] Digits 2 to 5 require both Hox d12 and Hox d13, but digit 1 requires only Hox d13. In birds, it can be shown that development of the first digit of the wing is controlled by Hox d13 only. Therefore, it must actually be digit 1 and homologous to digit 1 in theropod dinosaurs, and both birds and dinosaurs have lost digits 4 and 5, leaving digits 1, 2, and 3. This biomolecular evidence suggests that it is possible to argue for a genetic relationship between birds and dinosaurs without recourse to evidence from complex embryological events. In that case, zoologists have misinterpreted embryological events and the significance of physical characteristics of the hand since 1836. Embryology is a very difficult field of study and may be more prone to subjective bias than chemistry, but subsequent studies demonstrate that the interpretation of biomolecular data is also difficult and prone to subtle errors. In 2005, two papers appeared with contradictory results. Frietson Galis, Martin Kundrat, and J.A.J. Metz reported that the biomolecular evidence was not strong enough to support one hypothesis over the other [64], while a team led by Monique Welten reported a new theory from studies of the Sox 9 gene.[4] The latter's results support the existence of at least five digits in the embryonic hand of chickens and offer evidence for a sixth digit. If the primitive archosaurs, which gave rise to birds and dinosaurs, had more than five fingers, we may not need to worry about complexities like homeotic shifts because "the incongruity between digit domains and identities in theropods disappears" [226].

The digits in the foot have their own story of reduction and loss. Once again, digit 5 disappears. Digit 1 of the foot, usually called the hallux, typically points to the rear in birds, but it almost always pointed forward or to the side in dinosaurs. When the 10th specimen of *Archaeopteryx* became available to the scientific

---

3   The homeotic, or Hox, genes govern both overt and non-overt segmentation in the developing embryo.

4   Some Sox genes are involved in sex determination and others in neuronal development.

community, it was discovered to have the hallux mounted on the side of a foot, just like a dinosaur. In fossil footprints from the Cretaceous, the position of the hallux, when present, is a more reliable method of distinguishing between dinosaur and bird footprints.

In more modern birds, the primitive or ancestral arrangement appears to include a well-developed hallux at the back of the foot, where it can oppose the other three toes to grip a perch (Figure 1.9). The prevalence of this foot design supports the theory that birds originated with arboreal animals. In birds specifically adapted to walking, the hallux is often raised off the ground and appears on the back of the tarsometatarsus above the other toes. It may even disappear altogether. In swimming birds, the hallux is usually so small that in spite of a flat, finlike lobe, it is hard to imagine that it has any useful function.

Running appears to have been the preferred form of locomotion in many small predatory dinosaurs but it is rather uncommon in birds. Only a few flightless species, such as ostriches, rheas, and emus, are completely dependent on it, and all have specially adapted feet. The ostrich foot is reduced to just two toes, the largest of which looks much like a hoof.

The aptly named roadrunner has become an icon of American deserts. It is actually an unusual cuckoo whose habits are unique in a family whose members are mostly forest species. The roadrunner has a lean and streamlined profile that looks much like reconstructions of the *Archaeopteryx,* another bird that spent a lot of time on the ground. Although it is a skilled runner, the roadrunner does not have highly specialized feet. It uses the same zygodactylous foot, two toes forward and two back, as the forest cuckoos. The most unusual feature of its feet is the ability to fan out its long toes for enhanced stability on a flat surface. Long toes and long claws may be very useful for clinging to tree bark or perching on thin branches but they would seem to be inappropriate for life on the ground. They might drag or become entangled in surface litter, but perhaps the dry surface of the desert is relatively free of such debris.

For many years, naturalists thought that the roadrunner was yet another flightless species because there were no reliable reports of its flight. We now know that it rarely needs to fly in the treeless desert but it is adept, if slow, in the air. To some extent, it has replaced flight with its own unique form of locomotion. When it wants to get into a shrub, it is able to spring more than a metre off the ground without opening its wings.

In fast-running birds such as the ostrich, the hipbones are unusually long and narrow, somewhat like those of theropod dinosaurs. The roadrunner has retained wide hips with the boxlike sacrum typical of other cuckoos, but has developed bony flanges that extend out over the thigh, further exaggerating the area's width and lengthening the thigh muscles that the bird uses in running. Some slower-moving cuckoos such as *Carpococcyx,* or supposed cuckoo relatives, such as the turacos that spend a lot of time on the ground, have rather small flanges on their hips. The strictly arboreal cuckoos have none at all.

The ability of the roadrunner to make prodigious leaps illustrates a form of locomotion that is widespread among many small birds. Because the bird's femur rests more or less horizontally, its movement does not contribute much to the length of the stride. When it is rotated downward, however, it transmits a great deal of force through the lower limb bones to the ground. Many small species have added to that thrust by perfecting the toe-spring hop. As the femur rotates, they flick themselves into the air, off the tips of their toes.

This hop is an outstanding example of the use of inner-bird architecture. The muscles that flex the toes are attached to the femur, high on the leg, and the connecting tendons travel across the knee and ankle as well as the joints in the foot and between the digits of the toes. In very long-legged birds such as the cranes and storks, which typically walk rather than hop, this tendon is among the strongest structures known in nature. Although we usually do not think of them as being particularly long-legged, because their plumage hides the upper leg bones, many birds that hop about on the forest floor also have very long legs. The pittas (Pittidae) of the Old World Tropics and the unrelated antpittas (Formicariidae) of South America have a particularly long tarsometatarsus that is about half the length of the body. It holds them very high off the ground and makes them look a little like tiny storks. They also have very long toes and are able to spring into the air to escape predators or pluck a spider out of its web.

The toe-spring hop requires no special structures in the foot. The toes don't drag on the ground because the body is propelled vertically into the air. Not only do they not drag but the leverage created by their length increases the thrust needed for lift-off [217]. In the forest canopy, the toe-spring hop is often coordinated with a flick of the wings that aids in balance and slows the landing so that there is time for the toes to get a secure grip on a branch.

Hopping has become such a widespread habit among very small songbirds that walking is the exception rather than the rule. The ground-dwelling Ovenbird of North America rarely hops and this characteristic was so unusual that it was formerly known as the walking warbler. Hopping is also an important part of predator avoidance in small forest birds, which are most vulnerable on the ground and depend on a sudden vertical take-off for escape. Their short stubby wings create a great deal of excess lift but a toe-spring hop offers a useful quick launch into the air. It is unlikely to have been a useful trick in dinosaurs, even in the smaller species.

Although the roadrunner and other cuckoos share very similar feet, the other limb bones are more adapted to the varied lifestyles found in the group. William Engels was one of the first to compare arboreal cuckoos with their walking (but not running) relatives among the anis [57]. In general, among vertebrate animals, limbs used for speed tend to have long outer elements and short inner ones. All three types of cuckoos fit that pattern. There were only small proportional differences in the inner bones (humerus and femur) and somewhat greater differences in the next element (ulna and tibiotarsus), but the differences in the terminal elements (manus and pes) were quite large. As expected, the arboreal cuckoos, which flew regularly, had a relatively long hand while the roadrunner's hand was short. The tarsometatarsus, on the other hand, was very long in the roadrunner but short in the arboreal species. Comparable differences in the legs of advanced theropod dinosaurs suggest that some of them could have been very much faster than others.

Heads and Brains

The ability to move the long-range sensory receptors for smell, hearing, and sight into a useful position may be the head's single greatest value. It is the main reason that all of the most active forms of life have at least a distinctive front end. The ability to raise such an array of sensors high off the ground confers a distinct advantage on upright animals that want to look around for food, mates, and predators. They can detect such objects of interest much further away than crawling forms can. When we first make contact with advanced forms of extraterrestrial life, we can expect them to have a head carrying all of the appropriate sensory receptors – assuming that their ancestors also had to get off their bellies and live in three dimensions.

Vertebrates have protected their sensory receptors and brain by housing them within a bony skull. In primitive vertebrates, the skull consists of many individual plates. Advanced vertebrates have reduced its weight by fusing as many of its components as possible so that they form a close-fitting braincase. Because this structure fits like a glove, even a fossil can reveal much about the soft tissue of the organ it once contained.

Although the phrase "bird-brained" is often used as an insult, birds are one of the few animals to have a large brain, and it is often a useful character in separating fossils of birds from those of dinosaurs. We know from casts and from X-ray techniques such as tomography that the brains of a few advanced theropod dinosaurs such as *Tröodon* were as large, relative to body size, as those of modern birds, but most were much smaller. We also know that *Archaeopteryx*, in spite of its dinosaurlike body, had a very birdlike brain many millions of years before the advanced theropods appeared [2, 226]. *Archaeopteryx* depended heavily on vision, as do its modern descendants; consequently, its cerebellum is broad, about half the width of the whole brain. Its well-developed inner ear implies that it may also have depended heavily on hearing [2]. Taken together, these features imply that *Archaeopteryx* had the advanced neuromuscular integration needed for successful aerial flight. Whether it had sufficient muscular strength to get into the air is an entirely different question.

The reason birds have such large brains is that they need sufficient space for a very elaborate mapping system that evaluates and stores visual information [1]. The brain processes the flood of information from the eyes when the bird is in flight, but it probably developed long before the ability to fly. It goes without saying that an animal that moves at high speeds must have a good sense of distance to solid objects, and the bird's brain integrates the visual input as three-dimensional information. This is a feature found in many animals that live in complex environments such as the canopy of trees. There are no examples of specialized three-dimensional orientation in ground-dwelling animals except those descended from arboreal ancestors [13].

Early mammals also developed a large brain, and it may well be linked to the ability to maintain body temperature and increased activity. The operation and fine structure of the brain in mammals is very different from those in birds. Mammals have an elaborate set of visual maps in the grey matter of the neocortex but they connect to the rest of the brain through long fibres in the white matter. In a bird, there is a single large retinal map connected to the rest of the brain by the hyperpallium (from the Latin for overcoat, as in a pall of smoke), a blanket of star-shaped neurons whose interconnections take up less space than the white matter of the mammalian brain and may be much more efficient.

Interestingly, the brain of Sankar Chatterjee's controversial Triassic fossil *Protavis* is also very large [24]. If this highly controversial collection of fossils represents a bird, it is 75 million years older than *Archaeopteryx* and implies that the large brain developed very early in avian history. Endocasts of its braincase appear to show features consistent with *Protavis* being a more primitive animal than *Archaeopteryx*, but the nature of its relationships to other lineages is uncertain. The skull is fairly convincing as a bird, but the remainder consists of collection of bones that may be a chimera (an assembly of several different animals). Unfortunately, the Triassic has such a desperately poor fossil record that we may never find more informative material.

The skulls of the various lineages of dinosaurs tell a very different story from that of the birds. Most lineages of dinosaurs appear to have followed Cope's Rule in that they seem to have produced larger and larger animals with the passage of time. Among the predators, this trend meant that many of them developed fearsome jaws capable of ripping huge chunks of meat out of their victims. This required a very strong skull and a thickening of the vertical struts in front of and behind the eye. One result of this reinforcement is that the orbit of the eye has frequently been fossilized and is one of the best known parts of the face. The other is that the bony growth changed the shape of the eye, and many of the larger predators had relatively small eyes [83]. In *Compsognathus*, a small theropod that may have been somewhat similar in size to ancestral birds, the orbit was proportionately eight times larger than in an adult *T. rex*, but *T. rex*'s skull was 1,500 times stronger.

The skulls of most birds suggest that evolutionary pressures to reduce the size and weight of the head have been very strong. Most appear to have been reduced to an absolute minimum. Birds that live at sea, however – such as petrels, auks, loons, and some ducks – have gone against this trend and increased the weight of their heads by developing exceptionally large salt glands that sit on the skull, just above the eyes. The glands are often protected by thick bony ridges (Figure 1.2). Although Mesozoic seas are supposed to have been less salty than the modern oceans, the ancestors of modern marine birds also found that their kidneys could not cope with the inevitable accumulation of extra salt. Surprisingly, they all developed very similar osmoregulatory glands (salt glands) from tissues that may once have been responsible for the production of tears. The long grooves that house the salt glands are a distinctive feature in the skulls of all living marine birds. They are also obvious in the fossils of *Hesperornis* and *Ichthyornis.*

Life in a New Environment

Although the Triassic archosaur *Quianosuchus* lived at sea, we have yet to find a later dinosaur with unequivocal adaptations for swimming, let alone adaptations for life at sea. We know nothing about any glands that might have been situated on the heads of dinosaurs. As a group, however, they were so thoroughly terrestrial that it seems unlikely that they would have needed anything as large or specialized as the avian salt gland. The willingness of birds to head into the water is one of the clearest behavioural differences between them and the animals that appear to be their most recent ancestors.[5]

Four unrelated groups of birds have been especially successful in marine environments. The Procellariiformes are the most oceanic. They include petrels, storm-petrels, diving-petrels, prions, shearwaters, and albatrosses. The Charadriiformes include many groups, such as gulls, terns, and sandpipers, that prefer fresh water or return to fresh water regularly. They also include skuas, jaegers, noddies, and auks, which are more truly marine. The third group is the Sphenisciformes, or penguins, which are well known as a specialized group of successfully flightless marine birds. The fourth marine order is the Pelecaniformes, but it may be an artificial assemblage of families that share some similarities but are not closely related (see Chapter 6). Many members of the Pelecaniformes live on the continental uplands around large lakes and rivers. Although their coastal relatives have well-developed salt glands, they require reliable sources of fresh water and do not wander too far from shore. Other kinds of birds, including many ducks, loons, and grebes, also spend long periods, even years, on salt water, but they are not fully marine. They all return to fresh water to breed.

Many fossils of marine birds have been described as "pelecaniform," including the flightless *Plotopterus,* which once lived like a giant penguin in the North Pacific. Without clear criteria for membership in that group, however, it is difficult to understand what "pelecaniform" means.

Some families within the Procellariiformes and Charadriiformes have unique lifestyles. Others form pairs of rough equivalents that have converged closely in both appearance and their ecology even though their lineages are only remotely related. We will examine the lifestyles of marine birds in the closing chapter of this book, and see how they have become some of the most amazing animals on earth. They are so thoroughly oceanic that, like turtles, they come ashore only to reproduce. They have abandoned almost every aspect of their reptilian heritage, and it

---

5 The other giant marine reptiles of the Mesozoic were not dinosaurs. They appear to have been more closely related to varanid lizards and perhaps snakes.

is even difficult to see them as close relatives of the forest birds. They have turned the forest bird's fragile, gnome-like body into a robust, densely muscled power-house. Simple flapping flight has given way to unique flight styles like the elegant, dynamic soaring of an albatross or the projectile-like, high-speed propulsion of an auk. Such techniques enable these birds to travel great distances and carry amazingly heavy loads. They even propel themselves deep beneath the ocean's surface. They are living advertisements, not for the biomechanical constraints imposed by flight and the inner-bird structural strategy but for ease with which evolutionary processes have overcome them.

PART 2

What Kind of Bird Is It?

The Kinds of Birds

5

Classification and taxonomy

Chapters 3 and 4 compared dinosaurs with birds because comparison has been a major tool in biology since Aristotle first undertook the objective study of nature some 2,300 years ago. The underlying implication is that the more similar two animals are, the more closely related they are likely to be. It seems to have worked well in the sorting of some groups of animals, and their relationships are portrayed in well-accepted family trees. Birds have proven difficult to organize, however. Even though they are among the most intensely studied animals and modern ornithologists have access to all sorts of space-age technology, there is still no generally accepted family tree for the birds. This chapter and the next will discuss why similarity is such a slippery concept and why we continue to find it so difficult to understand the relationships among birds.

When people call an ornithologist to help identify a strange bird, they probably conjure up an image of some kind of museum specialist surrounded by trays of dusty specimens. For a whole host of reasons, that picture is out of date. It is almost Victorian, or at best left over from the 1950s. The portraits used to illustrate textbooks or histories of ornithology are one hint that the popular view of ornithologists comes from an earlier time. They are typically formal, black-and-white photographs of men in business suits. Many of the most important figures are wearing celluloid collars.

The stereotypical ornithologist puttering about among stuffed birds has become a rarity. Many museums have found that collections of bird skins were so expensive to maintain that they simply disposed of them and replaced curators of ornithology with less specialized collection managers. Universities have either dropped courses in ornithology or converted them from classical studies of avian taxonomy to surveys of regional birds and their habitats. Throughout North America or Europe, if you want a bird identified, your best option is likely to be the local natural history society instead of some government agency or research institution. The amateur ornithologist, or "birder," has largely replaced the professional academic as a source of expert knowledge – and is much easier to find.

The marketplace has responded to the abundance of birders by producing huge numbers of field guides and setting affordable prices for high-quality optical equipment, sophisticated cameras, and even the specialized equipment for recording birdsong. It also produces prestige items that signal an amateur's commitment to his avocation. The preferred models of binoculars may cost as much as a serviceable used car. Amateur birders have become so well equipped and experienced that wildlife conservation agencies regularly depend on their volunteer support for population surveys and research projects.

Within the professional biological community, classical ornithologists, in the sense of specialists in avian classification, have become a tiny minority. They have been replaced by ecologists, physiologists, geneticists, paleontologists, demographers, and other researchers who use birds to test general scientific theories. This wider scope for ornithology has led to a huge increase in the membership of the professional ornithological societies and an almost uncontrollable explosion of technical literature about birds. The flood of new information is so overwhelming that no one can keep on top of it without a complex infrastructure of indices, abstracting services, and search engines.[1] The combination of high volume, increased technicality, and specialized jargon has left modern ornithology with a split personality. Amateur ornithologists are deeply interested in the basic question of a bird's identity, while professionals are interested in how a bird makes its living.

The gap between amateur and professional ornithologists is largely a 20th-century phenomenon. Until the end the 19th century, both amateur and professional ornithologists participated in very similar kinds of activities, and it was often difficult to distinguish between the two. Both were deeply involved in attempting to understand God's plan at Creation by describing all aspects of nature. This passion dominated ornithology for the four centuries between 1500 and 1900 and led naturalists into the most remote corners of the globe in a quest for new and exotic species. In the process, they described and named most of the world's birds; in company with equally dedicated botanists and other naturalists, they laid the foundation for our modern knowledge of global biodiversity.

The thoroughness of the early ornithologists' search for novelty has contributed to one of the more peculiar characteristics of field guides. There are a great many versions of these books but they all contain very similar information. A new edition may include recent name changes and updates on ranges and distribution, but the bulk of the content was collected more than a century ago. Except for innovative approaches to illustration, one field guide is much like another, and only low prices and the birder's passion for the latest style keeps the market for these books alive.

---

[1] In 2004, SORA (Searchable Ornithological Research Archive) came online and made 19th- and 20th-century publications in *Auk, Condor, Wilson Bulletin, Bird-Banding,* and *Field-Ornithology* available without charge.

Historical Background

Classical Ornithology, 1500-1900

The scientific study of birds began 2,300 years ago with Aristotle. Its more recent history can be divided into a long period between 1500 and 1900, dominated by descriptive natural history, and the 20th century, dominated by the application of experimental sciences to the study of birds. Thanks to a hundred years of amazing scientific progress and an astounding flood of new fossil birds and birdlike animals, we may look forward to great advances in our understanding of birds in the 21st century.

As you might expect, the periods that mark the history of ornithology are closely linked to major events in global history that are far removed from the study of nature. We are all familiar with the social and technical turmoil of the 20th century, but events in the 1500s were just as dramatic and continue to have a great impact on our modern world. Three of the major events at the end of the 15th century were particularly important to ornithology and other natural sciences: Columbus discovered America, Guttenberg introduced the printing press, and Theodor of Gaza published the works of Aristotle in Latin.

If Theodor of Gaza had failed to rescue the last remaining copy of Aristotle as Salonika fell to the Turks, we might not have a modern version. Only a few disjointed fragments have been found since. Theodor was a bit of an entrepreneur and printed a popular Latin translation of Aristotle at the end of the 15th century. As a result, the next generation of naturalists was the first to be exposed to the complete works, and the first to have them as a readily available reference. The ideas of Aristotle were not completely unknown to the medieval scholars of the western world. Two hundred years earlier, Michael Scot had translated a version of Aristotle that had been in widespread use in Persia and Islamic countries since the time of Alexander the Great. By the time it became available to Scot, it included material from Arab and Persian philosophers that was far beyond anything known in Europe. Scot took the translation to the University of Paris in 1230. It was an intellectual bombshell that had a profound influence on Roger Bacon, Albertus Magnus, Thomas Aquinas, and other scholars of the 13th-century Renaissance. Without the printing press, however, it had limited distribution and is largely forgotten today.[2]

Ironically for a philosopher who preached the value of direct observation, the printed version of Aristotle's books made him an unassailable authority. Between 1500 and 1900, biological knowledge, right or wrong, was based on Aristotle's interpretation of natural events.

---

[2] Aristotle was never translated into Latin by the Romans, and knowledge of him in the West was lost during the Dark Ages. Copies of his works may have been carried eastward, however, during Alexander the Great's program to spread Greek culture. During the Islamic conquest of Persia, Aristotle's ideas had a major impact on metaphysical thinking, and he is still known as "first teacher" in the madrassas of Iran.

We usually think of the mechanical press as the first way to distribute written information rapidly and cheaply, but it had another significance for natural history that is generally unappreciated. The press also made it possible to reproduce images reliably and accurately. The medieval academic world learned to cope with the inevitable errors and omissions of hand copying. Words are discrete kernels of information with some internal logic, and transcription errors did not necessarily destroy the meaning of the text. Images presented an entirely different kind of problem. Each copied version was bound to contain less information than its predecessor. Detail was lost and every level of simplification increased the risk of significant misunderstanding. More than any other science, biology depends on communication through complex images filled with minute but critical details. The inability to copy images accurately must have limited progress and the spread of new information even in important biological subject areas such as anatomy and medicinal botany.

The scale of the problem is clear in the images that have come down to us in medieval manuscripts. Even though the copyists must have developed some basic artistic skills, familiar items like horses, dogs, and even trees are portrayed in a highly stylized manner. The images rarely suggest personal observation by the artist and are often mere cartoons, simple enough to be reproduced by other copyists. The cost of reproducing imagery must have been enormous, and it may be no accident that one of Aristotle's lost books, *The Dissections,* may have included his biological illustrations.

If we find his illustrations, they may be similar to Pierre Belon's famous representation of dangling skeletons (Figure 5.1). Aristotle's approach to understanding nature depended heavily on comparisons between similar objects, and Belon's illustration offers a perfect example of the kind of parallel construction that would have fascinated him.

Even with the guidance of Aristotle, biological discoveries from the New World overwhelmed Europe's tiny scientific community. Just a generation or two earlier, medieval naturalists had looked at nature as a static and unchanging phenomenon. They were completely confident in their position as God's most favoured creation. Their world did not extend much beyond the shores of the Mediterranean Sea, and it held only the few hundred plants and animals that God had created in the Garden of Eden, that Adam had named, and that Noah had saved from the Flood. There was no need for a complex classification system because the names of the few hundred familiar species could be easily remembered, while written descriptions of plants and animals dated back, literally, to the beginning of time. In such a world, it was inconceivable that anyone could discover a truly novel species; consequently, there was no need for a special protocol for naming new animals or plants, and no language to do it with.

For much of the 16th and 17th centuries, information simply accumulated in Europe. Detailed reports and a few mummified specimens, often collected at the

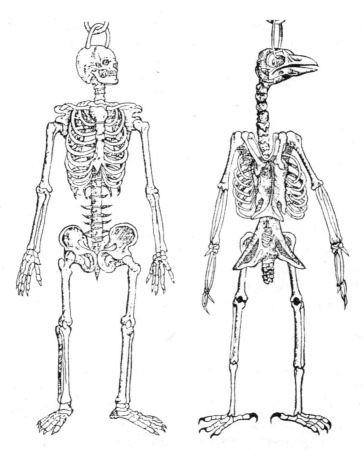

Figure 5.1 Pierre Belon published this classic comparison of human and bird skeletons in 1555, shortly after the first publication of Aristotle in Latin. It demonstrates how strongly Aristotle's ideas influenced early European natural history.

risk of life and limb, mouldered in forgotten state archives or royal collections. Some had potential as important marketable resources and were kept a closely guarded secret; others were seen as a threat to social or religious order. The rest disappeared into dust, simply because no one knew what to do with them.

It was not until the 18th century that the idea of global biodiversity became general knowledge and began to be celebrated as a wonder of God's creation. This change came about in part through the inevitable spread of rumour and myth, but improved printing techniques played an important role. Naturalists found patrons who could afford to produce books illustrated with colour images of newly discovered birds and other creatures.

Between 1731 and 1743, Mark Catesby published spectacular images of birds from the Bahamas and Carolinas. Each plate was printed in outline and painstakingly coloured by hand. Such labour-intensive production methods meant that they were available only to the profoundly wealthy, but they opened the public's eyes to the wonders of nature in the New World and began our long fascination with the outward appearance of birds.

In the same year that Catesby's first volume appeared, Eleazar Albin produced an illustrated book on the birds of Britain that called attention to wildlife "closer to home" and in some ways foreshadowed the development of the modern field guide. Albin was the first naturalist to include images of eggs with his birds. Compared with modern pictures, his drawings look rather stiff and even distorted, but his birds have a naive charm and the colours he chose are bright and perhaps overly imaginative. Like the authors of modern field guides, Albin drew his text from other sources.

Drawing was an essential skill for naturalists in the 18th and 19th centuries because they lacked two pieces of technology that we take for granted. They had no reliable method of preserving a specimen and there were no cameras. This presented particularly serious problems for the study of soft-bodied animals such as fish or amphibians, whose colours faded quickly after death. Fortunately, ships' officers were required to make accurate drawings of landscapes as part of the mapping process and many turned their talents to sketches of local wildlife. The better-organized exploration missions included dedicated artists for the production of accurate and detailed records. Images by W.W. Ellis from Captain James Cook's last expedition, and by Atanasio Echevarria and Jose Cardero, who accompanied Alejandro Malaspina, are well known for their portrayal of the people met by the early explorers. They are also a critical record of the wildlife that existed before the arrival of European colonists. In some cases, their drawings are the only evidence for the existence of species that subsequently died out.

The flood of new material presented the early naturalists with problems with which they had had no experience: how to name groups of animals and how to define those groups. The vernacular offered all sorts of names like finch, thrush, wren, or sparrow. These, and their equivalents in other languages, had been adequate for the few European birds since the beginning of recorded history but were of little help when it came to exotic types that were not quite one or the other. Aristotle did not offer much help either. He did state that one species does not change into another, so early naturalists were at least aware that they could be looking at something novel. To facilitate discussion, Aristotle had sorted birds into large groups according to their behaviour. Like most early naturalists, however, he knew only 100-200 species and never needed to set out a formal method of classification.

Faced with the growing diversity of new animals, two English naturalists made an early attempt to sort them out scientifically. Like Aristotle, John Ray (1628-1705) and his pupil and patron Francis Willughby (1635-72) began with a decision to depend on observations. They deleted all references to "hieroglyphics, emblems, morals, fables, presages or ought else pertaining to divinity, ethics, grammar, or any sort of humane learning" and restricted their comments to what "properly belongs only to natural history."

Aristotle had organized birds into three groups (terrestrial, shoreline, and aquatic), and by the 17th century, he had gained such stature as an authority that most scientists seemed unwilling or unable to look at alternative approaches. When Willughby and Ray sorted the whole animal kingdom according to physical features, it represented a revolutionary approach even though it was based firmly on Aristotle's concept of comparison. They used the presence of gills to separate fish from other vertebrates, and differences in the ventricles of the heart to distinguish amphibians from reptiles. They became the first naturalists to successfully sort out the confusing mass of invertebrates. It is surprising, but perhaps a sign of the times, that this highly successful team lost the courage of its convictions and left whales with fish, where everyone knew they belonged. Simply by looking at structure, the next generation of biologists could begin to sort out the puzzling array of creatures arriving from unknown corners of the globe and from even more alien continents under the microscope.

In 1676, John Ray published his friend's *Ornithologiae libri tres* posthumously. It was the last attempt to write a comprehensive ornithology along the lines of Aristotle's *de Animalia*. Even in the 17th century, the mass of ornithological material was far too large for any one person to digest completely. Perhaps the book's most significant contribution was a break with 2,000 years of tradition. It abandoned the attempt to classify birds by behaviour or activity, and relied instead on form or feature.

Like Aristotle, Willughby and Ray separated the birds that lived in the water from those that lived on land. Finer divisions were based on structure – beak shape, foot structure, and body size. It was the first truly scientific classification and became the basis of the concept of comparative morphology that carried ornithology into the 20th century. If they had written in English instead of Latin, their elementary descriptions would sound almost modern. Pigeons are terrestrial, neither the beak nor claws are large; both are straight or slightly curved. Gulls are aquatic with short legs; they have four toes, three face forward and are webbed, and one faces backward; the beak has a sharp tip and is narrow; the wings are long and narrow.

Willughby developed a particular interest in the problem of avian flight and used his profound appreciation of internal anatomy to develop some rather astute insights. After comparing human muscles with those of a starling, he recommended the use of leg muscles instead of arms for human-powered flight. Three centuries later, when the Paul B. McReady's *Gossamer Condor* took wing, it was powered by a bicyclist.

Ray's greatest contribution to science was the first really useful and meaningful definition of species. Aristotle and all his followers had worked from the top down, dividing big groups into smaller and smaller units. Ray began with individual animals and plants that looked alike and brought them together in groups. In Latin, "to look at" is *specere,* hence species in the sense of groups of individuals that can be looked at and seen to be the same because they share a suite of characters. More importantly for the biology of Ray's time, the members of his species mated with each other and produced young of their own kind. There was no room for mythical hybrids or the products of spontaneous generation. Later biologists criticized Ray's apparent emphasis on the fixedness of species. Actually, he says that they were only "fairly constant ... not invariable and infallible."

One effect of the switch to an emphasis on structural features was that the specimen became its own document of identification and therefore needed to be preserved in some form for future consultation. We call such items "type specimens" and make special efforts to ensure their preservation. In the 17th and 18th centuries, there were no reliable methods of preservation. In his Copenhagen museum, Ole Worm (1588-1654) laid specimens on open shelves or dangled them on bits of string. Other collectors found that small items could be pickled in distilled alcohol or even strong vinegar, but larger items were usually dried or "stuffed" with varying degrees of skill. Eventually, an enthusiastic bird collector, Jean-Baptiste Bécoeur (1718-77) invented arsenical soap to preserve his prizes.

Unfortunately, only modern technology can provide real security. Even ancient specimens loaded with toxic chemicals have crumbled to dust under the combined onslaught of moths, beetles, and fungi. In spite of a huge 18th-century collecting effort in the Americas, Asia, Australia, and Africa, only a tiny number of early specimens survive. In Europe and Asia, important collections survived into the 20th century only to go up in smoke during the two World Wars. Sadly, some of the survivors have since fallen victim to shortsighted cost cutting or conversion of state museums to free-enterprise operations. Marketing and the Disney mentality have no need of scientific collections and are more of a threat to our scientific heritage than any war.

Large-scale bird collecting began in the 17th century and quickly led to a variety of systems for sorting and classifying specimens. At first, ornithologists kept a few special skins or other specimens at home with their books. Gradually, these private examples were gathered together into larger representative collections. Perhaps the earliest mass collector was Sir Hans Sloane (1660-1753), who had been a friend of John Ray and sponsored the publication Catesby's bird illustrations. His collection of 1,172 ornithological items became the backbone of the British Museum when it opened in 1759. Big collections presented museum staffs with unprecedented problems of handling and organization. They had no instruction manuals for such an operation, and the organization of bird books of the 17th century was often a matter of convenience or economy of space. Publishers regularly used eccentric sequences and juxtapositions for species because they took the opportunity to sandwich a small bird in the space left between two large ones. Museums could not afford that luxury. They needed a cataloguing system, if only to plan the storage space. System also helped if you expected to retrieve an item once you had put it away.

Although Willughby and Ray had shown the way, European biologists continued to stumble along with classifications based on "philosophical principles" for another 60 years, until the accumulation of new organisms had brought the biological community to a state of crisis and confusion. In 1735, Carolus Linnaeus presented his *Systema naturae* to the world. At a stroke, it dismissed a huge number of esoteric philosophical problems as irrelevant and became the standard method throughout northern Europe. Predictably, the British objected to its use of Latin (a "foreign" language!) and the French objected to its lack of natural order. Unlike competing systems that were hurriedly cobbled together, however, it worked.

In 1788, the Linnaean Society was established in London and the entire biological community adopted Linnaean binomial nomenclature as the international standard. From that point on, all animals and plants were identified by just two names – the genus and the species.

Linnaeus had been strongly influenced by Aristotle's idea that there were "kinds" that differed from each other in fundamental or structural ways; within kinds, there were groups that varied by degree, what Aristotle called "the more and the less." Linnaeus used the word "genus" for the finest division of kind. Ernst Mayr, the dean of modern taxonomists, describes it as a category containing a single species or a monophyletic group of species that is separated from other taxa of the same rank by a gap. Within a genus, the individual species differ from each other by degree. They have longer or shorter or bigger or smaller versions of the same characters.

Like the family, the order, and other higher taxa, the genus is defined by comparison with similar groups. Only the species is defined by reproductive capability. A species is a group whose members can interbreed successfully and excludes those that do not share its courtship, anatomy, chromosome structure, or geography. Their progeny can also reproduce successfully. Sometimes members of different species can hybridize, but their offspring are typically sterile or cannot reproduce, and are therefore an evolutionary dead-end.

The adoption of a two-name system seems almost trivial, but standardized systems for identifying people by surname and given name were also rather new at the time. The oldest plant names date back to Linnaeus' first publication in 1735, but animal names date back only as far as the 10th edition of *Systema naturae,* which appeared in 1758. By international agreement, the official name of a new species is the first one published, and committees of taxonomists monitor the order of precedence.

In the 10th edition of *Systema naturae,* Linnaeus divided birds into six families:

- Accipitres – vultures, hawks, owls, shrikes, waxwings
- Picae – parrots, woodpeckers, kingfishers, hornbills, crows, grackles, hoopoes, hummingbirds
- Anseres – all swimming birds
- Grallae – bustards, ostriches, sandpipers, and other long-legged wading birds
- Gallinae – quail, grouse, peafowl
- Passeres – pigeons, thrushes, larks, tits, swallows, swifts, nightjars

None of the groups of birds in the *Systema naturae* are exactly the same as in modern classifications, but birds may have been outside Linnaeus' expertise. He was a botanist and had carefully constructed his classification of plants based on the arrangement of flower parts. His student, J.C. Fabricius, used the same approach to classify insects according to variations of their mouthparts. Birds did not receive the same degree of attention and Linnaeus failed to set out criteria for determining which kinds of characters were best to use in their classification. Birds do not offer a wide array of external features, and he appears to have used very few characteristics to distinguish the major groups. He should have consulted the works of Willughby and Ray, who do offer long lists of suitable characteristics.

Nonetheless, ornithologists followed other biologists in the application of the Linnaean method, and the number of identified species grew quickly after 1758. In that year, only 564 bird species had been described, but in 1760 Mathurin-Jacques Brisson added 1,000 species and in 1790 John Latham added 1,400 more from the voyages of Captain Cook. By the time Charles Darwin was writing, in 1841, 6,000 birds were known to science. The modern total exceeds 9,000 and may be growing as well-established species are split, thanks to studies of their DNA. The creation of new species, however, may not be keeping pace with the effects of deforestation or other industrial forces of mass extinction.

After the publication of the 10th edition of *Systema naturae*, there was a marked increase in the volume of biological literature. In earlier times, only a handful of important players wrote books and became known to history. Their extended support groups of friends and acquaintances who shared an interest in natural history were often anonymous or recorded only as names. Often these supporting actors included well-educated and expert individuals. Once the Linnaean Societies were established, there was a vehicle for them to document their own observations and contribute directly to scientific progress. From that point on, there has been a steady increase in the number of biological journals and the number of societies producing technical reports. Paradoxically, the great increase in communications and research activity has reduced the significance of each individual contribution. Only a few truly great men, and lately great women, have been able to gain any measure of fame.

It is not a stretch to suggest that biology became an academic industry after 1758. Scientific expeditions set out with specific goals and objectives based on prior knowledge or at least expectations. A host of naturalists was let loose upon the world in the later 18th and 19th centuries who made a good attempt to describe all its fauna and flora. The Linnaean system showed them how to put their findings in some sort of meaningful order. Merely building a huge catalogue of names and descriptions did not address Aristotle's main reason for inventing biology in the first place, however. He had hoped to discover the underlying cause of natural events. In spite of massive efforts, there were no great advances in that direction until 1859, when Charles Darwin published his theory of evolution by natural selection and gave biology an overarching theory that linked cause and effect. It swept away the remaining Platonist, non-scientific approaches to biology. The role of comparison in taxonomy also shifted subtly. For centuries, comparison had been the key to the identity of individuals; now it became the indicator of relationships among groups of organisms. Darwin's theory has been carried far beyond biology and has come to influence almost every aspect of human existence. It organized the science of life just as surely as Newton's ideas had brought order to physics and astronomy 200 years earlier.

The Victorian Basis of Modern Taxonomy

A little luck helped Victorian ornithologists in their attempt to organize groups of birds according to evolutionary principles. Just a few years after Darwin published *On the origin of species,* the world was amazed by the discovery of well-preserved fossils of very ancient birds. The first and most dramatic was the *Archaeopteryx* found in Europe, which was soon followed by the discovery of toothed birds in North America. *Hesperornis* was a large flightless bird with a superficial similarity to a loon; *Ichthyornis* may have looked somewhat like one of the large terns. Thomas Huxley, Darwin's most eloquent advocate, was quick to point out the reptilian features of these animals, and the fossils soon became icons for the theory of evolution.

In 1867, Huxley produced a family tree by expanding on earlier work by J.E. Courtnay, who described variation in the shape of bones in the roof of the mouth. He used the cumbersome word "dromaeognathous" to describe the palate of ostriches. In 1900, W.P. Pycraft suggested that it be replaced by "paleognathous," to reflect the "reptilian" nature of the ostrich's palate. The new word gained popularity and is still in use even though it implies, without any real justification, that the palate of the ostrich is somehow more primitive than that of other birds. Like many classifications based on a single character set, Huxley's attempt to reveal the evolutionary story of the birds was not as useful as its author had hoped. It did have one long-lasting effect: we still separate ostriches and their relatives from other birds according to the structure of their palate, and all other living birds are considered neognaths.

Many of Huxley's contemporaries were equally certain that there was a simple way to illustrate the evolution of birds from reptiles and attempted to develop phylogenies based solely on one body part or another. All failed, and the classification of birds remained one of the most difficult problems for ornithologists until the end of the century.

In 1888, Max Fürbringer demonstrated the weakness of family trees based on single characters when he published one based on 51. It represents an amazing achievement and required intense, sustained effort. He collected measurements and descriptive data from a large number of species and laboriously calculated the degree of similarity for each combination of characters. In the days before electronic computers, the process must have been incredibly tedious, but it produced a workable classification that lasted in one form or another for over a hundred years and is still very influential. With some changes to accommodate new information, Fürbringer's system appears in the organization of libraries and the sequence of species in textbooks and field guides. Although recent DNA-based genealogies suggest new sequences, most of the higher taxa, the families and orders, retain the memberships set out by Fürbringer.

Fürbringer answered all the taxonomic and evolutionary questions that his colleagues were in a position to ask, and said almost all that could be said about birds from the Victorian perspective. In subsequent years, avian taxonomy began to decline as the dominant field within ornithology. Physiology, behaviour, demography, ecology, and genetics had been given new impetus by Darwin's revolutionary ideas and came to dominate 20th-century ornithology.

The rigorous hypothesis testing that characterized these new sciences made comparative morphology look rather old-fashioned, and professors of avian taxonomy soon had trouble attracting resources while the best students naturally gravitated to better-funded projects in trendier fields. In many universities, traditional ornithology either disappeared completely or was relegated to some tiny and poorly maintained museum. Predictably, taxonomic problems left over from the 19th century began to plague ornithologists, but few researchers were able to acquire sufficient funding to visit rare and widely scattered specimens or collect new material.

During the centennial of Darwin's publication, Erwin Stresemann despaired of ever reconstructing the family tree of birds [203], but very few workers were able to tackle the problem [e.g., 197]. The same complaint was made by W.J. Bock, Robert Raikow, and Storrs Olson in subsequent years. By the 1980s, the failure of ornithologists to find a generally accepted family tree for the birds was impairing our ability to discuss avian evolution sensibly [171].

The problem has not gone away but there appears to be hope for resolution in the near future. Paleontologists have rekindled an interest in comparative avian anatomy by describing dozens of exciting new fossils, and it seems likely that we are close to a new understanding of the early evolutionary history of birds. In 2001, Bradley Livezey and Richard Zusi [117] infused new life into classical avian taxonomy by publishing a tree based on some 1,400 anatomical characters, 28 times the number available to Fürbringer. It was a preliminary effort but it demonstrated that progress was possible.

Relating Birds to Each Other

Although classical taxonomy may have become the most neglected field in modern ornithology, it had a long run as the richest and most important. Throughout the long period of exploration and colonization by Europeans, the classification of birds and other animals or plants was driven by the very practical purpose of organizing a great mass of unfamiliar information into something more digestible. It is a lot easier to begin the study of birds by contemplating the distinctive characteristics of two dozen rationally defined orders than by tackling 9,000 unsorted species. We use this kind of systematic classification to assess global biodiversity but the early naturalists had a second and more profound goal: they wanted to discover the meaning of life.

Aristotle took up the study of animals to uncover the mysterious causes of natural events, and this search for causation has remained the driving force behind all kinds of scientific inquiry. Unfortunately, causal links are extremely difficult to demonstrate, and in the centuries after Aristotle's death, only a few exceptional individuals took up his ideas. In most cultures, natural events of all sorts were attributed to the whims of the gods. Religious faith also simplified the world for medieval naturalists and, in later centuries, led many naturalists to look for the plan of God at Creation in their catalogues of species. It was sometimes a risky activity. Examining the plan of God could be perceived as an act of unbelievable arrogance, verging on blasphemy, and it occasionally attracted unwelcome attention from political and religious authorities. Even Linnaeus had reservations about the religious significance of his system of classification, and moved to renounce it before the end of his life.

Interest in causation returned in the early 19th century, when strange new philosophies began to undermine the influence of the church and the spreading industrial revolution introduced a more mechanistic view of nature. Interest in coal and its sedimentary geology pushed the age of the universe back into "geologic" time and changed the scientific perception of the world. Without an understanding of the great age of the earth, there would never have been enough time for the tedious, inexorable process of natural selection that drove Darwinian evolution. The age of the earth was as profound a discovery in the 19th century as its great size had been in the 16th. It set the stage for Darwin to propose the first testable theory for the diversity of life and causation in the natural world.

The concept of evolution gave a sense of direction to the classification process. Before Darwin, the categories of animals had no more connection to biology than the alphabet has to spelling or grammar. With the advent of evolutionary theory, biologists leapt at the opportunity to look for relationships among groups and fit them along trends from primitive to advanced. In the case of birds, those trends were supposed to run from reptilian to avian.

Victorians had lived through the industrial conversion from muscle power to steam power, and the evolution of nature was very much in keeping with their concept of progress. They eagerly set about making family trees and creating elaborate evolutionary stories to accommodate newly discovered fossils. Perhaps they should have shown more restraint. The theory of evolution had set biology in turmoil, and persons with more academic rank than information were able to propound their favourite theories with no real fear of effective criticism. Because both classification and phylogeny were based on a comparison of the same set of anatomical characteristics, there was no way to test the validity of any new ideas about evolutionary direction.

The Traditional Classification of Birds

The rhetoric of evolution's causal arguments was very compelling and versions of its ideas quickly took root outside biology. Evolution quickly became confused with the concept of progress in political and social institutions and offered spurious theoretical support for the evils of social Darwinism, eugenics, and racism. Most of its excesses are things of the past, but evolutionary ideas are now part of everyday life whether we like it or not.

Before 1859, it was sufficient to list a bird's physical characters and file it in the appropriate slot with the most similar types. If you were a knowledgeable biologist, you placed warblers with warblers according to bill shape, foot structure, and plumage characteristics. If you were an amateur enthusiast, you might set up a personal system and sort by colour or size without being ridiculed. Individual data sets were relatively small and the choice of filing system was largely a matter of personal convenience. There was no pressing reason to choose one system over another as long as you could communicate with your immediate colleagues and you put the names in a proper Linnaean format.

After 1859, Darwin's theory made it mandatory to organize animals according to perceived relationships with their ancestors. At first, ornithology seemed blessed with a remarkably clear evolutionary story. *Archaeopteryx*, a fossil bird, had been described in 1861 from Jurassic rocks in Germany. Its feathered wings made it a bird but its teeth and long tail were clearly reptilian. It was a perfect archetype, although, strangely, it was never embraced as an ancestral bird by Darwin. He left its interpretation to Thomas Huxley. Shortly after, Othniel C. Marsh found the fossilized remains of another toothed bird in much younger Cretaceous rocks from North America. The flightless *Hesperornis* helped confirm Victorian ideas about the reptilian nature of primitive birds, and its general similarity to living loons and grebes had a long-lasting influence on evolutionary stories for the birds.

Max Fürbringer was strongly influenced by *Hesperornis* and placed it near the root of his avian family tree. The loons and grebes looked so much like *Hesperornis* that it seemed only logical to place them and similar waterbirds nearby. They have been perceived as more reptilian and primitive ever since, even though we now know that they are unrelated to *Hesperornis* and have some very advanced characteristics. Hans Gadow produced a rather militaristic system (Table 5.1) as the English version of Fürbringer's classification. It organized the birds with keeled breastbones (carinate birds) into 2 large brigades, 4 legions, and 13 orders. He applied only 40 of the 51 characteristics used by Fürbringer, but all his groups shared at least 10 of those characters. None shared more than 34.

Victorian opinion was not wholly on the side of Fürbringer and Gadow. Alfred H. Garrod, Frank E. Beddard, and Elliott Coues decided that the apparent reptilian features of the loons and grebes were equivocal or at best irrelevant. They and other 19th-century ornithologists placed the woodpeckers and songbirds near the root of the family tree. These two groups are very similar and are often referred to as the Pico-passeres. It is still difficult to find unique anatomical characteristics in woodpeckers or songbirds, and many modern classifications have had difficulty separating them. Livezey and Zusi place them in a single group [117], even though DNA-based studies such as that by Sibley and Ahlquist suggest that they are not very closely related [192].

Gadow appears to have considered the birds without a keeled sternum (Ratitae) as the most primitive group. Perhaps he was influenced by their flightlessness. Since his time, the flighted tinamous have been placed within the Paleognathae and the keel has lost its importance as a distinctive feature. Gadow also set up a whole special division, Odontolcae, for extinct, toothed birds that modern ornithologists scatter among four subclasses, only two of which were known in Gadow's day. In the large group of keeled birds (Carinatae), he followed Fürbringer's lead and placed the loons and grebes at the base. Now that we have a great many avian fossils to study, it is difficult to understand why the flightless *Hesperornis* was so much more influential than the equally reptilian but ternlike *Ichthyornis*. Both were discovered at the same time and both were clearly toothed, primitive birds.

Gadow recognized one of the problem species that would continue to trouble ornithologists through the 20th century. He was unable to determine the close relatives of the Hoatzin, and even though it is represented by only a single species, he gave it its own order, Opisthocomi. Most ornithologists no longer see the Hoatzin as a relative of the chicken, but arguments about its relationship to turacos or cuckoos continue, and even advanced biomolecular techniques have failed to produce compelling evidence one way or the other. The issue will be discussed further in Chapter 6.

Placing *Hesperornis* at the base of the avian family tree was not an arbitrary decision. Birds appear to have evolved from more reptilian ancestors, such as *Archaeopteryx*, primarily by the gradual loss of unnecessary parts. Paleontologists refer to the history of such events as "reduction series." In birds, the primitive reptilian condition of five digits (or perhaps more) on each limb has been reduced to three on the hand and four on the foot. In some birds, the number of toes has been further reduced to three or, in the ostrich, to two. The loss of bones and their attached muscles has required that the remaining structures pick up new responsibilities and some specialized functions. The jointed thumb once grabbed prey in ancient reptilian ancestors, but its remaining bone supports the alula – the bastard wing that reduces the risk of stall at low air speeds.

The logic of organizing according to primitive and advanced characters is compelling when you are comparing an ancestral form to something much more modern, but it is less convincing when you are making comparisons within a group of advanced forms. It is not enough to state that a feature looks reptilian; it might well be a matter of convergence. For the past hundred years, most of the arguments in taxonomy have been about the most appropriate method for such comparisons.

By modern standards, the 51 characteristics used Fürbringer and the 40 used by Gadow seem to be rather small numbers with which to arrange some 9,000 species into two dozen orders and some 200 families. Recently, paleontologists have been

Table 5.1

Hans Gadow's 1893 classification of birds

Division	Brigade	Legion	Order	Membership
Ratitae				Ostriches, emus, rheas, cassowaries, kiwis
Odontolcae				Extinct toothed birds
Carinatae	One	Colymbomorphae	Colymbiformes	Loons, grebes
			Sphenisciformes	Penguins
			Procellariiformes	Petrels, albatrosses, shearwaters, prions, storm-petrels, diving-petrels
		Pelargomorphae	Ciconiiformes	Boobies, gannets, cormorants, frigatebirds, tropicbirds, herons, shoebills, storks, ibises, flamingos
			Anseriformes	Swans, geese, ducks, screamers
			Falconiformes	Hawks, falcons, eagles, vultures
	Two	Alectoromorphae	Tinamiformes	Tinamous
			Mesites	Mesites of Madagascar
			Turnices	Button quail
			Galli	Grouse, pheasants, pea fowl, etc.
			Opisthocomi	Hoatzin
			Gruiformes	Cranes, rails, coots, gallinules
			Charadriiformes	Sandpipers, gulls, auks, sand grouse, pigeons, doves
		Coraciomorphae	Cuculiformes	Cuckoos, turacos, parrots
			Coraciiformes	Kingfishers, rollers, owls, nightjars, swifts, hummingbirds, mousebirds, trogons
			Passeriformes	Woodpeckers, songbirds

Source: [63]

able to find more than 200 characteristics with which to classify dinosaurs, even though they have only skeletal features to work with. Nonetheless, in the 20th century, Alexander Wetmore, J.L. Peters, Erwin Stresemann, and many other tax-onomists relied on those same few characteristics to revise the avian family tree. Considerations of time and money, and the difficulty in reviewing hundreds of specimens, discouraged them from adding new characters to the basic data set that Fürbringer published in 1888. As you might expect from a system based on Victorian ideas, none of the characteristics used for classification are related in any way to flight capability or aerodynamic sophistication.

Characteristics Used in Classification

Several characteristics are used in the classification of birds:[3]

Characteristics of the nestlings. Are the nestlings naked or downy when hatched? Do they stay in the nest until fledged (nidicolous) or leave while still downy (nidifugous[4])?

Keratinous structures. Are the down and body feathers restricted to specific tracts (pterylae) or are they scattered over the body? Do the feathers have an aftershaft? Is there a tuft of down on the oil gland? How many feathers are there in the tail or the wing, and how are they arranged? How many separate plates make up the beak?

The organization of feathers into discrete tracts (pterylae) can be very distinctive, and illustrations of feather patches and gaps appear in most standard text-books. Often, a surprisingly small number of feathers is sufficient to cover a bird's whole body. In the heron, just two strips, of two feathers each, cover most of the upper and lower surfaces. In other birds, the tracts may not be very distinct and could be interpreted as being similar to several different groups. The Hoatzin has particularly indistinct feather tracts. Semiplumes are scattered over its body and blur the edges of the pterylae, contributing to the difficulty in classifying this strange bird.

Storrs Olson points out that pterylae appear to be a universal feature of birds, at least in embryos, and may therefore be a primitive character inherited from early reptilian ancestors. He feels that pterylae imply that even the earliest types of true feathers had an aerodynamic function, while simple hairlike structures were sufficient for insulation in flightless animals, as they are in the kiwi [151].

Feather tracts can be examined only in an everted skin. In other words, the skin must be pulled off like a sock or glove. Usually this means that ornithologists must have access to a freshly killed specimen because few museums retain specimens in this rather unattractive format. Feather tracts are therefore difficult or costly to

---

3 In this section, I have followed the plan that Charles Sibley and Jon Ahlquist [192] used for their history of ornithological taxonomy.

4 The term "nidifugous" is derived from Latin roots that mean "flees the nest."

examine first-hand, and there is a tendency to rely on previously published information about this character.

The absence of a fifth secondary feather (diastataxy) in the wing is one of the few characteristics that can be seen on a typical museum skin. It has been used as an argument for placing loons, grebes, petrels, and penguins among the primitive birds because the same feather is missing in *Archaeopteryx*. It is also missing in pelicans, herons, storks, flamingos, ducks, falcons, cranes, parrots, owls, sandpipers, gulls, auks, and rollers. The condition may simply be a feature of large birds or birds with particularly long, pointed wings. Significantly, ducks are the only member of the Galloanseriformes without a fifth secondary feather, and the only member of that group with strongly pointed wings.

The birds that always have a fifth secondary feather (eutaxy) tend to be small species with short, round wings, such as the songbirds, woodpeckers, kingfishers, trogons, mousebirds, and cuckoos. The characteristic also occurs in some relatively large birds, such as tinamous, galliform birds, the Hoatzin, trumpeters, and screamers, but this assemblage includes birds from several unrelated orders. As the only flying members of the Paleognathae, tinamous might be expected to have unusual wings. You might, however, expect them to have the supposedly primitive type rather than the more advanced one. All the other large eutaxic birds are round-winged representatives of the Galloanseriformes, except the Hoatzin, whose relationships to other groups remain unclear. Perhaps the fifth secondary feather appears to coincide with taxonomy only because body size and flight style also tend to be split along taxonomic lines. The split is especially clear in the Sibley and Ahlquist biomolecular phylogeny [192], which tends to place small birds near the base and large birds near the crown (Chapter 6).

The bony skeleton. Recent studies of birdlike dinosaurs suggest that a great many skeletal features in modern birds have potential value in classification. These need to be studied across the entire range of birds. Historically, only a few carefully selected examples have been used. The main questions have been: Are the nostrils separated by a solid septum (holorhyny) or is there a gap (schizorhiny)? Is there a basipterygoid process? Is the temporal fossa deep or shallow? Is the inner end of the lower jaw long and extended or short and simple? How many vertebrae are in the neck? Is there an extension (hypocleideum) at the bottom of the wishbone (furcula[5])? How many toes are in the foot, and how are they arranged in relation to each other (Figure 1.9)?

All modern classifications of birds recognize a fundamental description of the two main groups of living birds as paleognathous and neognathous. Initially, that division was based on the characteristics of the palate. Taxonomists found it convenient to use related characteristics to divide the neognaths into smaller groups.

---

[5] The term "furcula" is derived from the Latin for a small fork or crotch. It is related to the name for the peculiar yoke-shaped strut on which Venetian gondoliers rest their single oar.

Each group is based on the shape of the vomer bone in the palate and whether or not it touches other bones. The terms "paleognathous" and "neognathous" are used regularly in ornithological literature, but the minor descriptors of palate structure – desmognathous, schizognathous, or aegithognathous – appear only in specialized taxonomic and paleontological literature.

In Chapter 2, we saw that biomechanical studies suggest that such features of the palate vary according to feeding habits and are not particularly useful in classification [77, 78]. As such, they illustrate one of the basic problems in comparative morphology. The shape of a character has always been seen as far more important than its function. Consequently, highly variable structures, such as the hypapophyses on the thoracic vertebrae, have received very little attention.

Distribution and organization of limb muscles. The long history of birds as pedestrians is reflected by the absence of some muscles in the hind limb and variation in the arrangement of the neighbouring muscle groups. Over the years, these features have been used as the basis for various family trees by such notables as Thomas Huxley. Although none of those projects proved particularly useful in the long run, they left us a heritage of anatomical characters that are still considered important.

Alfred Henry Garrod also used the arrangement of hip muscles to classify birds in 1873 [65]. Some of the characters that he used were picked up by Max Fürbringer and are still used. Modern taxonomists have added data from eight more muscles, recording the information about their organization as yet another alphanumeric code. For example, the pelvic code of owls is "AD" but that of hawks is "A+," indicating that both have a caudo-femoral muscle (A) but only the hawks have an ambiens muscle (+).

Historically, the ambiens muscle has played a prominent role in avian classification. It is reported to help a bird grip a perch when it is asleep, but the supposedly automatic locking action of the toes was misunderstood until recently. The relationship of the ambiens to the tendon locking mechanism, discussed in Chapter 2, has not been described [169]. Garrod used the presence of the ambiens, along with the arrangement of the plantar tendons in the foot, to place parrots, cuckoos, and turacos together in a group called Homalogonatae. Woodpeckers, toucans, barbets, and their relatives, which lacked an ambiens, were placed in the Anomalogonatae. These names appear sporadically in the historical literature but are no longer in widespread use.

The use of foot tendons in classification dates back to the early 19th century. C.J. Sundevall, the Swedish anatomist, began with an analysis of muscle connections for the leg in 1835 but went on to erect a phylogeny based on the deep plantar tendons of the foot. Those are the tendons that move the toes when a series of small muscles anchored to the femur contract. Arrangement of these tendons is still widely used in taxonomy. Fortunately, modern taxonomists have recognized the pointlessness of long names derived from Latin or Greek, especially in a multilingual world, and refer to each of the eight basic categories by alphanumeric coding.

The tendons between the muscles and the toes pass through openings in a box-like structure at the top of the tarsometatarsus called the hypotarsus (Figure 2.13). The arrangement of the holes in this bone provides another useful characteristic for classification, especially in fossil species.

Features of the syrinx (voice box). Many birds have a complex syrinx associated with song. It may be located in the bronchus or the trachea or in between. It may also have an array of special muscles attached to particular points. All these features are used in both historical and modern classifications.

Arrangement of the carotid arteries. Birds may have a left and a right carotid artery, but sometimes there is only a left branch. There are also various elaborations on that basic structure. Garrod made use of the variation for classification but over the years the discovery of exceptions and special cases has reduced its reliability. When the information on the carotid artery is used, it is usually presented as an alphanumeric code.

Organization of the gastrointestinal tract. Many birds have a specialized tongue with a distinctive shape associated with their preferred type of food. Whole groups of birds are specialized for eating seeds or catching fish, but generally the tongue is too variable for use in classification. The fine structure of its skeleton, the hyoid apparatus, is more useful.

Periodically, attempts have also been made to classify birds using other parts of the digestive tract, especially the twists and turns of the intestines and the presence or absence of functional caeca (appendices). Modern analyses indicate that the characters are not reliable. There is a specific terminology for the contortions of the intestine, but the technical terminology (i.e., cyclo-, ortho-, iso-, anti-, plagio-, and peri-coelous) is rarely encountered outside Victorian texts and is of historical interest only [8, p. 24].

Species of parasites. By far the most intriguing indicators of relationships among birds are relationships among the lice that live on their bodies and among their feathers [194]. The lice found on birds are such complete homebodies that they have a great deal of difficulty surviving anywhere but on their chosen host. Consequently, they rarely get transferred to another species. In an ancient lineage such as birds, any distribution among clearly unrelated birds suggests that there was some significant evolutionary or ecological link earlier in the bird's history.

Such tiny creatures are rarely preserved as fossils but one has been found recently in exceptionally fine Eocene deposits near Manderscheid, Germany [223]. The specimen appears to belong to a distinctive group that parasitizes only waterbirds and shorebirds, but other species must have existed at that time. We might eventually find fossil lice on some exceptionally well preserved feathered dinosaur.

Sometimes there seems to be a simple mechanism that explains how closely related lice spread among a wide variety of birds. The wing louse *Anaticola* is found on ducks (anatids), and its relatives appear on other waterbirds as well as some gulls and shorebirds that share wetland habitats. Similarly, the louse *Brucelia* is found on a variety of unrelated small forest birds: 10 of its species occur on songbirds, 1 appears on trogons, another on motmots, and 3 additional species on woodpeckers.

Birds may also share lice for other kinds of ecological reasons. For instance, predatory raids by skuas on the nests of albatross may explain why both species share closely related lice. The wing louse *Harrisonetta* lives on the albatross while a close relative, *Haffneria,* is found on the skua. If transfers between the two groups were frequent, the lice would not have evolved into distinct species.

There are cases where the distribution appears to support an evolutionary relationship but other explanations may be equally plausible. Traditional taxonomy places the galliform birds close to the falconiform birds but biomolecular evidence places them far apart. The louse *Lipeurus* lives on chickens and the similar *Falcolipeurus* on falcons, suggesting either a genetic or a predator/prey relationship. Neither theory explains why another relative, *Archolipeurus,* is found only on ostriches. An evolutionary relationship is possible but seems unlikely, or perhaps very ancient. DNA studies place the division between ratites and galliform birds in the Cretaceous. Some ancestral lipeurid louse may have lived on their common ancestor.

Just as often, the distribution of closely related lice seems to defy rational explanation. *Goniodes* is a parasite of the galliform birds, most of whom live in the Northern Hemisphere, but the similar *Austrogoniodes* lives only on penguins in the Southern Hemisphere. Such exceptions are particularly frustrating to taxonomists because they suggest that theories based on the relationship between parasite and host could be fragile.

Ornithologists have struggled to identify relatives for the South American Hoatzin ever since it was discovered. It might have remained an obscure and unimportant species but the fingers on the wings of its nestlings caught the imagination of Victorian evolutionary biologists by hinting at a link to distant reptilian ancestors. Its anatomical characteristics are vague, however, and it even defies attempts to apply advanced biomolecular techniques to its classification. Not surprisingly, evidence from its parasites has also been equivocal. Some ornithologists believe that the Hoatzin is related to cuckoos; both groups share lice of the genus *Cucuphilus.* There are many cuckoos in the forests of South America, however, and the distribution of *Cucuphilus* could simply be the result of contact in shared habitats. The Hoatzin has a unique body louse, *Osculotes,* which has no close relatives and is therefore no help in classification.

Other features. Taxonomists have also made use of other characters, such as the shape of the bill, certain colour patterns in the feathers, the type of vomer in the palate, the presence of a pneumatic foramen in the humerus (an opening for the entry of the air sac into the core of the bone), the presence of supra-orbital glands (salt glands) on the skull, the presence of a crop or a penis, the arrangement of wing muscles, the type of nest, the structure of the egg, and even geographical distribution.

Adding New Characteristics to the List

Until very recently, only a few morphological characteristics had been added to those used in the 19th century, in spite of the many different attempts to redefine the relationships of the higher groups of birds.

One of the last of the classical taxonomists to tackle the problem of creating a new data set was Boris Stegmann [197], but even he was unable to complete such a project. In 1978, he published preliminary results in the form of an analysis of the muscles and tendons in the hand to test the usefulness of Gadow's Alectoromorphae (Table 5.1). His immediate goal was to determine the relatives of the sand grouse (Pteroclidae), a group whose characteristics place it somewhere between the pigeons and the sandpipers but not securely in either group.

Fürbringer had examined the pectoral girdle for usable taxonomic features in 1888 but found that there was little variation between groups. He took only a cursory look further out on the wing because he believed that the skeletal features would be too conservative to be useful in taxonomy while the wing's feathers would make too many accommodations for variations in flight style. Later taxonomists tended to concentrate on the reanalysis of Fürbringer's data sets, so potential taxonomic characteristics of the wing were never exploited. Stegmann was the first researcher in 85 years to develop new data for the avian wing from a fairly wide range of groups.

Based entirely on the sequence of losses for some muscles and the appearance of some specializations, he concluded that variations in wing structure generally supported the validity of Gadow's Alectoromorphae. He recommended, however, that the group be expanded to include the cuckoos, and also that it be split into two superorders, Alectoromorphae and Charadriomorphae, to reflect what he saw as divergent evolutionary trends.

Within the Alectoromorphae, the galliform birds occupied the centre, with tinamous and guans (cracids) near the base and Guineafowl as the most advanced type. Although Stegmann considered tinamous less advanced than other galliforms, he decided that the differences were not strong enough to place them elsewhere. Only recently have they been placed among the paleognaths with the ostriches.

Stegmann placed rails and button quail near the base of the Charadriomorphae and bustards and cranes nearer its middle. He also saw the pigeons as branching off from the main lineage near the base of the tree, and the parrots branching out from among them. As you would expect from a group with a name like Charadriomorphae, he considered the Charadriiformes (sandpipers, gulls, and auks) to be the most characteristic representatives of the group. He concluded that the sand grouse branched off from within the Charadriiformes, and not from among the pigeons.

Stegmann's project offers us some important lessons about the classical approach to avian taxonomy. First, it was his life's work and he simply ran out of time to address the whole of Class Aves. It took the effort of his whole career to investigate the wing structure of a relatively small number of groups and determine what information was needed for analysis. Second, although his analysis was based only on the peculiarities of wing musculature, many of his results agreed with classifications based on other, unrelated characteristics. This suggests that the various orders of birds are real biological entities. They are discrete and natural lineages that can be depended upon for other kinds of ornithological research. Third, his work reveals the underlying weakness of classical taxonomy. Where Stegmann's results disagree with the work of another researcher, there is no rational way to resolve the discrepancy. You either accept or reject his claims on the basis of your experience and knowledge.

When the prestigious Nuttall Ornithological Club published Stegmann's results, they included a foreword by Walter J. Bock, one of the most prominent evolutionary thinkers of the 20th century.[6] He set out the nature of classical comparative morphology in a very clear way and put Stegmann's work in historical perspective. His essay is very much an appeal for students to take up where Stegmann left off. Unfortunately, it was one of many that fell on deaf ears, and there was no renewal of taxonomy based on comparative morphology. Students in the 1980s and 1990s could not afford to risk their careers in a field that so obviously lacked resources and was so far from the mainstream of experimental science.

Although Stegmann's descriptions of the details of muscle attachments are highly technical and make very difficult reading, it is still worth examining his descriptions of how the muscle attachments affect the lifestyles of birds by enhancing their flight capabilities. His explanations of function help clarify the logic of his inclusions and exclusions in various categories. The enlarged *Musculus ulnimetacarpalis* on the wings of chickens has done more than augmented the offerings of the fried chicken-bits industry. Wild chickens and other galliform birds depend on an explosive take-off to escape predators, and the greatly enlarged muscle has strengthened their wrists. Apparently, the brief burst of deep but rapid strokes is not adversely affected by inertia generated by the weight of that muscle near the wingtip. That extra weight may be of more concern in other groups of birds, including the chicken-like tinamous, which have replaced the muscle with a light-weight tendon.

Applying System to Classification

Classification is a human activity and, like many human activities, it has been strongly affected by history. For much of the 20th century, it was influenced by the survival of Victorian scholarly traditions. Until recently, there were no formal criteria for decision making in comparative morphology; the whole subject depended heavily on learned opinion and experience. Consequently, it was possible for senior members of the scientific community to exert influence over research subjects and publications through what the modern taxonomist Robert Raikow calls "reverence over reference." As late as 1891, we find the eminent ornithologist

---

6 He was also an early proponent of the inner-bird concept.

William Kitchen Parker blithely dismissing the need for a uniform classification system with the comment, "Taxonomy is a very tentative science; it is only a sort of rough scaffolding, and not a Temple of Nature; moreover each ornithological artisan will use his own classification." At the time, he was expounding confidently on his theory that birds descended from the amphibians, or at least from dipnoan lungfish [155].

No modern biologist would tolerate such a statement but it was a product of its times. Victorians were working without definitive answers to many fundamental questions because they lacked the information to formulate such questions, let alone propose the answers. In spite of the lack of rigor in the Victorian methods, taxonomy far outstripped other aspects of biology, and further progress in the methods of classification had to wait until the other fields could catch up in the 20th century.

It took a long time. With the dawn of the 20th century, biologists lost patience with the previous era's rhetoric and became much more interested in clear communication (the sharing of intelligence) that required as stable a system of classification as possible. They also shifted their focus from the origin of large groups to the origin of individual species. It proved to be a productive field of investigation and led to the integration of ideas about evolutionary processes with all other aspects of biology [126]. Eventually, evolutionary approaches to subjects like physiology and ecology promoted the reinterpretation of dinosaurs as active animals and as the potential ancestors of birds.

Although taxonomists gradually became a very small proportion of the biological community, select groups of experts continued to debate specialized taxonomic issues. In 1985, Robert Raikow was able to list seven competing systems for the classification of birds [171]. All were based on very similar sets of morphological data but each represented a novel approach to the issues of comparative morphology, paleontology, and zoogeography: Mayr and Amadon (1951); Stresemann (1959); Wetmore (1960); Storer (1971); Morony, Bock, and Farand (1975); Wolters (1975-82); and Cracraft (1981).

None of these classifications was a trivial effort. Each was a carefully considered analysis by a well-known and reputable ornithologist with an extensive knowledge of evolutionary theory and avian zoogeography. Each researcher had devoted a significant portion of his career to the dogged examination of a confusing mass of characters. It was a thankless effort, however, and Raikow was not the only ornithologist to complain about the lack of progress towards a taxonomic consensus. Some of his colleagues were losing patience with the whole process. Storrs Olson caught their mood when he derided attempts at avian classification as "little more than superstition ... [bearing] about as much relationship to a true phylogeny for the Class Aves as Greek Mythology does to the theory of relativity" [171]. He blamed many of the problems of taxonomy on unsubstantiated speculation and a lack of effort by taxonomists to tackle difficult questions with large-scale research programs.

By the mid-20th century, most ornithologists had lost interest in the challenge and had made the wholly practical decision to settle for the classification offered in their favourite field guide. Much of their research on variations at the species level and the issues surrounding the higher taxonomic levels seemed to have lost their relevance. If possible, the link between birds and dinosaurs was even less important. It was only when advances in biomolecular chemistry created the opportunity to base taxonomy on gene sequences that avian classification began to make front-page news in the leading scientific journals. Coincidentally, paleontologists began to discover a whole new suite of dinosaurs and two whole subclasses of birds at about the same time.

Problems in Comparative Morphology

Polarization

One of the more difficult problems in comparative morphology is the establishment of a particular polarization for an individual character. Victorians described one end of the polarity as reptilian and the other as avian to reflect their belief that the evolution of birds from a more primitive form left identifiable traces. More modern taxonomists recognize that the dinosaur ancestors of birds were themselves a sophisticated and highly evolved type of animal, and use terms such as "primitive" versus "derived," "ancestral" versus "advanced," or "basal" versus "crown." The objective is to identify the condition of the characteristic that appeared in the shared ancestor without making assumptions about its link back to ancient reptiles.

As soon as a branch appears in a lineage, evolutionary pressures cease to act on the characters of the two descendant lineages in exactly the same way as they acted on the ancestor. If the ancestral structure was robust and continued to function well in the descendant lineages, it might change very little. It will change, however, if only slightly, because it is part of a thoroughly integrated organism. It will be dragged in new directions by more dramatic changes in other characters. For instance, among the petrels, the wings have undergone elongation and changed shape, and the beaks have been modified to cope with a variety of prey items. The legs and feet have changed hardly at all, however, and may be very similar to those of ancestral petrels.

It is difficult to see hints of a common ancestor that bridges the gap between the wings of an albatross and the flipper of a penguin, but the legs and thighs retain such close similarity in muscular and skeletal features that the two groups seem likely to have shared a common ancestor [117]. These features are so conservative that they suggest that both shared a common ancestor with the loon.

The difficult question for comparative morphologists is: Which characteristics of the wings in these three groups reflect the condition in the common ancestor? Without understanding the primitive condition, you cannot determine whether loons evolved from the petrel/penguin group or vice versa. When the question involves arranging over a hundred families in evolutionary sequence, the challenge is insurmountable without a well-understood ancestor. Resolution of the problem requires a very sophisticated protocol that describes and assesses the importance of hundreds of individual characteristics, and then develops compelling arguments for their polarization. The scale of the task is the main reason that the 1888 analysis of Max Fürbringer survived as a powerful influence on avian classification for a hundred years, until electronic computers became widely available.

Evolutionary Convergence

No problem confounds the analysis of morphological data more than evolutionary convergence – the tendency of animals to look alike if they share aspects of their lifestyle and habitat.

When HMS *Beagle* passed through the Straits of Magellan in 1838, Charles Darwin was amazed to see small black-and-white birds that looked and behaved just like the Dovekies or Little Auks of the North Atlantic [44, pp. 259-60]. They turned out to be diving-petrels, members of the Order Procellariiformes and completely unrelated to auks. Later, the similarity between these birds would be recognized as a classic example of convergent evolution. It is a concept that fascinated Darwin, and his theory of evolution by natural selection offered the first rational explanation of its causes. When creatures of different origin share a similar way of life, they come to resemble each other due to the forces of natural selection. Both the diving-petrels and Dovekies are small colonial seabirds that use wing-propelled locomotion to pursue plankton and small fish underwater. Both nest in underground burrows on isolated islands, lay a single large egg, and carry whole food items back to the nestling. Recognition of their membership in separate orders did not come easily. Discussions lasted until the middle of the 20th century and demonstrate the value of rigour in analyses.

There are many trivial examples of convergence among birds, especially among the numerous families of forest birds. Some of the most obvious are reflected in the choice of common names. The shrikes (Laniidae) are a family of predatory songbirds whose habits and a distinctive appearance have made them a particularly popular choice for names in unrelated birds. There are shrike-bills in the Monarchinae; shrike-tits and shrike-thrushes in the Pachycephalinae; butcher birds (another name for shrike) in the Cracticidae; helmet-shrikes in the Prionopidae; cuckoo-shrikes in the Campephagidae; peppershrikes in the Vireonidae; shrike-tyrants in the Tyrannidae; ant-shrikes in the Formicariidae; shrike-tanagers in the Thraupinae; and shrike-babblers in the Timaliinae. In every case, the shrike-like characters are limited to superficial features of bill shape, plumage, or behaviour, and the owner's true heritage remains obvious.

Convergence can be a powerful force and extends across higher taxa. Several major lineages of birds have come to resemble each other so closely that at various times some taxonomists have lumped the pairs together as members of the same family or order: grebes look like loons, hawks look like owls, and swallows look like swifts. Nevertheless, careful examination of the embryology and structural details or consideration of major features of behaviour or life history have led to the conclusion that even the most striking of these similarities is actually superficial. Loons and grebes, for instance, seem particularly difficult for taxonomists and frequently end up in the same order. Both loons and grebes have a very distinctive and large cnemial crest at the top of their tibiotarsus. It is the anchor for powerful swimming muscles from the hip. The structures look very much alike, but in loons the crest is derived from the patella or kneecap, whereas in grebes it is part of the leg bone. The function of the structure is the same in both groups, but the disparate origin of the component parts signifies that the knee structures of loons and grebes are analogous to each other, not homologous. They are not evidence that the two groups are related.

Convergence has even been able to bridge the largest taxonomic gap in living birds. Modern classifications recognize the split between paleognaths and neognaths as one of the earliest and most important divisions in the modern birds. One group of paleognaths is almost indistinguishable from a group of neognaths, however. The tinamous of South and Central America have been included with kiwis and ostriches in the paleognaths only after a lengthy debate. Although tinamous appear to have the distinctive paleognath palate, they are the only modern members of the paleognaths to have a keel on their sternum and the power of flight. It makes them seem very different from the ostriches and other living paleognaths, but the fossil lithornithids show that there were once other flying members of that group [93].

The paleognath tinamous bear a stunning resemblance to neognath ptarmigans and grouse among the galliform birds. Their skeletons appear almost identical, and the trailing edges of the sternum are deeply bisected in similar ways in both groups. The similarity between the two groups of birds even extends to the shape of their feather parasites. All the similarities are superficial, however. The structures of sternum that appear so similar are very different when examined in detail, and are derived from different embryonic components. The feather parasites are from unrelated groups. Both kinds of birds have rounded wings and use rapid take-off to escape predators but even their wings have subtle structural differences. Boris Stegmann found that tinamous have a relatively primitive wing musculature and are not nearly so well adapted for the kind of explosive take-off that characterizes the escape flight of grouse [197].

In the case of the diving-petrels and auks that attracted the attention of Darwin, the similarity of the skeleton is so strong that it has challenged opinions about the relationship between the birds. Most ornithologists consider the specialized bill of the diving-petrels to be sufficient evidence that they belong with other tube-nosed birds in the Procellariiformes and cannot be members of the Charadriiformes. There have been arguments to the contrary, however, and confusion between the two groups has a long history.

During the Spanish exploration of Pacific North America, Jose Mariano Moziño described many birds, including the Marble Murrelet, in his *Noticias de Nutka*.

Unfortunately, the illustration is labelled *"Diomedea exulans,"* a name now assigned to the Wandering Albatross. It might be a simple clerical error but at the time the Linnaean system was in its infancy and there were not many choices for names; it is possible that this was a simple misunderstanding of descriptive criteria. After all, in Europe the murrelet was later classified as a grebe.

If the label was not an accident, Moziño may have been influenced by the murrelet's peculiar tubular nostrils. They open to the sides but otherwise are similar to those of the Potoyunco or Peruvian Diving-Petrels, which open upward. He could easily have come across the diving-petrel during his undergraduate studies at the University of Mexico, before he ventured northward into habitat occupied by the Marbled Murrelet.

In 1958, R. Verheyen turned the whole story on its head by stating that it was the differences in bill structure that were superficial and merely the result of recent divergence within a single group of birds [216]. The proportions of various bones in the two groups were so often similar that he believed that both diving-petrels and auks belonged to a single order, Alciformes. In his eyes, the diving-petrel was a living representative of the ancestor of both the auks and the procellariiform birds. The fact that Verheyen was able to publish these seemingly extraordinary ideas in a reputable journal suggests that the theoretical basis of taxonomy had not advanced very far by the mid-20th century. Publication did not imply that Verheyen's ideas were widely accepted within the ornithological community, but without some form of independent test, any criticism of them was simply a contrary opinion.

Shortly after Verheyen published his theories, the Japanese ornithologist Nagahisa Kuroda began another study of auks and diving petrels [112]. His activities at Tokyo's Yamashina Institute for Ornithology (Yamashina Chôrui Kenkyûjo) marked the revitalization of Japanese ornithology after the catastrophe of the Second World War. He did not have access to a Little Auk from the North Atlantic so he used a specimen of the Ancient Murrelet that a friend had found in the Sea of Okhotsk. The Ancient Murrelet shows the same broad suite of proportional similarities that had fascinated Verheyen, but Kuroda noted significant differences in structural detail suggesting that any such similarity was merely superficial.

The skulls share similar proportions but the individual bones vary in shape and size. The nares, or bony nostril, of the murrelet is schizorhinal (open through the septum), not holorhinal (separated by an uninterrupted sheet of bone); the nasal bone itself is a small fused attachment, not the large independent structure found in the diving-petrel. The structure of the palate in the murrelet is more open and the shape of the postorbital process is distinctive. The skull of the diving-petrel also has prominent structures on its sides, called squamoso-parietal wings, that are absent in the murrelet.

The bodies of the two birds are constructed in significantly different ways. The furcula is deeply curved in both; in the diving-petrel it is fused to the sternum, but in the murrelet and all its relatives, it is an independent bone. Kuroda also discovered a significant muscle in the diving-petrel that is absent in the auk.

Diving-petrels have a thin middle layer in the pectoral muscle that occurs only in petrels and appears to be a diagnostic characteristic of the Order Procellariiformes. It is particularly large and distinctive in long-winged species such as shearwaters and albatrosses, where it consists of white fibres that are most suitable for brief bouts of repeated contraction. The longer-winged petrels apparently use it to manipulate the outer wing. Its presence, even as a small slip, in a short-winged species like the diving-petrel suggested to Kuroda that diving-petrels might represent the ancestor of the larger, more specialized aerial types.

Both auks and diving-petrels have unusually strong wing bones to accommodate the stress of underwater locomotion. In general, the diving-petrel's bones are a little more robust than those of the auk, but both have a distinctive flattened and arched humerus that is significantly longer than the ulna. This seems to be a special adaptation for underwater wing-propelled locomotion, although the exact purpose of its architecture is not fully understood (see Chapter 10). Species of shearwaters that use their wings to dive underwater have a similar but less distinct modification to their humerus, while the shearwaters that do not dive have unspecialized wing bones.

The modified humerus architecture is widespread in wing-propelled diving birds, including penguins and extinct flightless groups such as the giant Plotopteridae and the Mancallidae once found in the North Pacific. The only wing-propelled diving bird without a highly specialized humerus is the tiny dipper, or water ouzel. It is also the only aquatic songbird and the only wing-propelled diving bird of any type that regularly lives in fresh water. Its humerus is longer than its ulna but it is neither flattened nor arched like the humerus in an auk.

Perhaps the most significant characteristic of Kuroda's work is that it exhibits a very high level of expertise. Like the work of Boris Stegmann, it represents the accumulated personal experience of many years of study and a knowledge of the anatomy of many different species. It only demonstrates, however, that murrelets are unlikely to be related to diving-petrels and vice versa. It does not tell us about the relationship of either bird to other groups.

Many of the differences between diving-petrels and auks described by Kuroda involve soft tissues or fragile small bones that seem unlikely to be preserved after death. They suggest that it could be very difficult to distinguish between fossils of species as convergent as the diving-petrels and auks. The London clays of Britain and similar deposits in other parts of Europe, however, include some beautifully preserved specimens whose early dates make them important in understanding the recent evolution of birds.

In 1977, C.J.O. Harrison realized that it would be necessary to look at fine details to unravel convergence among such fossil forms [79]. Birds might have solved some large-scale problems that they shared in similar ways, but they have resolved smaller biomechanical issues in unique ways because they began at different starting points. According to Harrison, "given a single bone, or half of one of the larger limb bones, it should be possible ... to determine the family and order."

He found that the head of the humerus in both auks and diving-petrels was very similar but the structure of its inner surface was so distinctive that a single piece might include sufficient diagnostic characters. The elbow provided another set of features. Because water is so dense, both birds needed to strengthen the elbow joint and limit its tendency to shift backward under stress. Auks use a strong ridge of bone on the humerus to prevent such slippage, whereas the diving-petrels appear to have enlarged small irregularities that are present but barely visible in the elbow of other petrels. A prominent boss on the ulna of the diving-petrel fits snugly into a hollow on its humerus when the wing is extended. The lip of that hollow performs the same function as the strong ridge on the auk's humerus even though the two structures are entirely different in origin. Such small and apparently minor characteristics have become the basic tool of modern paleontological studies, and have been especially important in working out the phylogeny of dinosaurs from fragmentary remains.

The arguments put forward by Harrison and Kuroda are not really complete by modern standards. They accomplished only half of the job by telling us where birds do not belong taxonomically. Neither demonstrates that the individual characters are actually diagnostic of a group. For instance, Harrison does not compare characteristics of the elbow in the auks and diving-petrels with those in the pelecaniform birds or loons. Without an understanding of how such characters are distributed among other birds, it is impossible to use them to assign a fossil to a modern group with confidence.

Loss of the Reptilia

Recently, biologists have come to realize that a concept as familiar as "reptile" is actually based on an artificial assemblage of unrelated animals that have converged. They broke up the Class Reptilia and its name no longer has official status. It was not enough that all living reptiles were cold-blooded and scaly, because they had different kinds of skulls, and skull structure is fundamental to the history of vertebrate animals [158, p. 22].

Terrestrial vertebrates are now divided into three basic groups according to arrangement for the passage of muscles from the top of the skull to the lower jaw. In turtles and some extinct lineages of primitive vertebrates, the muscles pass smoothly over the outer surface of the skull to the lower jaw. All turtles were formerly placed among the reptiles on the grounds that they were cold-blooded and scaly and laid eggs. Mammals, and the extinct mammal-like reptiles that were their ancestors, have a single passageway on each side of the skull, and the muscles that move the jaw pass through it. Living mammals have never been thought of as reptiles because they are warm-blooded and furry and rarely lay eggs. The supposed position of the mammal-like reptiles is borne in their name. Crocodiles, snakes, lizards, dinosaurs, and birds have two openings in the skull for the passage of jaw muscles. Some members of this group are cold-blooded but others are warm-blooded. To varying degrees, they have scales. All lay eggs or produce eggs that are hatched internally.

A skull with no opening is "anapsid" and its possessor is a member of the group Anapsida. One with a single opening is "synapsid" and its owner is a member of the Synapsida. One with two openings is "diapsid" and its owner is a member of the Diapsida. The only remaining use of the word "reptile" is to imply that an animal is cold-blooded, sluggish, and a crawling type of creature. It is not a term that can be applied to mammals or birds, and modern reconstructions of dinosaurs suggest that it should not be applied to them either.

Typology

Some of the current difficulties in setting out a phylogenetic classification for the birds were inherited from the problems that Darwin set out to resolve. In the early part of the 19th century, there were many other explanations for the diversity of animals, each of which had its own historical foundation. The best-founded theories came from Richard Owen, the most knowledgeable Victorian anatomist. Although he had spent a lifetime studying the comparative morphology of fossil forms, he was unwilling to accept the great age that geologists assigned to the globe and so he put forward a theory of development from archetypes. He saw modern animals as modified representatives of the small number of archetypes found in the fossil record. The existing modern animals were the models for certain levels of organization and supposedly represented actual stages in the history of animal development. It was an influential point of view that gave special biological significance to higher taxonomic levels, such as families or orders.

The use of the "type" was a very popular idea with Victorians. Detective novels of the time often depended on the recognition of specific "criminal types." A sophisticated version is still a popular tool in criminology, under the modern name of "profiling." Unfortunately, "typology" has also had a powerful influence on evolutionary thinking and has led to all sorts of misconceptions, especially about animals known only from fossils.

Typology crept into the discussion of evolution when some of Darwin's supporters adopted very simplistic ideas of evolution as progress from the primitive to the advanced. At the time, such ideas were reasonable because no one understood how the characteristics of an animal could be passed on from one generation to the next. It was clear only that such a mechanism existed. After all, farmers used it to improve stock. Eventually, Gregor Mendel's experiments with peas were rediscovered at the end of the 19th century,[7] and genetics was able to provide a mechanism for the process of evolution. As a result, evolution is now reinterpreted as the transformation of characters gene by gene, and not the transition from type to type.

Dinosaurs have been the most prominent victims of typology. They have been seen as a type of reptile since 1825, when Gideon Mantell decided that *Iguanodon* had lizard-like teeth. The whole history of paleontology would have been different

---

[7] Gregor Mendel published in 1866 but the significance of his work was not understood.

if he had interpreted them as a type of giant bird. Dinosaurs remained sluggish, slow-moving lizards in everyone's imagination until Robert Bakker challenged traditional views with his *Dinosaur heresies* in 1986.

In spite of centuries of careful observation, birds have also been the frequent victims of typology. Typology, not phylogeny, made the loon more reptilian than other birds. Its general appearance was so much like the fossil *Hesperornis* that Max Fürbringer placed it near the base of his evolutionary tree.

Observations by W.K. Parker in 1891 express typological views about avian intelligence that are still widely held: "The raven is a modern type; for not only has he acquired all the highest accomplishments of which a bird is capable, but as the head of a long list of Families, he has an ornithological following of more than six thousand species" [155]. Most of us would agree that the raven is remarkably intelligent and it is one of the few birds to play games. It is difficult, however, to imagine the raven as the Victorian patriarch of House Aves. Parker was absolutely certain that birds (at least male birds) evolved like humans towards intelligence and that the brightest type represented the top of the heap. I doubt that he made himself popular with colonial colleagues when he went on to explain that primitive and distinctly "small-brained" types could be found in Australia.

Typology is so ingrained in biology that rooting it out is a bit like rooting sexism out of language. There are many examples from the 20th century. When Sibley and Ahlquist applied the name "Eoaves" to the ostriches in their biomolecular phylogeny, it implied that the ratites came from the "dawn" of time [192]. Such an idea flies in the face of logic, however. When a branch occurs within a group, the two resulting lineages are exactly the same age because they shared a common ancestor. If they eventually look different, it is only because each has responded to discrete sets of evolutionary pressures over millions of years. There is absolutely no meaningful evidence that the "Eo"-aves or "paleo"-gnaths represent a more primitive or older form than the "Neo"-aves or "neo"-gnaths. The living examples are more analogous to the two faces of the same coin than to the root and flower of the same tree.

When it was difficult to see relationships of any sort, Victorian typologists readily filled embarrassing gaps with "missing links." Occasionally, the less scrupulous conjured up physical evidence to illustrate a popular theory. The success of the famous "Piltdown Man" was assured by its fit with preconceived notions of appropriate types of fossils for human ancestors. Evolutionary thinking has moved on, and now you rarely hear of missing links, even in the popular literature.

For a while, there was some concern that one famous 19th-century bird fossil could be the "Piltdown Man" of avian paleontology, although by accident rather than devious intent. *Ichthyornis* was found in 1872 by O.C. Marsh. He described it from some headless fossils that he had found in Cretaceous rocks in Kansas. At about the same time and perhaps in the same place, he also found some small toothed jaws that he described as a reptile, *Colonosaurus mudgei*. Only later did he decide that the jaws were well enough associated with the *Ichthyornis* material to belong to that bird, in spite of the teeth. He had already discovered the very toothy *Hesperornis* and eagerly accepted the possession of teeth as a characteristic of a "type" or organizational level that he called "Odontornithes." Shortly afterward, the original fossil material was plastered into a display for the Peabody Museum of Natural History at Yale University, where it remained, inaccessible to scientists, until the end of the 20th century.

In 1881, Marsh published his *Monograph of the extinct toothed birds of North America,* which listed 14 species of toothed birds, only a few of which were actually known to have teeth in their jaws. Some of his contemporaries soon added several new fossils, such as *Laopteryx, Scaniornis, Enaliornis, Neogaeornis,* and *Cimalopteryx,* which were supposedly more toothed birds but also without the evidence of teeth in their jaws. By 1893, there were so many extinct toothed birds under discussion that Hans Gadow found it necessary to recognize them as a fundamental part of the Class Aves and inserted the Division Odontolcae, between the Ratitae and Carinatae (Table 5.1). He could not place these distinctly primitive types ahead of the ratites because the palate of the iconic toothed bird, *Hesperornis,* was clearly neognathous.

The special group of toothed birds remained an unquestioned part of bird lore for almost a century, until 1952. In that year, J.T. Gregory shocked the ornithological community by suggesting that the supposed jaws of *Ichthyornis* belonged to an unusually small mososaur called *Clidastes* [74]. The challenge caused paleontologists to re-examine the available evidence and the result was embarrassing. Pierce Brodkorb published a compilation of all the pertinent information in his *Fable of the toothed birds* in 1971, which made it clear that *Archaeopteryx* and *Hesperornis* were the only known fossil birds with an unchallenged claim to teeth [15]. Since then, additional specimens of *Ichthyornis* have shown that it also belongs on this short list, and other legitimate lineages of toothed birds have been discovered. The Confucius birds lacked teeth, but the early forms of the Enantiornithes (ball-shouldered birds) had some in the front of their jaw.

For many years, the only accepted specimens of *Ichthyornis* were those collected by Marsh, but in recent years, additional material has been collected from a wide range of sites across North America, Europe, and Asia. Gregory's conclusion was eventually rejected on the grounds that it was extremely unlikely that individuals from the same species of mososaur died near so many different fossils of *Ichthyornis*. Although Marsh's original work was vindicated, there is a strong lesson to be learned here about leaping to conclusions from fragmentary fossils. At least the controversy helped reduce the use of "types" in avian paleontology.

Recently, the *Ichthyornis* material was freed from its plaster display. When Julia Clarke examined it, she was able to confirm that the material supports Marsh's description of two species (*Ichthyornis dispar* and *I. minus*) [28]. As many as seven species had been proposed at one time or another. Seventy-eight of the bones belonged to *Ichthyornis,* two belonged to other known birds, and one belonged to a previously undescribed species. This important piece of work is discussed further in the next chapter.

As the 20th century came to an end, taxonomists were beginning to reassess the higher taxa in the classification of birds based on the century-old analyses of Max Fürbringer. The 200 or so families seemed fairly well founded, but there was less confidence in their distribution among the two dozen or so orders and some major changes were made. With a push from early biomolecular work, the Galliformes and Anseriformes were separated from the other neognaths and placed in their own major lineage, the Galloanserinae. A little later, the New World vultures were placed among the cranes. Although these shifts were generally applauded by the ornithological community, they did not really clarify the evolutionary story of the living birds, which was still unclear at the end of the 20th century.

Paleontologists had added Subclass Confuciusornithes and Subclass Enantiornithes to the extinct forms and reinvigorated the debate on the origin of birds with a host of other fossil birds and birdlike dinosaurs. At the same time, biomolecular geneticists began to decode the chicken genome and introduce new analytical techniques that helped demonstrate underlying weaknesses in comparative morphology as it had been applied.

In the next chapter, we will look at the checkered history of the use of such techniques in the study of birds. We will also look at the use of powerful new molecular tools that help us understand living species but are not directly useful in the study of fossils. In theory, however, the genome is another record of past events that we may be able to read.

Biomolecular techniques appear to have overwhelmed the more traditional approaches to avian taxonomy, but they have their own inherent problems (see Chapter 6). They should eventually help us link the morphology of birds to an evolutionary story for the group. That story may not be the one implied by the taxonomic characters currently in use. Perhaps those 40 or 50 characters represent an unintentionally biased sample. They appear to be sufficient in number to identify the clusters within such a large group but insufficient for arranging those clusters in a meaningful way.

One aspect of avian biology that might offer a clearer framework for an evolutionary story is the history of flight. Each of the three major lineages of living birds appears to have followed discrete evolutionary strategies that use flight in different ways. Flight capability is not a feature usually associated with the paleognath birds because most of the living species are pedestrian residents of open country that either cannot fly or rarely choose to. The tinamous are the only living paleognath to have anything like a functional wing, and they are the only paleognath that can fly at all. Among the Galloanseriformes, the galliform birds are almost as pedestrian as the paleognaths while the aquatic anseriform birds include many strong, long-distance fliers. The remaining neognath birds exhibit a wide variety of flight capabilities and flight styles. They also include 90% of the avian species. Understanding the significance of variations in flight capability must be the focus of any attempt to tell an evolutionary story for the whole class.

Flight has made birds a great success and may offer us a new opportunity to connect the physical features of a bird to its evolutionary history. At a simplistic level, it is relatively simple to measure a bird's wing and convert its characteristics into numerical values such as wing loading or aspect ratio. Those values are directly related to flight capability, flight style, and other kinds of behaviour. They are also reasonably distinctive in that they are usually fairly consistent within a family but vary between clusters of families and between orders. As we shall see in later chapters, flight capability is also related to egg size. Increased egg size is related to the structure of the hips and is a characteristic that separates modern birds from their early ancestors.

Further Reading

Few of the Victorian texts are still available, but Frank Beddard's book is still on the shelves of many public libraries. It also appears regularly in used-book stores and is available over the Internet. It provides useful insight into 19th-century thinking about birds:

Beddard, F.E. 1898. The structure and classification of birds. Longmans, Green, London.

Because taxonomy dominated biology for so long, descriptions of the contribution of Aristotle are often limited to the way he classified birds. If his interests had been that narrow, he would never have inspired early medieval naturalists such as Abu ibn Sina (Avicenna) or Frederick von Hohenstaufen. Recently, Aristotle's ideas about biology have been re-examined by James Lennox and reinterpreted as a research program into the life sciences:

Lennox, J.G. 2001. Aristotle's philosophy of biology: Studies in the origin of life science. Cambridge Studies in Philosophy and Biology. Cambridge University Press, Cambridge, UK.

Most histories of ornithology emphasize the 400 years between 1500 and 1900 when naturalists travelled the world looking for new specimens. That period certainly had the most interesting characters. Barbara and Richard Mearns tell the best stories, while Erwin Stresemann and Michael Walters provide more of the academic details:

Mearns, Barbara, and Richard Mearns. 1988. Audubon to Xantus; 1992. Biographies of bird watchers; and 1998. The bird collectors. Academic Press, San Diego and London.
Stresemann, E. 1951 [English trans. 1975]. Ornithology from Aristotle to the present. H.J. Epstein and C. Epstein (trans.), W.G. Cottrell (ed.). Harvard University Press, Cambridge, MA.
Walters, M. 2003. A concise history of ornithology. Yale University Press, New Haven, CT.

Charles Sibley and Jon Ahlquist include detailed histories of the technical aspects of traditional taxonomy among their introductory chapters:

Sibley, C.G., and J.E. Ahlquist. 1990. Phylogeny and classification of birds: A study in molecular evolution. Yale University Press, New Haven, CT.

The clearest ideas about the species concept and its role in evolution are probably those of the late Ernst Mayr. He wrote many technical and academic books, but the most recent was intended for a more general audience:

Mayr, E. 2001. What evolution is. Basic Books, New York.

6

That Bird Is Different from the Other One

Evolutionary relationships among families of birds

For much of the 20th century, ornithologists and other biologists turned away from issues surrounding the evolution of the major groups of organisms to questions about the origin of individual species. As evolutionary theory developed in the 1890s, biologists realized that it operated on individuals and that its effects were most likely to be observed at the species level. Questions about the evolution of species became the mainstream of 20th-century biology and helped link evolution to species-specific habitat or niche selection and interspecific variations in behaviour or physiology.

Although the early emphasis on experimental sciences in the 20th century led to a sudden decline in the importance of taxonomy, its questions and problems did not go away. As the depth of our biological understanding increased, those questions became more troublesome. Fortunately, intriguing new technologies were being developed in other fields that could be applied to biological problems, while a more sophisticated understanding of biological processes suggested that there were new ways to approach taxonomic problems.

Perhaps the single greatest advance in biology during the 20th century was the discovery, in 1952, that DNA was a double helix and capable of being the genetic material. It was probably more important than landing on the moon, and perhaps comparable to Columbus' discovery of the Americas. The new appreciation of DNA shifted the focus of evolutionary research away from changes in the whole organism to processes that occur at the molecular level. As with any theoretical breakthrough, however, it took a long time to develop the practical technology that would allow its full impact to be felt.

For years, simple biomolecular analyses required hours of tedious work in the laboratory. Today, machines do all the hard work and decode genomes or produce gene sequences to answer questions that would have been impossibly difficult just a generation ago. Still, if understanding DNA is like discovering a new continent, we have only begun to wander inland. Biomolecular genetics is incredibly complex. It has already produced some startling new ideas and it seems likely that it will produce many more in the future.

Using DNA to unravel relationships was still the stuff of science fiction in the mid-1960s, when biologists in several fields felt a pressing need for reliable family trees. They began to cast about for more rational ways to construct genealogies and resolve disagreements in taxonomy.

Numerical taxonomy appeared to offer the best answers at first. It was somewhat similar to the technique used by Max Fürbringer in 1888 but had received a great boost from the development of the electronic computer. For the first time, taxonomists could calculate large numbers of similarity indices to estimate "distances" between organisms. It was never clear, however, whether some characters should take precedence or exclude the use of others. There was also widespread distrust of computers, especially among traditional taxonomists. In spite of its use of rather simple indices, numerical taxonomy suffered from the perception of computers as an elitist tool, a black box whose results could be assessed only by those who understood FORTRAN or COBOL programming. In the end, however, it was problems with the definition and use of characteristics that drove taxonomists to look for a more sophisticated type of analysis.

Using the Power of Computers for Classification

Ornithology was not the only field of biology to have major problems with the traditional methods of taxonomy. Such problems also dogged the classification of other organisms, especially plants and insects. In 1966, entomologist Willi Hennig proposed a solution based on a rigorously logical protocol [84]. He called his approach phylogenetic systematics, but we know it as cladistics. The first step in a cladistic analysis is the conversion of all characteristics to binary code. The analysis then sorts the codes into a nested series according to the progressive sharing of similar characteristics.

The characteristics are divided into "apomorphies," a word derived from classical Greek for "shapes at the peak," and "plesiomorphies," meaning "shapes near [the base]." Apomorphy implies that the characteristic in question has moved away from, or is more derived than, the ancestral form. Plesiomorphy implies that the characteristic is ancestral. Each member of a group, or clade, shares a suite of synapomorphies (shared derived features) with all other members of that clade and their most recent common ancestor. The sharing of synapomorphies logically implies the sharing of a genealogy. An organism that does not have the full suite of synapomorphies cannot be a close relative unless it is possible to show that the missing feature was lost through recent evolutionary events. For instance, the absence of certain wing bones does not preclude the kiwi's being a bird.

In the days of desktop mechanical calculators, cladistic analyses were cumbersome and labour-intensive. Their numerous calculations are an ideal task for electronic computers, however, and today they can be carried out on most utility-grade computers. Cladistics is such a popular technique that several software companies have developed off-the-shelf programs for quick and easy analyses on laptop computers.

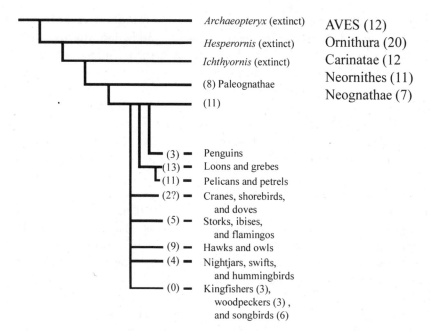

Figure 6.1 A family tree for the major groups of birds based on a 1981 cladistic analysis by Joel Cracraft [35]. The numbers in parentheses are the number of synapomorphies that unite the clade. There are no characters that unite the Coraciiformes and Piciformes with the Passeriformes in a monophyletic clade, and the numbers shown indicate the synapomorphies that support the monophyly of the individual clades. Note that the diagram leaves the 13 orders in the Neognathae as an unresolved group (polytomy).

In 1981, Joel Cracraft used a cladistic analysis to construct the classification of birds shown in Figure 6.1 [35]. As one moves up through the genealogy, changes in the suites of synapomorphies define discrete branches. We can see that the Neognathae and Paleognathae together form the large clade Neornithes, which includes all living birds. The Neornithes share 11 synapomorphies but within it are distinct subgroups. The Paleognathae, which includes only the ostrich and its relatives, share 8 synapomorphies. All the other birds, about 9,000 species, are members of the Neognathae and share 7 synapomorphies. Cracraft was able to distinguish seven large clades and several lesser clades within that group. At the highest level, the crown clade Passeriformes (songbirds) are distinguished by 6 additional synapomorphies that, by definition, occur in every songbird but are absent from cuckoos, woodpeckers, and all other birds.

Cracraft's decision not to separate owls from hawks or grebes from loons disagreed with the experience of most ornithologists, and his conclusions have not been widely accepted. His attempt reveals two potential weaknesses in what is otherwise a powerful analytical tool. First, cladistics requires a large set of characteristics; second, the analysis might be objective but the characteristics are vulnerable to subjective decisions by the analyst. Storrs Olson published a review that was very critical of Cracraft for not performing a fair test of the cladistic method by failing to follow the technique's basic principles and by introducing other kinds of arguments to make his point [150]. For instance, the technique is supposed to be based on similarities but Cracraft occasionally proposed differences as proof of the absence of a relationship.

Cracraft's inability to produce a convincing phylogeny for the living birds left ornithologists very suspicious of cladistics as a tool for classification. It made it very difficult for ornithologists to become excited about the subsequent success of cladistics in sorting out the dinosaurs. After all, if it is so difficult to construct a phylogeny for a well-known living group, such as the birds, how can cladistics be applied successfully to a group known only from fossilized, skeletal remains? As you might expect, the answer lies in lots of hard work to assemble a very large number of characters. Cladistics, unlike other forms of analysis, readily accepts new data. More importantly, it can provide a mathematical estimate of the confidence that scientists should place in the data. Each new fossil and each newly assessed characteristic make the results more reliable. It is necessary to proceed cautiously, however.

It was not until 15 years after Cracraft's analysis that Gerald Mayr and Julia Clarke made a cladistic assessment of the higher taxa of birds [131]. They were able to bring 148 characters to bear and had three fossil forms (*Apsaravis, Hesperornis, Ichthyornis*) as an outgroup. The did not have sufficient data for all the major groups of birds and left out the hummingbirds, swifts, nightjars, pelicans, anhingas, gannets, woodpeckers and their relatives, and mesites. They included kingfishers but left out all their supposed relatives.

Nonetheless, they were able to show strong support for the division of birds into two major lineages, paleognaths and neognaths. The living birds (Neornithes) shared 4 synapomorphies and were supported in 82% of the phylogenies created by bootstrap analysis. Bootstrapping is a statistical technique that repeats the analysis in different ways. The more frequently you get a particular result, the more likely it is to be the correct interpretation of your data. In this case, Mayr and Clarke could distinguish the Neornithes in 82% of the replicates. The Paleognathae shared 6 synapomorphies and had 76% support, while the Neognathae shared 10 synapomorphies with 88% support. Within the Neognathae, there were two main lineages: the Galloanserae (analogous to Galloanserinae or Galloanseriformes in other studies) and the Neoaves. A group that includes all the relatives of the chickens and ducks, such as the Galloanserae, is frequently identified in the biomolecular phylogenies discussed later in this chapter. Mayr and Clarke found that its members shared 9 synapomorphies but received only 57% support. Within it, the anseriform birds were recognized as a monophyletic lineage that shared only 4 synapomorphies but had 82% support. Neoaves represents all the remaining birds; it is sometimes called the Plethaves (majority of birds). Its members also share 4 synapomorphies, but perhaps it is no surprise that such a diverse grouping receives only 15% support from bootstrap analyses.

Gerald Mayr is the paleontologist who has proposed a relationship between grebes and flamingos [129]. In his work with Julia Clarke, the results show a very close relationship between grebes and loons, with 10 synapomorphies and 66% support. If the loons are removed from the data set, however, a relationship of grebes to flamingos appears in the dendrogram, with 12 synapomorphies and 79% support. Mayr and Clarke's conclusion is that a larger data set might reveal convergence between loons and grebes but will not indicate that they are related to each other [131].

Some of the synapomorphies shared by loons and grebes may seem fairly robust but they could be superficial similarities affected by common features of their feeding or locomotory techniques. Both groups have a zygomatic process in the skull, but this is also a characteristic of flamingos. Both have a smooth leading edge in their sternum, but this feature is shared with about half the groups studied. The tube for the spinal cord extends all the way through the pygostyle in both grebes and loons, but also in about half the other families. It may be related to loss of the terminal vertebrae when the tail becomes unusually short or the pygostyle begins to degenerate. The long, narrow pelvis is shared with other foot-propelled diving birds, including relatives of *Hesperornis,* cormorants, and sungrebes (a crane relative). Ostriches also have a long, narrow pelvis. Most of the other shared characteristics involve large, general features of the hind legs and feet, where convergence could be driven by similarities in locomotion. In contrast, features shared by grebes and flamingos often involve the fine details of bone and muscle structure, where convergence is unlikely.

Dinosaurs, Birds, and "Birds Are Dinosaurs"

It would be a lot easier to understand the evolutionary story of birds if we had a firm grip on a readily identifiable ancestor among the flightless animals of the Mesozoic. *Archaeopteryx,* the oldest generally accepted ancestor of birds, already had most of the characteristics of a bird – feathers, wings, lightweight skeleton, and a large brain. The highly controversial fossil, *Protavis,* described by Sankar Chatterjee [24] is much older but so fragmentary that it is never likely to displace *Archaeopteryx* from its central role in the early history of birds. Even the recent discovery of several feathered dinosaurs has not helped resolve that indistinct edge of ornithology where rather reptilian birds meet rather avian dinosaurs.

Within that grey zone, paleontologists have had some success in using cladistic analyses to set out the various diagnostic features of dinosaurs. Not all paleontologists agree with the details of the best-known dendrograms, but the remaining discussions revolve around facts and methods. The limited agreement on the overall pattern of dinosaur evolution is a far cry from the confusing muddle in ornithology.

Paleontologists succeeded where ornithologists failed because they were careful to work within the methodology of cladistic analysis and they were willing to do the work required to increase the number of characters being considered. In 1986, Jacques Gauthier used only 50 osteological characters to drive one of the final nails into the coffin of theories for a non-dinosaur origin for birds [69]. He was able to demonstrate that dinosaurs were a monophyletic group and that birds (i.e., *Archaeopteryx*) fit cladistically among the advanced theropods. How neatly and reliably the modern birds fit was another question, and one that is still actively debated. Within a few years, Mark Norell, James Clark, and Peter Makovicky had developed 205 characters just to sort out the 42 species of coelurosaurs [143], and a team led by David Pisani was able to construct a supertree that linked all 277 genera of the Dinosauria in one grand phylogeny [164].

Although paleontologists have been rather free with their use of words like "bird" or "Aves," they have not attracted much attention from students of the living animals. Although ornithologists have all heard the expression "birds are dinosaurs," it has had little effect on their day-to-day activities. The gap between the Cretaceous dinosaurs and modern birds is just too great for avian ecologists to consider. Anyone who insists on using "theropod" as a synonym for "bird" is likely to be dismissed as a quack. Ornithologists have a very clear idea of what constitutes a bird. It is a warm-blooded, bipedal animal with feathers that lays eggs and flies (in most cases). The existence of near-birds and feathered, flightless, bipedal theropods in the Cretaceous is intriguing but not directly relevant to the issues of conservation, endangered species, or wildlife management, which pay the bills of modern ornithology.

Paleontologists are faced with a much more difficult problem than that faced by students of living birds. The fossil fauna is more diverse and includes a variety of animals that are somewhat like birds but not quite the same as the living examples. In addition, fossils offer much less information than living animals and are spread vertically in geological strata that represent huge expanses of time. Unlike animals that exist in the present, fossils need to be arranged within a chronology and across a very different global geography.

One of the characteristics of the fossil history that still puzzles ornithologists is the casual way paleontologists can dismiss vast periods of time. *Archaeopteryx* is generally accepted as the ancestral bird, but it appeared about 25 million years before the advanced theropods that are supposed to be its ancestors. If you adjust the fossil stratigraphy, the difference may be reduced to a mere 10 million years, apparently a negligible error for paleontologists but one that still disturbs most ornithologists. Not everyone is content with the currently proposed timeline of evolutionary events. It disturbed Gregory Paul enough for him to work out an intriguing reconstruction of fossil history in which the birdlike advanced theropods of the Upper Cretaceous evolved as flightless descendants of earlier bird-ancestors [158].

Avian paleontology is only a little more than a hundred years old but it has already accumulated a thick overburden of technical terminology. Some of it is still meaningful but much has fallen into desuetude. In the case of birds, the technical terminology has become extremely confusing because some paleontologists have not been careful to use group names as they were originally defined.

To help paleontologists talk to each other and enable them to communicate their ideas to students of living birds, Gauthier and his colleague Kevin de Quieroz have proposed a new, more precise set of names for the ancestors of birds (Tables 6.1 and 6.2) [70]. The new names apply to the branches in their cladistic analysis and cannot be applied to other trees, such as Paul Gregory's "birds-first" proposal. The word "Aves" has become the technical name for birds a kind of organism. It does not refer specifically to any key feature or defining characteristic. Each of the proposed names refers to a specific structure that is a synapomorphy so that the name reflects the group's history and the organisms that it includes or excludes.

In 2002, Luis Chiappe used a similar argument when he defined the group "Pygostylia" as all birds with a pygostyle as distinct from *Archaeopteryx* and *Rahonavis*, which had long reptilian tails [25]. This group is roughly equivalent to the Carinatae in Table 6.1. In fact, "Pygostylia" may be a more useful term because the sternum is often poorly preserved and a keel may be difficult to detect. In the Confucius birds and early forms of ball-shouldered birds and modern birds, the flight muscles are attached to the furcula (wishbone); in some cases, the furcula's hypocleideum

Table 6.1		

Proposed names of clades to help resolve confusion around the use of the term "Aves" as it applies to both fossil animals and living birds

Name	Definition	Included groups
Panaves	The most inclusive clade that can contain Aves (e.g., *Vultur gryphus*)[1] but not crocodiles.[2]	Aves and all dinosaurs, plus such Triassic forms as *Pseudolagosuchus, Lagosuchus, Lagerpeton,* and *Scleromochlus;* also the Mesozoic pterosaurs.
Avifilipluma	A clade that includes the first member of the Panaves with avian feathers.	Aves, *Caudipteryx, Protoarchaeopteryx, Sinornithosaurus, Archaeopteryx,* Confuciusornithes, Enantiornithes, as well as *Ichthyornis* and *Hesperornis.*
Avialae	A clade stemming from the first panavian with wings homologous to those found in Aves and used for powered flight.	Aves, *Ichthyornis, Hesperornis, Apsaravis, Patagopteryx,* Enantiornithes, Confuciusornithes, *Rahonavis, Archaeopteryx.*
Carinatae[3]	A clade stemming from the first panavian with a keeled sternum homologous to that found in Aves.	Aves, *Ichthyornis, Apsaravis,* Enantiornithes, Confuciusornithes. (The Ratitae, *Patagopteryx,* and *Hesperornis* apparently evolved from keeled ancestors.)
Ornithurae	A clade stemming from the first panavian with a tail shorter than the femur that ends in a pygostyle, homologous to that found in Aves.	*Ichthyornis, Hesperornis, Apsaravis,* and all living birds.

[1] The Andean Condor (*Vultur gryphus*) was the first member of the Class Aves listed by Linnaeus in 1758.
[2] Crocodiles are considered to be the most primitive of the diapsid animals and are often used as an outgroup for birds in both morphological and biomolecular studies.
[3] Approximately equivalent to the Pygostylia [25].
Source: [69]

Table 6.2

The clades of living birds and their immediate relatives that survived the end of the Cretaceous

Name	Definition	Included groups
Aves	A clade including the most recent common ancestor of the Paleognathae and Neognathae.	All living birds and some fossil groups, but neither Confucius birds nor ball-shouldered birds.
Paleognathae	The crown clade stemming from the most recent common ancestor of ostriches, tinamous, and their relatives.	Tinamous, ostriches, rheas, emus, kiwis, cassowaries, the recently extinct moas and elephant birds, and the fossil Lithornithidae.
Neognathae	The crown clade stemming from the last common ancestor of *Charadrius pluvialis* and all extant birds sharing a more recent common ancestor than that of ostriches, tinamous, and their relatives.	Galloanseriform birds (chickens and ducks) and Neoaves or Plethaves that include the remaining living birds.

Source: [69]

extends along the sternum to create a keel. All these birds, however, have a large pygostyle that is frequently fossilized.

Cladistics based on large numbers of features has enabled paleontologists to place extinct animals in context with each other in a very convincing way, but we should not get carried away with the rhetoric of cladistic analysis. Defining birds as members of a clade that shares a common ancestor with *Archaeopteryx* usefully defines Class Aves and "birds." Extending the argument to claim that birds are dinosaurs is not so useful. Humans share a common ancestor with apes but humans are apes only in a specialized technical sense. They also have synapsid skulls without being mammal-like reptiles in any meaningful way. The statement "birds are dinosaurs" is helpful when it expands the connotation of the word "dinosaur" by extending its meaning from huge sauropods to tiny arboreal animals with feathers. It is unhelpful and mischievous when its intent is to expand the connotation of the word "bird." The ultimate purpose of both vernacular and technical classification systems is supposed to be effective communication – shared intelligence.

Paleontology and taxonomy are, by their nature, dependent on precision of language. It is particularly unfortunate, then, that theropods have come to be referred to as "avian" or "non-avian" according to their apparent relationship to modern birds. The term "non-avian" tells us what something is not without telling us what it is. Perhaps it would be better to use the phrase "reptilian theropods" to describe the groups that are not directly related to birds and "avian theropods" for flightless animals that are. Adherents of the theory that birds evolved from ancient thecodont reptiles could signal their obscurantism by using "theropodan avians" for birds that they see as only looking like theropods. If not for the unlucky termination of the

Cretaceous, various non-avian dinosaurs might well have continued to dominate the earth and mammals might still be an inconsequential part of the biota.

The 21st century has started off very strongly for avian taxonomy based on physical characteristics. Bradley Livezey and R.L. Zusi published a preliminary tree based on several hundred characteristics [117]. They have been able to take advantage of the recent increase in the volume of osteological work on dinosaurs to develop new characteristics for birds, and they are exploring some data sets that were collected earlier in the 20th century but have never been fully exploited by taxonomists. There are several Mesozoic fossils, besides *Archaeopteryx*, for which there is enough material for inclusion in a family tree. *Eoalulavis, Confuciusornis,* and the recently re-analyzed *Ichthyornis* have joined *Hesperornis* as useful examples of the basal groups. In addition, there are recently extinct groups such as the dodos, solitaires, and moas, as well as early Tertiary forms such as diatrymids, plotopterids, teratorns, pseudodontornithids, and others, for which there is sufficient fossil material for effective analysis.

Eventually, Livezey and Zusi hope to be able to exploit over 1,400 characters, but their current tree is based on only a quarter of the material – 359 characters from the skull and vertebrae. As with almost every phylogeny based on character-istics of the skull, the earliest branch of the tree separates paleognaths from the rest of the modern birds. A second branch takes off the galliform and anseriform birds as a discrete lineage. This branch has been reported regularly from biomo-lecular studies but not from previous analyses of comparative morphology. Livezey and Zusi also bring New World vultures into the storks; as we have seen, some tax-onomists have been recommending this since the 19th century. Some of the other placements are so startling that Livezey and Zusi are unconvinced by their own results (e.g., opisthocomids [Hoatzin], steatornithids [Oilbirds] phoenicopterids [flamingos], and many of the smaller forest birds). These placements may also be dependent on skull characteristics, and a larger, more varied data set may generate a different arrangement.

Livezey and Zusi's tree also shows an "alliance of grebes, loons, and penguins with tubenoses (petrels)." The relationship between loons, penguins, and petrels appears in several other phylogenies, including Sibley and Ahlquist's biomolecular phylogeny of 1990 [192], and seems to have achieved a level of consensus among ornithologists. A very close relationship between grebes and loons is difficult to ac-cept, however, in spite of their generally similar appearance. Perhaps the characteris-tics of the skull are indeed very similar, but the grebe's unique feet, specialized locomotion [103], unusual feathers, small eggs, large clutch size, fused thoracic ver-tebrae, distribution in the Southern Hemisphere, and other non-taxonomic charac-teristics suggest that any similarity to the loon is coincidental. Perhaps taxonomic characteristics from other parts of the body will eventually distinguish between these groups, or biomolecular studies will confirm that grebes belong in the Metaves and that the loons are members of the Coronaves (see page 196). Until the analyses are complete, it seems appropriate to reserve judgment on that particular relationship.

Livezey and Zusi make three important points about the state of affairs in modern avian taxonomy [117]. First, much of the progress in thinking and methodology is the result of interest by paleontologists in fossil birds and birdlike dinosaurs. Second, even if they finish their work and achieve a major breakthrough, it could be extremely difficult to publish the results along with the data set. Modern scientific publishers discourage the production of long papers, and many authors are forced to place large data sets in special archives. Third, there is still far too little funding for morphological studies of birds. This is not to say that there is no money for avian classification. Quite a lot has been spent on biomolecular approaches that have had some noteworthy successes. In many cases, however, molecular classifications have failed to provide convincing answers. Livezey and Zusi express dismay that such an "uninspiring record has neither slowed the flow of resources to molecular practitioners nor the growing preponderance of systematists who are experienced only in molecular techniques." We will examine some of the products of molecular studies of birds later in this chapter.

Many ornithologists have also found the record of cladistic analysts "rather uninspiring" [110]. We need a family tree that resolves the relationships of all the birds and does not include large polytomies. The relationships also need to be convincing. If owls and hawks or grebes and loons are genuinely closely related, then we need to understand the biological meaning of the obvious differences between those pairs.

The poor performance of cladistics in the past is most likely attributable to the sample of characteristics applied to the problem. Max Fürbringer was incredibly lucky to correctly identify so many of the orders and families with only 51 characters. What we need today is a large enough sample of characters to have a reasonable chance of detecting events throughout the whole history of birds; in this sense, Livezey and Zusi certainly appear to be on the right path even if their preliminary results are not entirely satisfying.

Birds are an incredibly ancient and conservative group compared with mammals. If dinosaurs had left us written records, we would have difficulty recognizing the mammals that they knew. In the past few million years, mammals have stopped laying eggs and developed placentas; learned to fly, glide, and swim; and figured out ways of going without food or water for very long periods. Mammalogists might disagree, but this series of rapid evolutionary developments has left a relatively easily interpreted record.

When, if ever, did comparable revolutionary changes occur in birds? Rather modern-looking birds began to appear over 100 million years ago, and many living families flew among the dinosaurs. That great age makes it difficult to understand which characters separate birds from pedestrian dinosaurs or from failed lineages of near-birds, and which separate the groups of living birds from each other. When you add the intense evolutionary demands of aerial flight to that incredibly long period of time, it is little wonder that birds are difficult to sort out taxonomically or genetically.

At the moment, many of the characters used in avian classification appear to be those that are important in anchoring the avian family tree in the distant past. The earliest ancestors of modern birds, *Archaeopteryx* and the theropod dinosaurs, represent architectural and ecological strategies that are very different from those followed by living forms, however. Perhaps we should worry less about the deep roots of the family tree and concentrate on animals that look less like dinosaurs. Some of the earliest evolutionary changes appear to have involved the head and the hips, leaving a very long time for additional changes to occur to those structures and obscure the historical record. It might be easier to find characteristics that distinguish modern families from one another by examining parts of the body that have been adapted to modern lifestyles and to advanced forms of wing-propelled locomotion.

C.J.O. Harrison laid the foundation for this kind of approach when he cleverly chose minor features of the elbow to distinguish between two intensely convergent lines [79]. The features reflect very different solutions to a simple biomechanical problem. Auks are able to maintain a stiff wing underwater, in part with the help of a bone protrusion that fits behind a ridge. Diving-petrels have a cusp that fits into a pit. Both structures can be traced back to less-developed features on non-diving relatives. It seems likely that other kinds of birds would also need a stiff elbow and those that flapped their wings very vigorously might well have distinctive supporting structures of their own.

Similarly, foot-propelled diving is an advanced form of locomotion, at least in the sense that none of the Cretaceous dinosaurs appear to have been extensively adapted for an aquatic existence. Two such divers, loons and grebes, have been repeatedly placed close together in phylogenies and they do, indeed, share many characteristics. As mentioned above, they have similar skulls, hip musculature, and elongated hipbones. Their ecology and behaviour suggest, however, that they are very different kinds of birds. Loons are large, lay a few rather large eggs, and spend much of their lives on the ocean. Grebes are small, lay many small eggs, and include many species that never go near salt water. Both have rather narrow wings and fly with rapid, shallow strokes, but they have a very different profile in the air and grebes use a unique kind of lift-based, hydrofoiling propulsion underwater [103].

Either the flight style or twisting and turning underwater appears to place their thoracic spines under some stress and both have stiffened them, but with very different structures. The grebe has fused a short series of the anterior thoracic vertebrae and reinforced them with a rigid brace of simple hypapophyses. The loon has chosen limited flexibility. Its wide thoracic vertebrae remain independent, and beneath each centrum, a large, flared hypapophysis extends into the abdominal cavity, where it is attached to the sheets of connective tissue that surround the vital organs (Figure 2.7). It seems likely that the sum of the ecological, behavioural, and structural differences argues against a close relationship. As we shall see below, that relationship may be much more distant than anyone ever suspected.

Classification and Biomolecular Characteristics

Early in the 20th century, advanced analytical techniques helped basic biochemistry make many important breakthroughs. By the 1950s, scientists were examining the fine structure of the giant organic molecules that are typical of living organisms. It quickly became apparent that some large molecules were characteristic of specific groups of animals and even individuals. Each organism's unique set of genes generated so many small and large differences in structural proteins, enzymes, and other large molecules that the array of biomolecules offered a set of characters as unique as a fingerprint. When blood typing was found to be necessary for safe transfusion, scientists began to look at large molecules as tools in classification.

In 1952, Watson and Crick announced their discovery of the structure of deoxyribonucleic acid, and the stage was set for a revolution that made DNA a household acronym and led to the deciphering of the human genome by the end of the century. Long before biochemists could imagine working in terms of genomes and genetic profiles, however, Charles Sibley and Jon Ahlquist were using features of egg white proteins to distinguish between different groups of birds. They hoped to go beyond Aristotelian comparisons of physical features to find a set of practical chemical markers that would resolve the problems in avian taxonomy. If the variations in the proteins matched the divisions between the traditional families and orders of birds, the principles of traditional avian taxonomy would be upheld. If they did not, it would be a signal to taxonomists that the classification of birds needed to be re-examined. The results of the analyses of egg white proteins were interesting but not as robust as Sibley and Ahlquist had hoped. They decided that it would be more effective to work directly on the genetic material, and were among the first to attempt a measurement of genealogical distances based on the physical characteristics of the DNA molecule.

The use of biomolecular data is a two-edged sword. On the one hand, it removes subjective judgments about the nature of physical characteristics by going to the very source of variation. On the other hand, it gives answers that have no practical value until they can be related to the whole organism. A huge list of genes from a sample of DNA has no value to the biologist until it can be connected to some identifiable ecological entity, such as a particular red flower or brown bird with yellow feet. Because of this gap between molecule and organism, the field guide will always be with us. It is essential as a practical tool for identification. Recent studies of rapidly evolving genes [80] suggest that future field guides will need to explain that groups of apparently identical birds are, in fact, superficial assemblages of genetically distinct species.

Nuclear DNA and Relationships Among Birds

Today, everyone has heard of cases where DNA has been used to identify criminals or solve questions of relationships, and the plots of television programs such as the *CSI* series[1] revolve around the successful use of recombinant DNA techniques and gene sequencing. In real-life studies, biochemists have successfully tracked down the living descendants of noteworthy individuals who happened to have left behind a hair follicle or a tooth root. All of this is possible only because various types of genetic analysis and gene sequencing have become a fully automated industrial process.

Sibley and Ahlquist began their studies long before automated equipment was available [192]. They knew that DNA should hold the answers to questions about avian evolution but the only practical approach available to them involved the exploitation of the physical characteristics of the DNA molecule. They chose to use DNA from the nucleus (nDNA) because large quantities could be collected from the chromosomes, which contain all the genetic instructions for constructing an organism. DNA is easy to collect from birds because, unlike mammalian red blood cells, avian red blood cells have a nucleus and a DNA sample can be obtained easily, without killing the donor. Sibley and Ahlquist's basic assumption was that the physical properties of DNA molecules changed with the accumulation of mutations, so each lineage of birds should have structurally distinctive DNA. The closer the relationship between two lineages, the more similar their DNA molecules should be.

DNA consists of two complementary strands that separate ("melt") at a specific temperature when the molecule is gently heated. If you mix the separated strands with a similarly treated sample from another species, the strands rebond at specific points in an attempt to reconstruct their original complementary nucleotide pairs. The new double strand will then melt at a lower temperature because the rebonding cannot be perfect between strands from different species. The greater the genetic similarity between the two samples, the greater the frequency of bonding. Increased levels of bonding create a more stable molecule, which melts at a higher temperature. DNA from distantly related birds melts at a lower temperature because there is less bonding. The technique is known as DNA-DNA hybridization.

Sibley and Ahlquist used the change in melting temperatures ($\Delta T_{50}H$) as a single numerical parameter that indicated the degree of similarity between pairs of DNA molecules. It offered a simple method of calculating the genealogical distance among samples of donors. In effect, their technique reduces the complexity of the genetic material to a single physical feature that can be measured objectively. At the time, $\Delta T_{50}H$ appeared to be the logical basis for the single-character phylogeny that had eluded 19th-century naturalists.

Over 12 years, Sibley and Ahlquist conducted hundreds of tests with samples from 1,700 species and 168 of the 171 living families of birds. By comparing the value of $\Delta T_{50}H$ for various combinations and permutations of donors, they collected a huge data set, with which they constructed a family tree for all the living members of Class Aves (Figure 6.2). It identified almost all the traditional families

---

1 Crime Scene Investigation.

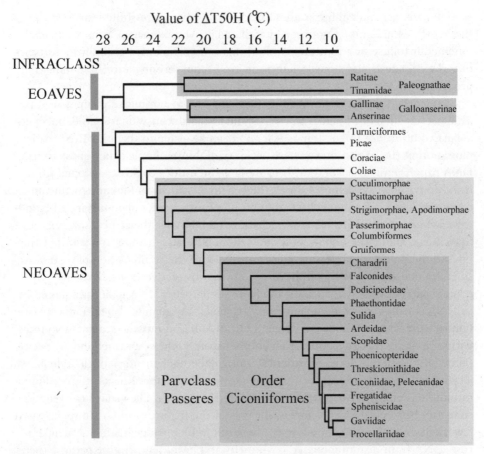

Value of ΔT50H (°C)

28 26 24 22 20 18 16 14 12 10

INFRACLASS

EOAVES

Ratitae
Tinamidae   Paleognathae

Gallinae
Anserinae   Galloanserinae

Turniciformes
Picae
Coraciae
Coliae
Cuculimorphae
Psittacimorphae
Strigimorphae, Apodimorphae
Passerimorphae
Columbiformes
Gruiformes
Charadrii
Falconides
Podicipedidae
Phaethontidae
Sulida
Ardeidae
Scopidae
Phoenicopteridae
Threskiornithidae
Ciconiidae, Pelecanidae
Fregatidae
Spheniscidae
Gaviidae
Procellariidae

NEOAVES

Parvclass   Order
Passeres   Ciconiiformes

Figure 6.2 The family tree of living birds derived by Sibley and Ahlquist from the results of biomolecular analyses [192]. This tree has been rearranged to present the groups according to increasing values of $\Delta T_{50}H$ (see text). Note the large numbers of clades included in Order Ciconiiformes and Parvclass Passeres. Note also the assignment of the anseriform and galliform birds to a single lineage and the assignment of that lineage to Infraclass Eoaves.

and sorted them systematically into an array of higher taxa; it was, in effect, the first independent test of comparative morphology. The biomolecular phylogeny did not presuppose the existence of the traditional taxa but derived them directly from comparisons of nDNA. Although some of the relationships among the higher taxa were predictable, others were very surprising.

Sibley and Ahlquist agreed with many traditional taxonomists when they put the ratites (ostriches, emus, and so on) together in a monophyletic clade and added the tinamous as a sister group to form the paleognaths. They also recognized the galliform birds and anseriform birds as monophyletic clades, but took the radical step of putting them together as sister lineages within a major basal branch, separating them from all other birds. Recent studies have supported the existence of this lineage but usually place it within the neognaths [76].

Sibley and Ahlquist also proposed controversial homes for troublesome birds such as the Hoatzin, sand grouse, and turacos, which had never been convincingly classified by comparative morphology. They were unsuccessful in finding a satisfactory place for the button quails, a difficult group that other workers have sometimes placed among the galliforms and at other times among the cranes.

To turn their phylogeny into a practical classification of birds, Sibley and Ahlquist assigned rank names to each of the lineages. Unfortunately, the branches appear to be randomly scattered over a wide range of temperatures without any obvious clusters or discontinuities. Sibley and Ahlquist simply assigned taxonomic ranks to arbitrarily selected ranges of the parameter $\Delta T_{50}H$. Most taxonomists make do with two basic ranks for the higher taxa – orders and families – but the scattered branches of the biomolecular phylogeny required a much larger array of names. Most of the roots of those names are familiar and they usually identify a branch that coincides with a traditional taxon, but the suffixes that indicate the rank of the branches are less familiar and are somewhat confusing.

Sibley and Ahlquist divided the birds into two infraclasses (Eoaves and Neoaves) and six parvclasses (Ratitae, Galloanserinae, Picae, Coraciae, Coliae, and Passerae). Within these large taxa, the major groups of birds were assigned to two superorders (Cuculimorphae and Psittacimorphae), four orders (Passeriformes, Columbiformes, Gruiformes, and Ciconiiformes). The order Ciconiiformes contained one suborder (Charadrii), one infraorder (Falconides), one parvorder (Sulida), and 12 families (Podicipedidae, Phaethontidae, Ardeidae, Scopidae, Phoenicopteridae, Threskiornithidae, Ciconiidae, Pelecanidae, Fregatidae, Spheniscidae, Gaviidae, and Procellariidae). Sibley and Ahlquist were unable to place the Turniciformes within any particular group. Two of the primary branches in the phylogeny are not named. One leads to the sister families Ciconiidae and Pelecanidae, near the top of the tree; the other leads to the sister superorders Strigimorphae and Apodimorphae, nearer the base.

Often the ranks in Sibley and Ahlquist's biomolecular phylogeny differ from traditional usage. For instance, the auks are familiar to most seabird biologists as the Family Alcidae, but the arbitrary criteria used by Sibley and Ahlquist reduce them in rank to a mere subfamily, Alcinae, within the family of gulls. The use of the subfamily rank appears to understate the many important biological differences between auks and gulls. Similarly, the use of the family rank to separate penguins, loons, and petrels appears to understate the fundamental nature of the differences in those groups. At the other end of the phylogeny, ancient groups such as the Galbulimorphae, Bucerotimorphae, and Coraciimorphae are placed in separate superorders even though they share many structural similarities and are often considered to be closely related families. Most ornithologists have not found Sibley and Ahlquist's approach very satisfactory, but assigning taxonomic names to the numerous branches of any biomolecular phylogeny is difficult.

The issue of rank remains a problem in both biomolecular and cladistic phylogenies. Both techniques generate phylogenies with a very large number of branches that do not necessarily coincide with taxonomic ranks. There are movements among taxonomists that propose to do away with ranks altogether. The cohesive character of a group like the auks or hummingbirds is self-evident and of practical use, even in everyday speech, and this utility has nothing to do with the group's rank as order or family or something in between.

Like any pioneering work, the phylogeny of Sibley and Ahlquist offers a great many intriguing possibilities but suffers from a variety of flaws and weaknesses. The chemical analysis was cutting-edge when the project began but it had already been superseded by more advanced techniques by the time it was completed. Most importantly for comparison with recent studies, the dendrogram was constructed without any outgroup. An outgroup puts the subject of analysis in perspective by anchoring it among other animals. Most recent biomolecular analyses of birds use crocodiles. They bear little resemblance to birds but are another ancient lineage of diapsid animals and are believed to be more closely related to birds than snakes or lizards. The tenuous relationship between crocodiles and birds is one reason for the serious inquiry into the existence of DNA in dinosaur fossils.

Sibley and Ahlquist's phylogeny has also been criticized for assuming that the molecular clock has moved at a steady pace. There is no justification for such an assumption, and evolution appears to have sped up and slowed down at various times in other organisms, if not in birds. Most ornithologists find some of the sister relationships difficult to accept and want much stronger evidence that falcons and grebes or frigatebirds and loons share a common ancestor. Major structural differences appear to make such relationships unlikely.

On the other hand, the broad significance of the phylogeny for the evolution of birds is intriguing. Its linear shape implies the possibility of trends in avian evolution. In particular, it appears to highlight a role for variations in flight capability. All the crown clades (Sibley and Ahlquist's Order Ciconiiformes) are aerial specialists of one sort or another that exploit aerodynamics in unusual ways.

Mitochondrial DNA

There are other kinds of DNA that can be used to construct phylogenies. A much smaller molecule of DNA is found in the mitochondria (mtDNA). Mitochondria are organelles that control the flow of energy in every living cell. They appear to be distantly related to chloroplasts, which convert solar energy into biological energy in plants. The existence of distinctive mtDNA suggests that the ancestors of these organelles were once free-living creatures that discovered the benefits of commensal association with other single-celled organisms and eventually became permanent residents of those organisms. A typical cell has just one nucleus and one package of nDNA, but each cell contains between 800 and 2,600 mitochondria, each with an identical string of mtDNA.

Peculiarly, only mtDNA from the female line is passed on to an organism's offspring because the sperm's mtDNA is not part of the package injected into the egg. It is lost on the outer wall of the egg, along with the sperm's cell wall and tail.

It is this characteristic that is the basis of the claim that all members of the human race can be traced back to a woman who lived in Africa a few hundred thousand years ago. So far, no one has been able to use mitochondrial DNA to replicate Sibley and Ahlquist's biomolecular phylogeny, but many laboratories are looking at pieces of avian phylogeny and testing the conclusions of previous work.

It may not be useful to use mtDNA to try to replicate a project like Sibley and Ahlquist's phylogeny because mtDNA is a relatively small molecule. As such, it is subject to change within fairly short periods and is an unlikely tool in the exploration of ancient events such as the origin of birds. It may, however, show traces of much more recent evolutionary events such as the origin of new genera or species. The small size of the mtDNA molecule offers some benefits, and researchers have been able to work out the gene sequences in mitochondria of a great many organisms. The first sequence for a bird was published in 1990.

The mtDNA of birds has some interesting features. In the whole animal kingdom, there are only three basic formats for the gene sequences in the mitochondrial molecule [170]. Almost all animals share the same type but lampreys and birds have their own unique sequences. In some ancestral creature, the mtDNA molecule twisted and broke. The loose ends rejoined but not in the original sequence. The result of such a catastrophic event should have been fatal but, surprisingly, the new sequence was passed on to future generations and the change must have included some hidden benefits that enabled its owner to survive. The change in sequence may have happened to the ancestor of birds, but it could easily have been a much more ancient event that occurred in the common ancestor shared by birds and dinosaurs. Unless someone succeeds in the improbable task of extracting mitochondrial DNA from a dinosaur fossil, we may never know.

So far, the only attempt to examine mtDNA from a dinosaur has failed. The scientists were extremely careful to keep the sample pristine but the results suggest that it was accidentally contaminated by the humans who collected or prepared the material.

Some other features of avian mtDNA are also special but more esoteric. Their examination is far beyond the scope of the amateur with a casual interest in birds, and even the literature on the subject is almost unintelligible to a non-expert. In a 1995 comparison of avian mitochondrial genomes with those of other animals, Thomas W. Quinn stated that "birds lack the hairpin structure that forms the light-strand origin of replication (O_L) in mammals and the *Xenopus* (a popular laboratory frog), and that the CO1 gene (Cytochrome c oxidase 1) has a putative GTG [guanine-thymine-guanine sequence] initiation codon that is unusual among vertebrates and is unique for this gene" [170]. His comments illustrate two important facts: avian mtDNA is distinct from that of mammals or other animals, and the science of mitochondrial genetics has quickly acquired its own technical language.

Perhaps the most unfortunate feature of the study of mtDNA is that it is dependent on complex and sophisticated technologies that make it a bit of a "black box" subject. Its technical literature is so full of specialized jargon that it is often opaque to non-specialists and certainly to the general public. Even review papers, intended for scientists in other fields, assume a detailed knowledge of molecular biochemistry. In one passage, Quinn explains an important point by saying: "several genes end with incomplete stop codons ... guanine is relatively infrequent at third positions of codons ... several genes ... overlap ... the control region includes a transcriptional promoter." The selected passage is a string of reasonably familiar words, but it is impossible to comprehend the statement without knowing the precise technical meaning used in this field. Non-specialists must be satisfied with only a general understanding of the conclusions and cannot hope to make an informed assessment of the shortcomings of one approach over another.

The inability to communicate clearly to a broader public was not a significant issue as long as the study of mtDNA was an esoteric subject, of interest only to biomolecular geneticists. It has recently been shown, however, to have a variety of practical applications in evolutionary and taxonomic studies. Some of these have involved the identification of bird species, suggesting that mtDNA can be a useful tool in the conservation of endangered species and will therefore play an important role in litigation about such species. We will need a useful bridge between biomolecular genetics and taxonomy.

Mitochondrial DNA is significant in the evolution of species because it is a small molecule that evolves rather quickly. The molecules of nuclear DNA are much larger and their genes appear to evolve much more slowly. One of the most useful mitochondrial genes is CO1, in which Quinn found the anomalous GTG base sequence (described above). This gene has 648 base pairs that vary so rapidly that there is often a measurable difference between individuals of the same species. Recently, Paul Hebert and his team have suggested a way of using it as a bar code to reliably identify species [80]. They found that the natural level of variation in this gene within a species was about 0.27% and that there was much more variation between species within a genus. In other words, there was a measurable gap, and perhaps an arbitrary value – for example, 10 times the level of variation found within a species – might be a useful way to distinguish one species from another. It worked well when they applied the technique to a sample of 260 North American birds (38% of the breeding species), and it even helped identify genetically distinct groups within four species that might qualify as new species. Hebert is so confident in the reliability of this approach that he suggests using physical characteristics to confirm biomolecular identification of species, rather than the other way around.

Bar coding works at the species level because the gene evolves very quickly. There are, however, much more conservative genes in mtDNA that show very little variation and evolve very slowly. One of the most conservative genes controls the production of ribosomal ribonucleic acid (rRNA), the molecule that translates genes into enzymes for use by the cell. The molecules of rRNA are a fundamental tool of cellular metabolism and are therefore very similar in every living organism. Almost any kind of variation in their gene sequence is likely to disrupt the function of the cell, with fatal consequences. The handful of existing variations represent very ancient mutations that can be used to distinguish between the major divisions of living things – between plants and animals, or between various kinds of single-celled creatures.

At the species level, the success of a criterion as simple as the bar code in giving a clear identification implies that the species concept is a real biological phenomenon. Generally, a species defined by biomolecular data complies with the traditional criteria for the same entity, as set by Ernst Mayr [126]. Separate species do not interbreed, or if they do, their young are sterile, and so on. Higher taxa are not defined by such absolute criteria and exist only in relation to each other. Bar coding and other forms of biomolecular fingerprinting that might apply to higher taxa are examples of the kind of classification that Livezey and Zusi [117] have set out to counteract with their massive collection of new physical features.

To identify major divisions among the higher taxa, we need to find a gene that consistently evolves at a slow rate. Rapidly changing genes, such as CO1, cannot be used because in the time it takes one species to become two, the pertinent segments of the gene may have changed several times. The evidence for genealogical relationships may simply have been lost because the trace has been overwritten by a series of mutations and has been completely lost. M.G. Fain and Peter Houde believe that they have found one useful, slowly evolving gene and are searching for more. Their results already suggest that we may need to take a very close look at the structure of Class Aves [59]. We will discuss this gene and its significance closer to the end of this chapter, when we look at the confusing history of attempts to classify the Hoatzin.

Most molecular studies have failed to find significant gaps or quantum leaps in genetic variation of the type that might mark the divisions among higher taxa. In nDNA, there appears to be an almost smooth continuum stretching from the gross differences between plants and animals to the nearly identical sets of genes shared by some species. There are no specific levels of either genetic or morphological variation that can be set as rational criteria to delimit classes, orders, or families. The higher taxa are like computer-generated fractal images. From a distance, they seem to have clear definitions with crisp boundaries, but as you get closer, the edges reveal ever-increasing detail as one shape transforms into another. That lack of a clear boundary implies that some of the higher taxonomic levels are arbitrary and not part of nature. They are nothing more than what they were designed to be – practical tools for organizing ideas.

There is a movement among geneticists pushing biologists to abandon Linnaean classification in favour of the molecule-friendly phylocode – no rank, just name and serial number. Such a system might be useful in the laboratory, but even if biomolecular classifications supersede systems based on physical features, there will always be a need for some kind of practical taxonomy for the field biologist. How else are students going to learn to identify birds in the wild? Some modern version of Fürbringer's classic application of comparative morphology will be with us for many years, although it may become an artificial construct – a practical but arbitrary tool for identifying birds without a meaningful theoretical basis. Most of us already use our field guides in exactly that way.

All this worry about the classification of birds may seem a rather esoteric subject and hardly worth the large amounts of time and money spent on it. Our knowledge of birds is becoming extremely deep, however, and we now understand far more about their ecology, physiology, and behaviour than we did just 20 years ago. It is becoming clear that knowledge gained about one species or family may not be equally applicable to another. If you are dealing with an endangered species or the spread of an avian virus, such discrepancies can be critical.

In the absence of a reliable avian genealogy, physiologists and biochemists have found it necessary to ignore potential differences between major taxa and treat all birds as equals. They often generalize from observation of a single species to the whole group. It is not difficult to find sweeping statements about birds as a group, based on data from typical laboratory species such as the chicken, pigeon, or starling. For instance, X-ray cinematography of a flying starling revealed the movement of the furcula (wishbone) [100]. The bone flexed with the beating of the wings and was interpreted as a spring that assisted with avian respiration. The original experimenters specified that their subject was a starling, but in the years since, the results have been applied across the whole class. The bone's interpretation as a spring has been criticized [5], but no one has ever expressed concern that the starling might not be a suitable representative for such a large and varied group.

The starling has a moderately large and strong furcula for a bird of its size, but many other birds lack a furcula entirely. Australian scrubbirds and bush birds are unusual as examples of passeriform birds without a furcula, but it is also absent in a wide array of birds in other orders and families. Some toucans and South American barbets, some owls, some parrots, turacos from Africa, and the mesites of Madagascar all lack a furcula but are able to fly. Among their close relatives, even when a furcula is present, it is often little more than a·thin strap of ossified ligament that loops down from the clavicles. It cannot support its own weight, let alone function as a spring. On the other hand, there are many groups of birds whose furcula is much larger than a starling's and too stiff or massive to act as a spring. In falcons, cranes, and other strong fliers, the arms of the furcula are large, hollow, and quite rigid.

Other kinds of observations from just a few representative species have also been extrapolated to the whole class, including statements about the distribution of various kinds of muscle fibres, the arrangement of nerve connections, or the nature of digestive and renal functions. The Class Aves is very large and diverse, making it risky to extrapolate from one or two species to the whole group. Sometimes it is even risky to generalize within a family, particularly since evolutionary relationships are so poorly understood that what seems to be a family may not be a collection of natural relatives.

Biomolecular Relationships and Morphological Characters

In their biomolecular phylogeny, Sibley and Ahlquist recommended many significant changes to traditional classifications [192]. Some had been anticipated by
earlier morphological studies but had never been acted on by taxonomists. For instance, they place New World vultures among the storks (Figure 6.2). These two
groups share a very similar skeleton, with quite long legs, and exhibit the strange
and distasteful habit of cooling themselves by defecating on their legs. J.D. Ligon
had made the same suggestion in 1967 [114], but he was not the first. In 1896,
while applauding a decision to put the owls in their own order, Frank Beddard
noted that all eight of the characters used to define the Order Falconiformes were
missing from the New World vultures. He bluntly asked: "What reason is there for
retaining American Vultures in this order at all?" For almost a hundred years, no
revision was undertaken even though evidence continued to accumulate. In 1946,
H.I. Fisher observed that the feet of New World vultures were very different from
those of Old World vultures: "No Cathartid vulture is capable of using the foot as
an organ for predation. They cannot even perch well on objects of small diameter
because the inner bones of the toes are too long to permit the flexing necessary to
grasp small objects and the muscles moving those digits are weak" [61]. The feet
of Old World vultures are more eagle-like.

The two groups of vultures also have distinct foraging techniques. New World
vultures use both their olfactory system and keen vision to find food. Old World
vultures have a very basic olfactory system and depend more heavily on their eyesight. Although all vultures look very much alike, especially in flight, the similarities
between the two groups are merely another product of evolutionary convergence.

Storrs Olson supported Ligon's hypothesis regarding the New World vultures
and wrote in 1979 that "the New World vultures ... are almost certainly derived
from a group of waterbirds that includes storks and pelecaniforms." He saw the
peculiar Shoe-billed Stork as a link between storks (and vultures) and pelicans
[149]. Sibley and Ahlquist also suggested placing pelicans among the storks and
vultures. Again, some early naturalists had drawn attention to the groups, and
John Gould had even described the shoebill as a long-legged pelican in 1850. A
century later, P.A. Cottam came to the same conclusion after detailed osteological
analysis [33].

Not everyone agreed with the suggestions for reclassification of such iconic birds
as pelicans and storks. In 1996, Alan Feduccia categorically denied that such a relationship was possible on the grounds that it flew in the face of embryological and
osteological evidence [60]. It is perhaps typical of historical problems in avian taxonomy that he made the pronouncement without presenting any pertinent evidence.

The two groups of vultures appear to be another example of the power of evolutionary convergence to drive changes in structure. Both groups share similar habitats and scavenge carcasses. They use large wings with slotted tips to search across
great distances without expending a lot of energy. There is even fossil evidence that

they once share each other's ranges and that their current distributions are the result of northern glaciations that incidentally pushed the so-called Old World vultures into Africa and the so-called New World vultures into South America. Old World vultures have been found in the La Brea tar pits of California, and the fossilized remains of New World vultures have been found in the Messel oil shales in Germany.

In hindsight, it might seem that the revision of the vultures should have taken place long ago. The lack of decisive action illustrates the central problem biologists have faced when using comparative morphology to erect a classification system. There were no criteria for clear decision making and the process depended heavily on narrative argument and opinion. Changes to an established classification represented a major expenditure of time and resources around the world and could not be undertaken lightly. Revisions needed to be supported by compelling arguments and shown to be necessary for the understanding of ornithology. To some extent, this preference for stability contributed to gradual decline in the role of comparative morphology and made the molecular projects seem all the more exciting and revolutionary.

Even in the case of the New World vultures, this level of caution may have been justified. In spite of numerous recommendations to reclassify them, not all the evidence was strongly supportive. Later studies of mtDNA found that although the link between New World vultures and falconiform birds was not strong, the evidence for a link with the storks was also weak [189, 229].

Sibley and Ahlquist also broke with tradition by placing the pelicans among the storks. In 1995, Douglas Siegel-Causey used mtDNA data to examine the different families of totipalmate birds, traditionally placed together in the Order Pelecaniformes, and found some support for Sibley and Ahlquist's decision on pelicans and for other aspects of their treatment of the Pelecaniformes [193]. According to Sibley and Ahlquist, anhingas, cormorants, and boobies should be placed in a discrete group named "Sulida," while the frigatebirds and tropicbirds belong in their own clades [192]. Siegel-Causey found molecular evidence for both a close relationship between the pelicans and the Shoe-billed Stork and separate status for both the tropicbirds and the frigatebirds. He concluded, however, that frigatebirds were probably closer to cormorants than suggested by their position in Sibley and Ahlquist's phylogeny.

Siegel-Causey did not leave the mtDNA data to stand alone but revisited the morphological characters used to define the Pelecaniformes. The order has long been known as the "totipalmate birds" because every member of the group has a web connecting all four toes. Some of the other characters that are supposed to demonstrate relationships are not so reliable, however. In the traditional diagnosis, totipalmate birds are supposed to lack a brood patch and therefore use the feet to incubate the egg, but incubation behaviour is variable within the group. Frigatebirds of both sexes do have a brood patch and use it for incubation. The tropicbirds lack a brood patch but use body heat in incubation stints that can last longer than two weeks. The other groups use their feet, but the egg sometimes sits on the top of the feet and sometimes the feet are wrapped around the egg. Among pelicans, the choice of incubation technique may vary from day to day. Perhaps the exact type of incubation is not very important to these birds; it is clearly not a useful taxonomic characteristic. Many of the totipalmate birds nest in the tropics, where shade may be more important to the egg than heat from the parent's body.

The absence of a character such as a brood patch has always been a particularly weak form of argument in classification. It is much better to use characteristics that are present but variable, as long as they are also homologous and not the products of evolutionary convergence. A character is homologous if it is derived from the same tissues and in the same way during development. All the totipalmate birds have a gular pouch but, as Siegel-Causey points out, Aristotle was the last biologist to examine the pouch and similarities in its structure may be superficial rather than homologous. The pouches of the tropicbirds and frigatebirds certainly look very different from those of their supposed relatives. Similarly, the webbed toes may look similar but their homology has not been demonstrated. The only other special character shared by the whole group is a bony nostril that is either absent, reduced, or indistinct, but this character is not unique to the totipalmate birds and appears in other groups.

Pelicans are also noteworthy for the extensive pneumatization of the skin, especially near the base of the neck. Erwin Stresemann suggested that it might act as an "air mattress" and protect the bird from the shock of impact when it makes a plunging dive out of the air. Tropicbirds and boobies are more vigorous plunge-divers and have even larger and better developed air mattresses, while cormorants and anhingas, which do not plunge out of the air, have relatively little pneumatization in the skin. The variation is what you might expect across a group of apparently related birds, but again the presence of the structure is not absolute proof of a relationship. The pneumatizations may have evolved independently in each group, and similar structures appear in other unrelated birds. For instance, the South American screamers are not related to any of the totipalmate birds but they also have strongly pneumatized skins. What use they make of this feature is unclear, but screamers do not plunge-dive.

The weakness of the long-accepted unifying characters, contradictory osteological evidence, and consistency between nDNA and mtDNA analyses suggest that the "Pelecaniformes" is indeed an artificial assemblage and that the phylogeny proposed by Sibley and Ahlquist may be close to the truth when it places the pelicans and the New World vultures close to the storks. The recent analysis of 359 skull characteristics by Livezey and Zusi [117] strongly contradicts Siegel-Causey's findings, however, and supports the traditional classification of the Pelecaniformes.

Mayr and Clarke did not include pelicans in their study but they found support for a close relationship between cormorants and frigatebirds [131]. The two groups shared 10 synapomorphies but three characters – the hooked beak, the gular pouch, and the globules of calcium phosphate on the eggshell – are found in other birds.

Four other characters are the absence of features. Frigatebirds and cormorants do share a bony canal for the toe tendon on the tarsometatarsus and a pectinate claw on the third toe. The "pelecaniform birds" are clearly a group that warrants more detailed study.

The general patterns of the biomolecular phylogeny have significant impact on the interpretation of extinct or fossil lineages. If the "Pelecaniformes" are to be broken up and the pelicans placed among the cranes, the description of dozens of fossils as "pelecaniform birds" has little meaning. It has always been difficult to relate some fossil groups to living birds because the fossil forms appear to have characteristics of more than one taxon. Sometimes they have been described as mosaics, with the implication that they might represent an ancestral line of one or both of the similar modern groups. *Presbyornis* was a long-legged wading bird with a very duck-like bill. In 1996, Alan Feduccia described it as mosaic of anseriform and charadriiform characters [60], but biomolecular phylogenies and recent cladistic analyses put ducks and plovers in widely separated lineages [131]. A recent osteo-logical analysis by Gareth Dyke [53] places *Presbyornis* among the Anseriformes. Its habit of wading instead of swimming might have contributed to characters that converge on those of modern shorebirds. Convergence may contribute to other mosaics among ancient birds, such as *Juncitarsus,* a supposed shorebird/flamingo mosaic, and *Rhynchaetes,* a supposed shorebird/ibis mosaic. In these cases, however, the DNA evidence suggests that the pairs of modern representatives may be closely related, and it is more reasonable to expect a fossil to show characteristics of more than one closely related group.

In the days before DNA-DNA hybridization, while Charles Sibley was looking at egg white proteins, other teams of researchers studied variations in the many complex chemicals produced by birds. Jürgen Jacob and Vincent Ziswiler examined the sebum and constructed a kind of multi-dimensional fingerprint based on dif-ferences in its complex alcohols and waxy acids [99]. They were pleased to find that their results were not too controversial and agreed with many features of classical avian systematics. Interestingly, they are also broadly consistent with Sibley and Ahlquist's biomolecular phylogeny. The most complex combinations of sebaceous products occur in grebes, penguins, petrels, and some of the "pelecaniform" birds, groups that all live in aquatic environments and are among the higher clades of the biomolecular phylogeny.

Like the results of many other studies, the distribution of feather waxes is con-sistent with the division of birds into many familiar orders and families but does not provide enough detail to support a particular phylogeny. Kiwis and galliform birds have similar waxes and both are representatives of the earliest divisions. Their waxes are very different from those in representatives of the Plethaves, such as cuckoos or cranes. The waxes of cuckoos seem almost identical to those of wood-peckers, while owls, swifts, and hummingbirds share similar types of wax with song-birds. Pigeons, parrots, and kingfishers have their own distinctive waxes. Waxes from the Shoe-billed Stork and the ibis are similar to those from plovers, and waxes from flamingos are similar to those from both ducks and shorebirds. As mentioned above, plovers are not far from flamingos in the biomolecular phylogeny, but they are far from ducks. Since there is additional support for the position of the ducks as a discrete group, it seems likely that the similarities in their "wax fingerprint" are examples of convergence.

In 1998, C. Jeffery Woodbury attempted to add to the list of characteristics available to taxonomists by examining structures that could be seen in a cross-section of the spinal cord [232]. In the process, he unintentionally demonstrated just how difficult it is to find a characteristic that varies consistently with the divisions of any form of classification.

The spinal cord is a very ancient and conservative structure because its role is so fundamental to the function of the organism. Spinal cords from all vertebrates are, in fact, very similar. In cross-section, they show a pair of rounded structures that point upward and outward, called dorsal horns. These appear to be responsible for receiving sensory information from the skin.

Although birds are an ancient lineage, the basic structures of the spinal cord may represent the condition in ancestors hundreds of millions of years older. You might expect all birds to have very similar spinal cords but, surprisingly, there are two distinct types. Sometimes the dorsal horns have a smooth, or "leiocerate," shape, which is shared with most other vertebrates and is probably the ancestral condition. In other cases, the cord is partially split lengthwise, or "schizocerate." Such a basic character should distinguish between major groups of birds such as the paleognaths and neognaths, but sometimes one family within an order will be leiocerate and another schizocerate. Such variation at the family level implies that the characteristic, like many others, is probably not very useful.

Sibley and Ahlquist's biomolecular phylogeny suggests resolutions for some of the peculiar distributions of this character [192]. It places the schizocerate sand grouse among other schizocerate families in the charadriiform birds, and the schizocerate turacos with the schizocerate owls. Recently, Janice Hughes and A.J. Baker have placed the schizocerate turacos and the schizocerate Hoatzin in their own order, Opisthocomiformes [95]. The biomolecular phylogeny does not offer resolutions for all the discrepancies, however, and it creates new ones. Its linear structure requires the schizocerate condition to arise independently in ducks, kingfishers and rollers, parrots, and cranes. It also implies that the woodpeckers, cuckoos, and songbirds reverted to the ancestral condition from schizocerate ancestors. Such switching back and forth is not consistent with our understanding of evolutionary events, and we can only conclude that the two sets of data are incompatible. Perhaps a clearer understanding of the function of this feature in the spinal cord would lead to an explanation of its distribution among the various groups of birds.

The Biomolecular Phylogeny as an Evolutionary Story for the Birds

The shape of Sibley and Ahlquist's phylogeny surprised many ornithologists. Most phylogenies look like a shrubby root system (hence the name "dendrogram"), but the phylogeny in Figure 6.2 looks more like a vine. In the jargon of dendrograms, it is extremely unbalanced. The version in Sibley and Ahlquist's book appears a little more shrublike but it has been folded so that the songbirds are at one end [192], where Fürbringer had placed them in 1888. About half of all birds are songbirds, and this group's story is just as complex and confusing as the story for the whole Class Aves. Sibley and Ahlquist knew that they were going to devote a large portion of their book to them and simply placed them at the end so that all of the lesser groups could be dealt with in the first half.

It may sound like a trivial reason for such a major decision but it made the book much more accessible and had no effect at all on the information content of the phylogeny. You can legitimately twist a dendrogram into any shape you want as long as you do not break any of the branches. It's better to think of a phylogeny as a shrubby plant with a complex, three-dimensional structure. You can flatten a plant on a page in several different ways. Some arrangements are more esthetic than others or emphasize certain structures, but none changes the relationships among the branches. Similarly, folding changes the information content of a dendrogram without changing the data. In Figure 6.2, Sibley and Ahlquist's dendrogram is unfolded and its branches are arranged in strict sequence of decreasing values of their physical parameter, $\Delta T_{50}H$ [192].

Although there is only a tenuous theoretical connection between the structure of the DNA and the characters used to define the traditional orders and families of birds, each lineage coming off the main trunk of the Sibley and Ahlquist vine is very similar to one of the well-known taxa set out by Fürbringer in 1888 [62] or Peters in 1934 [163]. In other words, comparative morphologists and biochemists have independently divided the birds into very similar groups.

The Hoatzin

Perhaps the clearest illustration of the problems surrounding the use of biomolecular information comes from the attempts to determine the relatives of the Hoatzin. Ornithologists argued about its relationships throughout the 19th and 20th centuries without coming to a satisfactory conclusion. In 1990, Sibley and Ahlquist used evidence from DNA-DNA hybridization to place it among the cuckoos. Most ornithologists thought that their evidence was weak, but attempts to apply information from base sequences in mtDNA and nDNA have also been inconclusive.

The Hoatzin lives along the quiet backwaters and *madre viejas* (oxbows) of rivers in northern South America [75]. It looks somewhat like a very scrawny chicken. Its ancestry has puzzled biologists for over 200 years, and a suite of suggestive anatomical features and peculiar behaviour have led only to dead-ends and failed speculations. In 1776, Müller decided that it belonged among the galliform birds, and it was placed there in 17 later classifications. At various times, it has also been placed in the cuckoos 8 times, 4 times in the turacos; once each in the cranes and pigeons; and 12 times in its own order, Opisthocomiformes. Thomas Huxley thought that its hips were similar to those of the button quail, another bird whose relationships have been difficult to determine. In fact, many of the groups proposed as relatives of the Hoatzin are themselves taxonomic problems. That is why the recent placement of the Hoatzin and most of its supposed relatives in Fain and Houde's Subclass Metaves is so intriguing (see page 196) [59].

Victorian ornithologists were puzzled by the presence of fingers on the wings of young Hoatzins and some adults. The free digits were suggestively reptilian and hinted at a link with the fossil *Archaeopteryx*. If baby Hoatzins had had teeth as well, ornithological history would have been rewritten. In 1891, W.K. Parker published a series of detailed drawings to show that the forelimb digits were true fingers, homologous in structure and development to the digits of other animals and not some disguised or misinterpreted facsimile [155]. Parker believed that the Hoatzin was primitive, and carefully described the bones in the palate, the structure of the legs and wings, and the shape of the vertebrae. He was not looking for reptilian characteristics; rather, he wanted evidence to support his theory that birds arose directly from amphibians or dipnoan lungfish.

Parker's contemporary Frank Beddard also saw the Hoatzin as a primitive bird but based his conclusions on less dramatic features [8]. In his view, "advanced" birds had their feathers organized in well-developed tracts, but the Hoatzin's pterylae were weakly formed and the spaces between the feathers were covered in sparsely distributed filoplumes. He thought that this might represent remnants of an unorganized distribution of feathers such as you might find in an animal whose scales were evolving into proto-feathers.[2] He also saw a parallel between the weak pterylae in the Hoatzin and narrow pterylae in the equally enigmatic African turacos. Like the Hoatzin, turacos produce young that are densely covered with dark down. Their nestlings also clamber about in the branches of the nest tree, assisted by claws on the wings. They lack only the Hoatzin's independent digits. In both groups, growth of the primary feathers is delayed so that the terminal claws can be used more easily by the young.

Although the Hoatzin's fingers look primitive, it is difficult to demonstrate their close relationship to the fingers of primitive birds. They could be a new feature convergent on the old. Other features, such as the Hoatzin's unique and highly specialized digestive system, appear to be a mix of the new and the old. The gizzard is unusually small and unspecialized, but the crop has a peculiar, deeply ridged interior lining. Its enlargement as a fermentation chamber is unique among birds. The middle sections of the gut are relatively unspecialized and somewhat similar to those of a cuckoo or pigeon. There are a series of paired caeca, which are sites for further fermentation. The hindgut and rectum are long and coiled like those of an ostrich, which Victorians would have considered to be a primitive type.

---

2   Feathers are not derived from scales; see Chapter 7.

Most of the important 19th century anatomists who expressed an opinion on the subject saw the Hoatzin as a galliform bird with links to cuckoos or pigeons, or perhaps to turacos. According to Sibley and Ahlquist, Erwin Stresemann, Alexander Wetmore, J.L. Peters, Ernst Mayr, Dean Amadon, Robert Storer, and other avian taxonomists of the mid-20th century placed the Hoatzin either among the galliform birds or in a closely related group, in spite of considerable evidence to the contrary [192]. Verheyen [216], de Quieroz and Good [48], and others have argued that the structure of its skeleton placed it elsewhere, perhaps among the cuckoos. Stegmann came to the same conclusion from his analyses of wing muscles [197].

The muscles of the Hoatzin's hip are supposed to be primitively organized and a particularly important factor in the bird's classification. Both Garrod [65] and Beddard [8] give the formula as ABXY+ and suggest that the Hoatzin bridges a gap between the galliform birds and the cuckoos. The "+" symbol in the muscle formula indicates the presence of the ambiens muscle, which is supposed to assist perching in arboreal birds. Because of its absence from reptiles, the muscle's presence in the Hoatzin is generally considered a more advanced, and perhaps cuckoo-like, characteristic. The character is inconsistent in the Hoatzin, however. Even when present, the ambiens is very small.

Unfortunately, the Order Cuculiformes does not have a clearly defined set of features and its supposed members may not be closely related to each other. Significant differences in skeletal structure, muscle arrangement, and pterology (feather tracts) suggest that the African turacos could be split off from other cuckoos, as a separate Order Musophagiformes. This ongoing uncertainty about the place of the Hoatzin among other birds is typical of the inability of classical morphological analyses to resolve problems [192].

DNA and the Hoatzin

In 1990, Sibley and Ahlquist used DNA-DNA hybridization to place the Hoatzin among the cuckoos, and the turacos among the owls and nightjars. The decision agreed with their earlier work using protein electrophoresis but was not widely accepted by ornithologists. In 1994, John Avise and colleagues examined the sequence of 961 nucleotides in the mtDNA coding for Cytochrome b in an attempt to determine whether the Hoatzin was closer to the galliform birds or to the cuckoos [4]. He found that all three had discrete lineages and suggested that the separation of the Hoatzin was too ancient to be resolved with this technique unless longer sequences could be used. Later, Kevin Johnson and colleagues used 429 base pairs from Cytochrome b and 522 from a gene called ND2 (NADH dehydrogenase) to find additional support for placing the Hoatzin outside the cuckoos [104].

In 1999, Janice Hughes and Alan Baker announced that they had successfully used a combination of mitochondrial and nuclear genes to resolve the relationships of the Hoatzin [95]. They also found that it was neither a galliform bird nor a cuckoo but that it belonged in the same order as the turacos. Because the name "Musophagiformes" is younger than "Opisthocomiformes," the latter has precedence and turacos joined the Hoatzin in that order.

At the same time, Hughes was working on osteological characters for the same birds [94]. She found that only 5 of the 32 characters traditionally used to define the Order Cuculiformes were actually shared by all cuckoos. To resolve the problem, she examined 54 families of birds to derive 9 more shared characters, and was able to base her cladistic analysis on a total of 135 characters. The resulting phylogeny pulls all the parasitic cuckoos from both the Old World and the New World into a single Subfamily Cuculinae. The strategy of reproducing by brood parasitism therefore needed to evolve only once. Earlier classification schemes had it evolving on as many as three separate occasions.

Three structural features of the ulnare, a small bone that affects the function of the wrist, were synapomorphic for cuckoos, turacos, and the Hoatzin, suggesting that the three groups formed a large monophyletic clade. Most significantly, 19 such shared characteristics supported a sister relationship between the turacos and the Hoatzin. In other words, Hughes' morphological information matched the results of her biomolecular analyses.

Unfortunately, this was not the last word on the subject. In 2003, Michael Sorenson and his team threw Hughes' conclusions into doubt [195]. They applied a 5,000–base pair sequence from mitochondrial DNA and 700 base pairs from nuclear DNA to conclude that the Hoatzin was not related to either the turacos or the cuckoos. Instead, there was a weak association with pigeons, as once suggested by Thomas Huxley and Max Fürbringer. Worse still, they detected "pervasive errors" that biased Hughes' work towards a sister relationship between the Hoatzin and turacos.

A year later, Matthew Fain and Peter Houde provided further support for a relationship between the Hoatzin and the doves [59]. It was just one result in an attempt to examine all the basal divisions in the phylogeny of birds, but it led to a proposal that might eventually resolve the most difficult taxonomic anomalies. It will also require, however, a complete reorganization of the Subclass Neoaves.

Fain and Houde based their analysis on "intron 7 of the b-fibrinogen gene" (FGB-int 7) from nuclear DNA. It is a moderately conservative gene that had already been used successfully to identify memberships in various families and orders. In 2005, Heather Lerner and David Mindell used it to demonstrate that eagles are a monophyletic group [113]. They believed that it could be useful at even higher levels because it showed a much lower rate of variation than other mitochondrial genes; evidence of evolutionary history was less likely to be overwritten and obscured by series of multiple mutations. It also had the same kind of "indel" features (insertions and deletions of molecules) that had provided "compelling evidence" in other genes to support the Galloanserinae and the Apodiformes as monophyletic clades.

Fain and Houde [59] were not able to base their study on as large a sample as Sibley and Ahlquist [192], but they had the advantage of working in a well-established

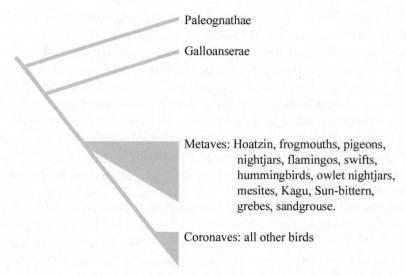

Paleognathae

Galloanserae

Metaves: Hoatzin, frogmouths, pigeons,
        nightjars, flamingos, swifts,
        hummingbirds, owlet nightjars,
        mesites, Kagu, Sun-bittern,
        grebes, sandgrouse.

Coronaves: all other birds

Figure 6.3   The phylogeny of birds based on variation in the b-fibrinogen gene from nuclear DNA [59].

field and were able to focus their attention on 79 families in the Neoaves (73 of the 88 nonpasserine families and 6 of the 59 or so families in the Order Passeriformes). They anchored their study with 8 outgroups. The analysis of their results produced a phylogeny that began by dividing birds into four main clusters (Figure 6.3). As expected, the paleognaths split off first and then the Galloanserinae, but the next split was surprising. It divided the remaining birds into two large monophyletic clades. The proposed division met the demands of several different measures of statistical confidence. Most significant, the split required the removal of some members from long-established orders of birds.

If additional evidence had not supported the split, it might have been dismissed as some sort of procedural artifact, but the two new clades have some unusual characteristics. The larger of the two has been dubbed "Coronaves" (literally, "crown birds") and contains a great many widely distributed birds. The smaller (19 families), has been named "Metaves" ("almost birds") and tends to contain families that are endemic to southern continents (8 of 19), consist mostly of southern species (8 of 19), or are absent from the fossil record in the north (4 of 19). Most of the members are highly specialized types that are represented by only a few species. The humming-birds, swifts, and grebes are the strongest exceptions to this rule. Each of these groups is represented by a number of species and is fairly widely distributed, even though they are concentrated in the Southern Hemisphere. There are 328 species of hummingbirds, mostly in South America; 92 species of swifts (plus 4 tree-swifts), mostly in the tropics; and between 17 and 20 species of grebes, mostly in the Southern Hemisphere. There were 22 species of grebes but between 2 and 5 species became extinct in the 20th century.

In addition to the grebes, Metaves contains other families that have been problematic for taxonomists, including Hoatzins, turacos, mesites, sand grouse, and flamingos. A seemingly unlikely but close relationship between flamingos and grebes has been suggested by other biomolecular studies [26, 213]. Such a relationship would appear to make more sense if the members of the Metaves were lonely survivors of an early, post-Cretaceous radiation that was all but extinguished by a later radiation of more advanced Coronaves.

Fain and Houde make an inspired comparison of the Metaves/Coronaves split to the separation of marsupial from placental mammals and their subsequent convergence as ecological and behavioural types. Many families in the Metaves have imitators in the Coronaves, and this convergence may well have combined with the biomechanical constraints of flight, the bipedal stance, and other basic avian features to contribute to an extreme degree of morphological similarity that subsequently confused many taxonomists. Once again, however, some 19th-century experts were able to find morphological evidence that anticipated modern biomolecular results. Thomas Huxley suggested that the Hoatzin was related to the doves in 1888; Edward Bartlett linked the sunbittern to the mesites in 1877; and W.A. Forbes linked both birds to the Kagu in 1882. Michael Evans, Bradley Livezey, Gerald Mayr, and other 20th-century anatomists have also reported evidence that supports some of the relationships within the Metaves.

The limitations of the method are amply illustrated by the membership of the Coronaves. Fain and Houde were unable to settle on relationships for many of the major groups and, as in Joel Cracraft's cladistic analysis, they were forced to include a large, unresolved polytomy. Significantly, it shows a close relationship between the songbirds, woodpeckers, and kingfishers that is typical of most biomolecular and morphological classifications. Some of the other sister relationships in their phylogeny suggest, however, that there is significant convergence or that some genetic information has been "overwritten" by repeated mutation. Although it appears unlikely, a close relationship between falcons and mousebirds is at least conceivable. Both are terrestrial birds. A comparable sister relationship proposed for cuckoos and loons is incomprehensible and can only be the accidental result of molecular convergence or overly strict adherence to the rule of parsimony within the analysis.

Overall, the content of the Metaves is consistent with the notion of that group being the remnants of an early radiation of Neoaves that has largely been replaced by a more recent radiation of the Coronaves. Fain and Houde are currently looking for further confirmation of their results and examining 11 other genes with somewhat similar characteristics. If additional studies uphold the division of Neoaves into Metaves and Coronaves, they will have a profound effect on our view of avian taxonomy and may make it easier to find that elusive evolutionary story. If nothing else, Fain and Houde have presented us with a totally new solution to the puzzle of the Hoatzin's relationship to other birds. Perhaps there is no longer any excuse for describing it as an enigma.

A Last Word on Avian Genetics

In 2004 the International Chicken Genome Sequencing Consortium published the first complete genome of a bird [98]. A host of cultural and biological reasons led them to choose the best known and most important of all birds – the domestic chicken. A great many of the protein-coding genes in chickens are directly compa-rable (i.e., orthologous) to the genes found in other vertebrates, in spite of the vast amount of time since terrestrial animals split from fish, or diapsids (e.g., birds) split from synapsids (e.g., mammals). About 60% of chicken genes are orthologous to human genes, but there are two special families of proteins not found in humans and 21 families not found in puffer fish (*Fugu* sp.).

Just as you might expect, there are no genes in the chicken that control the pro-duction of either tooth enamel or the casein proteins of milk. Among the chicken's typically avian genes are 166 for the production of the eggshell and 150 for the special keratins used in the production of feathers, beaks, and claws. None of the chicken genes produces the high-sulphur type of keratin found in mammalian hair. Chickens also have a gene for indigoidine synthase, an unusual enzyme involved in the production of blue pigments. Mammals cannot produce blue pigments and lack a comparable gene. Although chickens and other birds excrete a paste of uric acid rather than fluid urine, the genes that control the full urea cycle appear to be intact and orthologous to those found in mammals. Perhaps they are involved in some other function in birds.

Shortly after the consortium presented its genome, Luis P. Villarreal published an equally far-reaching synopsis of recent advances in virology [219]. His central theme was the frequency with which viruses have colonized the genomes of various organisms and become an integral part of their genetic inheritance. This ability of viruses to attach themselves to chromosomes has profound significance for natural selection as an evolutionary process. Variation need not wait for the tedious accu-mulation of point mutations to effect change when a colonizing virus can attach hundreds of alien nucleotides to the genome in a single event.

Viruses are parasites and their DNA may be redistributed by processes other than genetic inheritance. Like the feather lice found in birds, virus particles may move among susceptible animals that share the same habitats, eat the same foods, or are in a predator/prey relationship. They may even be passed on by wide-ranging macro-parasites such as mosquitoes, or picked up casually as migrants move into an area previously occupied by an infected group.

The consequences of viral colonization can be profound, and Villarreal discusses evidence linking viruses with such fundamental structures as the nucleus of cells, the flowers of plants, and the placenta of mammals. The ease with which viruses colonize mammals may explain the vast amount of non-coding DNA in their genomes. There may be tens of thousands of such repeated units, but they are not "junk" DNA. They record successful invasions of the mammalian genome by retroviruses that have had dramatic impacts on the evolution of that group. As endogenous retroviruses (ERVs), they played important roles in the evolution of higher mammals and may even have had a significant effect on the development of higher social behaviour in humans.[3]

The genome of birds is about 60% smaller than that of a mammal and appears to lack significant colonization by retroviruses. Some ERVs that may antedate the bird/mammal split were found in the chicken genome but none are known from wild birds. Instead, birds have SINEs and LINEs – short and long insertions of nucleotide elements – which appear related to historical virus activity. In the chicken genome, there are indications that no SINEs have been active as genes for the past 50 million years. It will be interesting to see whether highly specialized birds such as the Hoatzin, or very unusually shaped birds such as the penguin, have evidence of novel LINEs or SINEs in their genomes.

If ERVs are related to the spectacular structural diversity of mammals, their absence from birds and the inactivity of avian SINEs and LINEs might account for the general conservatism of birds as a group. Most of the higher mammalian taxa appear to have arisen in the Eocene but did not achieve their modern appearance until the Miocene or Pliocene. This makes mammals relatively young compared with any of the higher avian taxa. There is some evidence for loons and other modern groups of birds in the Cretaceous. Certainly, birds achieved the major features of their modern form long before the extinction of the dinosaurs.

Crocodiles, another conservative group, also have a small genome. Perhaps a small genome has always been a characteristic of diapsid animals and provides an explanation for the high degree of structural conservatism within the dinosaurs. In spite of the elaborate frills and spikes that decorated some, the group was restricted to a rather small number of long-lived body plans, whereas mammals have produced everything from whales to bats and humans in a few million years.

Recently, avian viruses have been at the centre of international concern over the spread of influenza. Our fears may be justified because humans are no longer typical mammals. Mammalian viruses evolved as relatively slow-acting residents because mammals were sparsely distributed, territorial animals. In general, the surest method of transmission from one host to another was through sex. Avian viruses evolved in populations of highly mobile animals that often gather in large flocks or live in dense colonies. It benefits them to be highly infectious and to have acute effects on their host. Many avian viruses are transmitted by mosquitoes. Humans have

---

[3] Retroviruses may be important factors in human evolution. The separation of apes and humans may have been related to colonization of the primate genome by SIV (simian immunodeficiency virus) millions of years ago. More recently, human speech and advanced social function may be related to another successful colonization. If a retrovirus gave Cro-Magnons an advantage that contributed to the extermination of their Neanderthal neighbours, we should be interested in the possibility of new species arising from among the many millions of modern children being born to parents infected by the human immunodeficiency virus (HIV).

long suffered from a host of avian pox and influenza viruses that have crossed species lines, sometimes into pigs, cows, or other domestic stock, on their way to infecting people. From the viruses' point of view, they have been a great success because humans live like birds – in large, dense masses where highly infectious diseases are quickly passed on.

Conclusion

At the beginning of the 21st century, ornithologists are still in the embarrassing position of not having a generally accepted phylogeny for their subject group. They have been unable to put together a convincing evolutionary story for living birds, whereas paleontologists have made great strides in sorting out the long-extinct dinosaurs, using little more than osteological evidence.

In the past few decades, paleontologists have also introduced us to a great many new fossil birds, including the Confucius birds and the ball-shouldered, or enantiornithean, birds. These groups represent two entire subclasses whose existence was completely unsuspected less than 25 years ago.

Fortunately, biologists are beginning to learn how to read the biomolecular clues left on the chromosomes by evolutionary events, and may eventually fill in the enormous gap between the modern birds and their Mesozoic ancestors. Surely we cannot be too far away from some real breakthroughs and a new understanding of avian evolution and classification.

Further Reading

The subjects covered in this chapter are relatively new forms of science. I have not come across books on mitochondrial DNA or other current aspects of biomolecular genetics that are accessible to non-specialists. Similarly, there are no commentaries in a popular format on Luis Villarreal's ideas about viruses or the decoding of the chicken genome. All these topics are discussed on various websites, however, as well as in magazines such as *New Scientist* or *Discover*. Information in the magazines is likely to be more reliable than that on the Internet.

Although they do not deal directly with the material in this chapter, three books by Matt Ridley use everyday language to discuss the deeper significances of evolution with wit and insight: *The red queen: Sex and the evolution of human nature* (1993), Penguin Books, New York; *The origins of virtue: Human instincts and the evolution of cooperation* (1998), Penguin Books, New York; and *The genome: The autobiography of a species in 23 chapters* (2000), Perennial (HarperCollins), New York.

PART 3

How Does a Bird Fly?

Feathers and Feathered Dinosaurs

7

Flight in early birds and birdlike animals

The previous chapters have described how various aspects of avian anatomy have been used to answer the questions "What is a bird?" and "What bird is it?" In the last four chapters, I want to examine the bird as a flying animal. As a first step, we need to look at the plumage because there is so much more to feathers than the bird's outward appearance.

We know quite a lot about the feathers of birds. For centuries, they have been valued for softness, warmth, and bright colours, and their fine detail was one of the first natural phenomena investigated under the microscope. Each one is an incredibly sophisticated object; arrayed together as plumage, they create the most complex body covering in nature. Feathers would be amazing enough if they only covered the body, but they also contribute to locomotion and communication and make the "inner-bird" strategy feasible. Without such an effective body covering, it would impossible for a creature as fragile as a bird to withstand the world's harshest environments.

It had always been safe to assume that feathers were unique to birds, but recently they have been showing up on the fossils of many different kinds of dinosaurs. These fossils come as no surprise to supporters of the theory that birds evolved from dinosaurs, but feathered dinosaurs raise as many questions as they answer. Why did dinosaurs need feathers? Why use something as complex as feathers instead of something as simple as fur? If feathers are such an ancient phenomenon and so useful to birds, why did dinosaurs not produce other success-ful feathered animals that followed their own version of the inner-bird strategy? Where are the fossil remains of all those dinosaur feathers? One would expect dinosaurs to have relatively large feathers that would have been cast off as they wore out. Why, then, are there so few fossils of feathers, even in amber or the fine-grained rocks that have preserved the colour patterns of things as delicate as but-terfly wings?

Feathers and the Body

Care and Maintenance of Feathers

In many ways, the success of birds depends on their ability to get the most out of their feathers. The plumage of birds is the most complex and dynamic body covering in the animal kingdom [196, 198]. It simultaneously provides its owner with protection, insulation, communication, and locomotion. Muscles flap the fleshy part of wings but it is the broad areas created by specialized feathers that provide lift, thrust, and control. Those same feathers often carry contrasting colours that flash signals to other birds. Simple contour feathers cover the body and cushion it from the shock of unexpected impact while helping to fend off sharp objects, parasites, and disease. Tiny muscles in the skin control the position of each feather, slowing or speeding the radiation of body heat and enabling the bird to maintain its temperature within a very narrow range. Feathers are able to cope with so many responsibilities because they are fully integrated with living tissues and supported by the elaborate maintenance and renewal system operated by the inner bird.

The inner-bird strategy, which concentrates most of the muscles and vital organs in a central mass, leaves the bird's rather fragile body dependent on two thin layers of non-living keratin for all of its protection. The outer layer is the plumage, visible as a continuous blanket of feather tips, laid out smoothly like overlapping shingles on a roof. The inner layer is little more than a thin film of keratinized skin cells. This sclerotic layer is not unlike the outer layer of our own skin, but it is much tougher and covers much more of the body. It even extends into the air sacs, lungs, and hollow bones so that the inner spaces in a bird are lined with the bird's outer covering. In a strictly topological sense, the thin sclerotic layer makes the hollow spaces within the body of a bird part of the external world.

The feathers that form the outer protective layer are rooted in arrays of small follicles that are one of the few interruptions in the continuity of the skin's sclerotic layer. Feathers are non-living extrusions of keratin, an inert compound very similar to the material that makes up our own fingernails. Like fingernails or claws, the tips of feathers wear away naturally from abrasion. They are so important, however, that if they deteriorate too quickly, the loss of integrity threatens the survival of their owner.

The main threat to feathers comes from the bird's life in the open. Other animals spend much of their lives under rocks or in burrows, but birds usually seek shelter only during the few weeks each year when they incubate eggs. At other times, they are exposed to the sun, whose ultraviolet radiation greatly accelerates the deterioration of feathers. Surprisingly, evolution has not found a chemical fix to protect keratin from sunlight, and birds must discard old feathers and replace them with a new set during a regularly scheduled moult. A bird cannot risk having feathers that are shredded or worn out, so it delays the onset of that catastrophe by expending a great deal of energy on feather care. It preens obsessively, keeping its plumage soft and flexible with regular applications of restorative oils produced by its body.

The most important of these oils is the sebum produced by the skin. It is a complex mixture of large molecules, including dozens of different waxes, fatty acids, alcohols, and complex lipids that work together to delay the effects of ultraviolet radiation. Birds may have 150 genes involved in the production of keratin for feathers, and have another 210 that control the activities of the skin and fatty tissues that produce the sebum. Mammals also produce an oily mixture in their sebaceous glands but it is a relatively simple product responsible for both sleek pelts and greasy hair.

Most of the sebum used by the bird seeps through its skin and is gradually transferred to the feathers by the bird's persistent preening activity. Dead cells from the keratinized layer over the epidermis become coated with sebum and transfer it to the feathers as they slough off. Preening with the beak then works the oily cells into the spaces between the barbules.

The exact combination of feather-care products in the sebum varies from species to species, and some birds have used them for exotic purposes. Five songbirds in the genus *Pitohui* and another called *Ifrita kowaldi,* all from Papua Niugini, have gained a certain notoriety as the only known poisonous birds. Homobatrachotoxin, better known as a dart poison distilled from frog skin, appears in the feathers and on the skin of these birds. The source of the poison appears to be toxic beetles that are part of the birds' diet [51]. The toxin simply moves from the skin to the plumage with the sebum carried on dead skin cells. The *Pitohui* must be immune to the poison if it is to survive preening its own feathers.

Birds never leave proper maintenance to chance and preen busily for long periods every day. The beak prods and strokes each of the feathers in turn and makes sure that it is lying in its proper position. This would be a very tricky operation if birds still had teeth, and feathered dinosaurs probably found it easier to use their hands for grooming. The beak would appear to be an overly simple tool for an operation as complex as feather care, but its hard covering disguises its internal sophistication. Tiny sensors line its tip and may be able to taste the amount of sebum already on the feathers and help the bird determine which areas of the plumage need the most attention.

Preening usually redistributes sebum secreted by the skin, but birds can also collect a supply from a specialized oil gland, the uropygial gland, on the top of the body, near the base of the tail. This oil gland is the only sebaceous gland found in birds except for a pair of small and poorly understood cloacal glands. The openings for these glands are the only breaks in continuity of the skin besides the feather follicles.

The oil gland is often rather large. One is present in almost all wild birds, even though it usually provides less than 10% of the total amount of sebum used in preening. Ostriches and some of their relatives lack an oil gland, but the large, fluffy feathers of an ostrich may not need more chemical protection than that provided by the secretions of its skin. Oil glands are also missing in some domestic races of pigeon, suggesting that they are a useful option but not an absolute necessity in feathered animals. They have not been found in any fossil birds or dinosaurs. Like other soft tissues, sebaceous glands are poor candidates for fossilization.

· The oil gland is probed frequently during the preening process and its accessibility is important. In mammals, large sebaceous glands that produce milk are often located on the breast or under the forelimb, where they are convenient for suckling the young; their position does not need to be convenient for direct use by their owner. A bird, however, must be able to use its own sebaceous gland. If it had to bend its head down to reach a gland on its breast or under the wing, it might lose track of nearby predators. In addition, such a location would be extremely difficult for waterbirds. Ducks and other swimming birds already undergo complex contortions while preening on water. They roll over onto their sides and stretch their limbs into unusual postures that look risky in terms of escaping from sudden attack. They would, however, be at a much greater disadvantage if they also had to dip their heads below water to collect sebum from the oil gland. A site at the base of the tail is safer and within convenient reach of the bird's long neck.

Although birds cannot prevent feather wear, they can slow the process. The colour of a feather has a strong effect on the rate of deterioration from sunlight. Dark pigments, such as melanin, protect the keratin in feathers just as they protect tanned or dark human skin. White feathers, like the skins of fair-skinned people, are quickly damaged by ultraviolet light and break down very easily. In sandpipers, the white parts of the feathers erode so quickly that the birds appear to change colour in seasons when they are not undergoing a moult. The white tips and edge markings disintegrate and fall away to reveal the darker feathers underneath. Usually, these white areas are very small and their loss does not interfere with the function of the feathers. Many white birds have dark wingtips so that their melanin can reduce the rate of deterioration in the feathers that are most exposed to wear and tear during flight.

Birds also produce specialized feathers for unusual functions. Some grow a special kind of feather called "powder down," which may help maintain the condition of the rest of the plumage. The delicate tip of the powder down barb breaks off as it grows, producing a fine talc that spreads over the bird. This dusting of feather fragments gives herons and some pigeons a characteristic soft or matte appearance. Powder down may have more than a cosmetic role, however. Herons have been seen using it to mop up fish slime from their feathers. Mixing powder down with the slime thickens it, so that the heron can more easily preen it away with its beak.

Protection of the body includes defending it from attack by microscopic organisms, but feathers would seem to be a poor design for such a function. They create a warm, moist space next to the skin that could be an ideal incubator for spores. Wild birds rarely suffer from skin diseases, however. The chemicals in the sebum include an array of antibacterial and antifungal agents that help the skin stay healthy.

The useful properties of sebum are quickly destroyed by contact with petroleum and other oils. The rehabilitation manuals at animal rescue centres refer to "recovery of waterproofing" as one of the main goals of treatment. Unfortunately, the role of sebum in feathers is much more complex than just waterproofing, and rehabilitation specialists have yet to discover a practical way to restore the integrity of chemically damaged plumage. In some cases, they have kept an oiled bird in captivity for months before it appeared normal. In spite of the best efforts of rescuers, such patients may never make a full recovery. There are no records of oiled seabirds rejoining breeding colonies after a lengthy period in care.

The natural renewal of the plumage during moulting is a relatively simple cyclical phenomenon controlled by hormones. Old feathers drop out and new ones grow from the same follicle on a schedule that is tied to other cyclical events, such as reproduction. In the Northern Hemisphere, most birds moult just after the breeding season or in the autumn so that the insulating layer is new and complete for the winter. Many birds change colour at this point, as though they were adopting some sort of camouflage. Ptarmigan become white to match the snow; marine birds, such as murres, guillemots, loons, and grebes, become a uniform grey above and white below. Late in winter, the moult begins again and birds grow feathers that signal their reproductive status, in time for the breeding season.

Feather moult is more about wear and tear than colour change, however. Near the end of the useful life of its wing feathers, a bird is forced to expend extra energy in flight because the capacity of its wings to generate propulsion and lift decreases as the tips of the feathers wear away. Many forest birds and soaring birds make a smooth transition from old to new plumage by moulting a few feathers at a time, but waterbirds and diving birds usually drop all their flight feathers at once and become flightless for a few weeks. Ducks and geese are so safe in the water that they schedule their flightless period to match the development of wing feathers in their young. The family group gains its flight ability together, in time to migrate before winter sets in. Adult loons, grebes, and sea ducks delay the flightless period until they return to the sea for the winter.

Auks use their wings underwater as well as in the air, and their feathers must withstand the combined onslaught of solar radiation and salt water. When the complete wing moult makes them flightless, the remaining stub may actually become more efficient underwater because the long primary feathers are no longer there to generate drag. In effect, the auk's wing becomes a simple penguin-like paddle.

Feathers and Mechanical Protection

No single responsibility of the plumage is more important than another. They are all fundamental to the bird's existence. Protecting the body from physical damage is perhaps the most straightforward and least complicated function. It is also probably the benefit conferred on ancient birds by feathers before they became useful for flight.

Plumage protects the body by forming a springy cushion that reduces the impact of sudden blows and may even prevent the penetration of sharp objects, such as the beak or claw of a predator. All feathers point backward, making it easy for a bird to slip between thorns and twigs as it dives through shrubbery. Even if the plumage is not dense enough to protect against the talons of a hawk or the claws of a cat, a predator must learn to make an accurate attack on the central core of a bird if it is going to come away with more than a few inedible feathers.

The resilience of feathers is particularly useful in a fall. Air resistance or drag prevents a light object from picking up speed in a long drop, and the halo of feathers decreases the bird's overall density and reduces the impact of landing. This is especially important for young birds that cannot fly. The downy nestlings of waterfowl that nest on cliffs or in tree cavities depend on the ability to bounce. They are expected to leap from high nests within a few hours of hatching, with only a thick layer of down and soft, flexible bones to protect them from injury. A soft landing is also important to young forest birds during their first experiments with flight.

Feathers and Temperature Control

Warm-blooded animals depend on body heat generated by metabolic processes to maintain a high level of activity regardless of environmental conditions. Unfortunately, this heat diffuses into the external environment unless the animal is properly insulated. Birds are entirely dependent on their plumage for such insulation. It is so effective that they can afford to spend long periods incubating a clutch of eggs, and some species are able live most of their lives swimming in ice-cold water. One theory for the development of feathers among the non-flying ancestors of birds is that there was a benefit to be gained from the effective insulation provided by plumage.

Birds are rather small for warm-blooded animals and enhance the insulating effect of feathers with other features. Perhaps the most important is the distinctive shape of the body. The muscle mass has been centralized to create a nearly spherical body. A ball is the most efficient shape for heat retention because it slows heat radiation by minimizing the surface-to-volume ratio. It also makes it easy for the slightly curved feathers to form an uninterrupted insulating layer. Birds enhance the effectiveness of their shape by adopting characteristic resting postures that include pulling the neck, wings, and legs close to the body. Many species also tuck the head and feet against the body when they are in flight. Usually, none of the exposed parts are fleshy. They are either armoured, like the beak, or sticklike and scaly, like the toes and lower leg.

You might expect simple physics to force such a small animal to be tolerant of temperature changes but, surprisingly, birds are very sensitive to variations in body temperature. A bird typically operates at a higher body temperature than a mammal of similar size but it tolerates a much narrower range. If the body temperature varies, critical physiological processes such as brain function may cease. Too much or too little heat and a bird will collapse until conditions improve or it dies. For some, collapsing into torpor is a regular part of their daily lives and a successful survival strategy. Very small swifts and hummingbirds find it particularly difficult to maintain body temperatures during periods when they cannot feed. They enter torpor during the cold of night, with just enough energy in reserve to wake them to the warmth of the sun in the morning.

In the 1960s, military and civilian researchers tried to exploit the bird's sensitivity to temperature change to reduce the number of birds that collided with high-speed aircraft. They had seen that birds passed out when irradiated with microwaves and speculated that feathers acted as miniature radio crystals, discharging small but disabling amounts of electricity into the bird's muscles. Experiments with shaved chickens eventually determined that the microwaves actually warmed the bird's brain to the point where it simply ceased to function.

The plan was to render birds unconscious with the beam from on-board radar. Birds would simply fall out of the flight path of the aircraft and, it was hoped, reawaken before they hit the ground [10, 207, 208]. Unfortunately for the engineers, radar transmitters strong enough to disable birds were too heavy for use in aircraft and the project was abandoned. Designers turned to the construction of stronger compressor blades in the jet engines and housed the engines in external sponsons that were isolated from vulnerable areas of the aircraft's body. Governments all over the world, however, have built their biggest airports on estuaries where flocks of large, heavy geese and other migrants stop for a rest. Birds remain a major threat to jets and turboprop aircraft at such sites, and local hazard reduction programs often kill them in large numbers.

Effective insulation and a compact body do not answer all of a bird's temperature control requirements. Living within a cloud of feathers poses serious problems for many species. Fortunately, plumage is not a simple, impermeable shell and is able to do more than insulate. It is part of a dynamic, integrated system that has other tools beyond the basic insulating effect of dead air, including an elaborate heat control system that responds to the bird's internal temperature. Each feather has its own set of muscles and nerve endings, which enables patches of feathers to be consciously raised or lowered. This adjustment changes the effective thickness of the plumage and varies the rate at which heat radiates from various parts of the body. In the heat of summer, urban sparrows look lean and scrawny in spite of abundant food. In the winter cold, they fluff themselves up for warmth and look chubby even when food is scarce.

The ability to increase apparent body size is also a useful defence tactic. Owls are particularly well known for dramatic displays that involve erect feathers, open wings, and unnerving hisses. They are dangerous birds and are treated as a threat by many other species. The American Bittern has a trick that exploits that reputation. When threatened, it bends over backward to look like the face of a giant owl by exposing large eyespots on its wing.

Although the physiology of a bird's central nervous system appears to require a high degree of temperature stability, the feathers on the head and along the backbone appear too small to be effective insulators. A layer of air trapped between the microscopically thin plates of bone that surround the brain and in the hollow spaces along the vertebrae may help protect the most sensitive areas. Long neural spines on the thoracic vertebrae also increase the distance between the spinal cord and the bird's outer surface. Most birds lack neural spines on the sacrum, and that segment of the spinal cord often lies close to the surface. Eagles, owls, and a few other birds that lurk in the cold for long hours have hipbones with a series of greatly inflated hollow chambers over the spinal cord that submerge the nerves deep within the body (Figure 2.12).

Birds live in a variety of climates, and it is often as difficult to cool down as it is to keep warm. Birds cannot lose heat by sweating, and even in cool climates, intensive muscle activity during flight may generate more heat than mere feather manipulation can deal with. They typically solve this problem by modifying existing structures to serve as radiating surfaces.

Hawks and loons use an energetically expensive form of flapping flight, and the action of their large breast muscles generates a great deal of excess heat. They are, however, able to transfer the heat into air sacs that run through the pectoral muscles and into an exceptionally large and hollow furcula. Heat from the muscles radiates into the air spaces and is carried out of the body as the bird exhales. Other birds exploit their featherless legs and feet as radiators. If the surrounding air or water is cool enough, they can pump blood into their feet, where the naked skin allows heat to radiate from the body.

Birds are also able to change their behaviour if they get too hot. Like other animals, they seek shade, pant, expose patches of bare skin, or excrete body fluids. Some overheated birds flutter the skin under their throat to cool down by increasing airflow over moist tissues. New World vultures and cranes apparently decrease their internal temperature by defecating on their legs to take advantage of the cooling effect of evaporation. Some waterbirds also use evaporation, and intentionally wet patches of their plumage in an effort to cool down.

Many ostriches live in tropical deserts, where keeping cool can be a serious problem. Nonetheless, they are able to withstand long periods of very high temperatures. The ostrich has a unique physiology that enables it to pant for long periods without being overcome by acidosis. In acidosis, the concentration of electrolytes in the blood changes to the point where it interferes with normal physiological processes.

An ostrich also gets help from its fluffy plumage. The large feathers are ideal for catching the heat of the sun and keeping it away from the bird's body. They also help increase the animal's profile. Ostriches are one of the few birds big enough to cool off in their own shadow.

The ability to cool off in your own shadow is not a trivial trick. It enables penguins to nest on equatorial islands such as the Galapagos. Penguins share much of their breeding range and some of the breeding sites with nesting albatrosses, another very large seabird. Both leave their young for long periods, during which the nestlings must keep control of their body temperature. In Antarctic and Sub-Antarctic regions, it is a matter of staying warm; at the equator, it is a matter of staying cool. The young of both birds are covered in an exceptionally deep layer of loose, ostrich-like down that makes them look more than twice their real size. The down effectively prevents heat transfer through the body surface. In the polar regions, it prevents heat loss; in the tropics, it prevents the radiant energy of the sun from overwhelming the body's other defences against overheating. If there is shade from shrubs or rocks, the young will move into it. Usually, however, the chicks simply sit still and do as little as possible.

Young albatrosses have one big advantage over the neighbouring penguins. Their webbed feet are huge and effective radiators. The chicks simply rock back on their heels (their real heels, the joint between the tarsometatarsus and the tibiotarsus) and lift both feet into the air. Only the small patch of heel is in contact with the hot ground, and both surfaces of the web are exposed to such air movements as there might be. Penguins use the same trick to raise their toes off the ice in the brutal cold of the Antarctic winter.

In polar and temperate regions, birds usually face the more familiar problem of retaining heat. One simple method is to have a large body mass. Mammals in cold climates increase their mass with heavy layers of insulating fat. Penguins, which live at temperatures below those tolerated by most mammals, are also able to store extravagant layers of heavy fat, because they are flightless and do not need to worry about being too heavy get into the air. The extra weight may be of help in over-coming buoyancy when penguins dive to great depths beneath the sea. Although fat floats in water, it is much denser than the air that occupies the internal spaces of most birds. Fat also insulates against the extreme cold of ocean depths, and con-tains additional myoglobin, which acts as an oxygen source during long dives. Auks and loons of the Northern Hemisphere look somewhat like penguins, and they also store fat in layers beneath their skin. They have retained the ability to fly, how-ever, and their fat reserves are much less extravagant. They must eat regularly even in winter, and depend on reliable supplies of food.

Feathers and Communication

All animals need to be able to communicate with their own kind, and often with potential predators or competitors. Communication in birds and mammals is a very complicated process, and the two groups have very different strategies.

Mammals are typically much more secretive than birds, and are usually not very colourful. When a specific signal is needed, it usually consists of a white patch of fur that contrasts with the animal's darker background coloration. Mammals tend to advertise their presence with relatively subtle modifications of their environment that reflect their close interaction with it. They litter their sur-roundings with chemical markers, urine, feces, shed hair, scratched trees, wallows, beds, and dens that are often intended as messages to members of their own species.

Birds tend to spend less time in any one place and are often irregular in their habits. They rarely mark their territory. They may advertise their personal presence with showy displays of bright colours but tend to be very circumspect about where they have been in the recent past. Birds drop large numbers of feathers over the year but do not appear to use them for communication. Perhaps feathers blow around too easily to be useful as calling cards. Birds never use chemical or scent markers; often, the nest is the only physical evidence of an occupied territory. Even the obvious excavations in tree trunks left by woodpeckers seem more interesting to squirrels and insects than to other woodpeckers. Colonial birds may be one of the few groups to advertise by using great splashes of white wash around breeding areas to attract new recruits. Solitary species are more circumspect. They are careful to deposit fecal material far from the nest site, but otherwise seem very casual about the disposal of their personal waste. When you are dropping it from a great height into someone else's habitat, there is little value in being careful.

One plausible explanation for the differences between mammalian and avian communication strategies is that the plumage wraps the inner bird in a kind of portable habitat. Most species have little need for a fixed den or abode outside of the nesting period, and frequently wander huge distances during migration. Displays, calls, and other forms of direct interpersonal communication are more reliable for such highly mobile animals.

The elaborate displays of birds have long attracted human interest and help explain our fascination with the group. Humans enjoy spectacle, and the combination of visual and vocal displays is the bread and butter of the burgeoning bird-watching industry. Mammal spotting, except at sea, has not become a widespread pastime. Humans are ill equipped to appreciate the sensory subtleties of urine trails, scent marks, and other features left by voles or other small mammals in their natural habitat. Small mammals also tend to be nocturnal and nearly invisible, whereas birds, in colourful dress, are conveniently active during daylight hours. Humans now control so much of the world's habitat that a bird's charisma might be a significant evolutionary advantage when it attracts the attention of conservationists. Avian displays intended to attract their own species may have become even more valuable because of their unintended effect on a totally unrelated class of animal.

Feather Growth

Feathers share some characteristics with the hair of a mammal but they are constructed from beta-keratin and are not derived from primitive scales. Mammalian hair, on the other hand, is constructed from alpha-keratin and is apparently derived from scales. Feathers are also far more elaborate than any keratinous structure in other animals. Their production is an outstanding example of the complexity that can be achieved by processes working at the molecular level [167].

The 150 genes involved in the production of keratins [98] control complex activities that take place as the feather grows in its follicle. The follicle is not a simple pit. It begins as a small cluster of cells on the skin when the embryo is only 12 days old, and becomes a deep pit with the fibre-producing cells in its base. Each fibre is secreted by one cell in a ring of special cells that form a collar in the base of the follicle. The individual fibre-bearing cells are pushed upward, away from the nutrients in the base, and eventually out of the follicle altogether. The keratin is left behind as the cells die.

The collar is a complex structure with several layers. Its outer layer produces the thin sheath that is shed as the feather matures but houses the feather while it is growing. The middle layer is organized in a series of longitudinal ridges. Differential development of the cells in these ridges produces barbules, barbs, and the tubular rachis in the central rib of the feather. The cells are not arranged directly above one another but slant towards a midline on what will become the outer surface of the feather. As the cells divide, they are pushed along a helical track, screwing outward in a gentle curve to their eventual death. The keratin from cells nearest the midline fuses to form the central tubular rachis. Cells that arrive later contribute to the other components of the feather.

Because fibres of different length have the same age, it is easy for the follicle to apply simple bars and stripes of colour across a feather. A similar process may incidentally record other events in a bird's life, and feathers frequently have "wear bars" running straight across their surface. These patches represent discontinuities in the fibre's development during a period of stress, such as a food shortage. If the event affects a nestling, the wear bars in the tail will line up because all the feathers grew at the same rate. Such alignment is very useful later, when bird-banders need to identify full-grown birds as young-of-the-year. In adults, the wear bars are staggered because the feathers grew at different times during the moult period.

The working surface of a feather consists of flat vanes on either side of the central rachis. In the contour feathers that cover the body, the left and right halves are usually roughly equal in width. The same can also be said of most of the tail feathers (rectrices) and the large feathers on the inner part of the wing (secondaries). The outer wing feathers, or primaries, are closely associated with the power of flight and are decidedly asymmetrical. The inner (aft) vane of a primary feather is much broader than the outer (forward) part, and the whole feather is slightly curved. This gives each primary the appearance of a small wing in itself. The trailing edge of the primary may be smooth and uninterrupted, but in many species, part of the outer edge is cut away so that the tip is narrower than the main body. A cutaway shape usually indicates that the bird is very fast or capable of exploiting some special technique, such as thermal soaring. The curved asymmetry of the primary feathers is usually attributed to the demands of flight. As we shall see below, the primary feathers of other feathered but non-flying animals may also be curved.

Evolutionary pressures have ensured that the vanes of feathers are sufficiently springy and the rachis sufficiently strong to avoid collapse under the day-to-day stresses of a bird's life. Flight is an extremely demanding form of locomotion. Although feathers drive the bird through the air, their shapes are finely tuned to the demands of efficiency and saving energy. Occasionally, we see birds with a few miscoloured feathers, but misshapen, wrongly sized, or poorly fitting feathers probably lead to an early failure in situations that challenge the bird's ability to escape predators and are rarely seen in the wild.

In modern birds, there are only four basic types of plumage. The flightless ratites make do with basic body coverings that include very few feathers with specialized shapes. Nonetheless, their feathers range from the simple hairlike structures of the kiwi to fluffy plumes in the ostrich. The cassowary also has somewhat hairlike feathers, but its primary feathers are highly specialized. They lack vanes and have become long, robust spines that protect the bird's flanks.

Flying birds offer a sharp contrast to the ratites in having a rather large number of specialized feathers. They come in two basic types. Soft and flexible plumage is usually found in terrestrial birds, including both small forest species and larger soaring birds that inhabit open areas. Typically, it has a small number of relatively large feathers. Ducks and seabirds tend to have densely packed and firm plumage that usually consists of a very large number of small feathers. The use of this type is carried to an extreme in penguins, where the feathers are so tightly packed that they create a waterproof surface that looks much like the scaly covering of ancient reptiles.

The similarity of feathers among both modern birds and their distant ancestors suggests that evolutionary constraints on feathers have always been strong. A few species have developed showy plumes to advertise sexual prowess but these usually fold away so that they do not interfere with flight. In exceptions, such as the peacock and widowbird, exaggerated tail feathers that greatly increase the costs of flight would be a severe evolutionary disadvantage if they did not enable the bird to show off its strength to a potential mate. The Pennant-winged and Standard-winged Nightjars of Africa have elongate plumes that trail from their wings, but this display is unique. Once birds were committed to flight, there appears to have been little room for experimentation with the basic shape of feathers that had important locomotory duties.

The production and maintenance of feathers consume more of a bird's resources in terms of time and energy than any other activity, with the possible exception of reproduction. Researchers, who go to great lengths to determine the scale of parental investment or lifetime reproductive effort, rarely give it a second thought, however. As a result, there is very little information on how the moult cycle affects the bird's life, and feathers were rarely discussed in avian evolution. All that has changed with the discovery of fossils of feathered dinosaurs.

Feathered Dinosaurs

In 1997, two specimens of a bizarre and spectacular animal were discovered in China. They showed just how difficult it could be to distinguish between birdlike dinosaurs and dinosaurlike birds. This creature was not just another dinosaur with hints of integumentary structures. It had strongly vaned plumes on the hands and tail and was named *Caudipteryx zoui* meaning "Vice Premier Zou Jiahua's feathered tail" [101]. Mr. Zou had provided much-needed support for the scientific paleontological excavations in Liaoning, China.

The feathers caused considerable controversy. Was this animal a feathered dinosaur or an early bird that had abandoned flight? Although the feathers on the "wing" of *Caudipteryx* appear robust enough for use in flight, they are far too small to have lifted such a large animal off the ground. The tail feathers also appear too small to be of much use. The form a frond, like a shortened version of the tail in *Archaeopteryx,* and appear to be striped. It is unclear whether the marks were due to the preservation of pigments or to the effects of localized structural variation, analogous to the wear bars found in the tails of modern birds.

Although *Caudipteryx* was originally classified as a maniraptoran dinosaur, the presence of vaned feathers encouraged the view that it might be a flightless bird. There were teeth in its jaws, but this is not uncommon among ancient birds and it appeared to have a whole suite of other seemingly avian characteristics. Some of these features were poorly preserved and could be interpreted in several ways, while others were shared by both birds and advanced theropods. In the end, it was the comprehensive logic of a cladistic analysis that placed *Caudipteryx* near the equally strange *Oviraptor,* among other maniraptoran dinosaurs. There are now 10 specimens to clarify the less certain features, and there appears to be little reason to look at *Caudipteryx* as a potential bird.

Oviraptor had enjoyed a brief period of notoriety as an unusually birdlike dinosaur that specialized in the predation of dinosaur eggs. Egg eating was an inviting interpretation of its strange, toothless jaws, and the first fossil was discovered in the remains of a nest that it had apparently been raiding. Later, the fossil scene was reinterpreted as the tragic death of an incubating parent, and the kind of food handled by the well-muscled, horny jaws remained a mystery. It has been suggested that they could crush mollusks, but modern, beaked birds that eat mollusks swallow them whole, and no likely mollusk remains or those of any other prey have been found within its fossils. Except for its peculiar jaws, *Oviraptor* is relatively unexceptional as a maniraptoran theropod.

The position of the featherless oviraptor suggested a possible purpose for the short plumes on *Caudipteryx.* Thomas Hopp and Mark Orsen suggested that they could have had an early importance in care of the young [87]. Modern birds "brood" their young by spreading the wings and tail as a protective shield. In one simple action, they help the nestlings retain body heat, deflect the wind, and hide the young from passing predators. It is a posture that takes advantage of the characteristic ability of both birds and theropods to fold their wrists to the side (protraction). Some distant feathered ancestors may have enhanced the protection by increasing length of the wing feathers and developing of a flap of feathered skin (propatagium) between the shoulder and the wrist.

The posture of the oviraptor that died in its nest supports such an idea [142]. The eggs are cradled between the forelimb and the body, just as in a modern bird. Oviraptors are now among the theropods believed to have had feathers, although none were discovered in that particular fossil. Without wing plumes, its eggs would have been quite exposed, while even short plumes could have provided a useful amount of cover.

Caudipteryx is not much more birdlike than *Oviraptor,* in spite of its feathered "wings," and even those feathers are not exactly like a bird's. In *Archaeopteryx* and all modern birds, the outer vane of the wing feathers is narrower than the inner, and the feathers are aerodynamically curved. In *Caudipteryx,* the feathers appear to be more symmetrical and straight. They are also organized on the arm in a very non-avian way. The longest is closest to the body and the shortest is the furthest out on the hand. This peculiar arrangement suggests that the most important duty of the feathers was to protect the flanks from abrasion, or perhaps insulate them from heat loss when the animal was roosting. The tail feathers are also simple, as they are in most birds, but it is difficult to suggest a likely function for them in a large flightless animal like *Caudipteryx.* Perhaps they were simply for show, but in a world where all parts of most animals are fleshy and vulnerable to attack, a distracting flag at the least vulnerable end could offer a unique and effective defence.

The skeleton of the tail is perhaps *Caudipteryx*'s most birdlike feature. It has 22 vertebrae, the same number as *Archaeopteryx* and 8 less than most other oviraptors. The individual caudal vertebrae do not become longer towards the tip, as they do in oviraptors. The vertebrae support 11 pairs of tail feathers, as in many modern birds. In one specimen, about two-thirds of the distal vertebrae form a stiff rod that looks somewhat like an incipient pygostyle, but the bones are not fused.

Other parts of *Caudipteryx* are more typical of maniraptoran theropods. The clavicles are fused into the broad, boomerang-shaped furcula found in *Archaeopteryx, Tyrannosaurus rex,* and the theropods. The wall of the abdomen is supported by pairs of slender floating bones called gastralia. Similar bones appear in the abdominal walls of all non-avian theropods, crocodilians, and many early birds.

Although the front limbs of *Caudipteryx* are long, they are shorter than those in other coelurosaurs. In modern flying birds, the ulna carries a series of nodes associated with the wing quills. There are none in *Caudipteryx,* but there are none in *Archaeopteryx* either. The pectoral girdle does not seem particularly well designed for a flapping stroke, and there is a pair of separate sternal plates, instead of a fused sternum, on the chest. The scapula is longer than the humerus, but it is always shorter than the humerus in flying birds.

When *Caudipteryx* was first discovered, it was exceptional because it had well-preserved and undeniable feathers, and the feathers led to debates about it being a bird. At about the same time, however, paleontologists discovered a number of dinosaur fossils with the remains of suggestively featherlike "integumentary structures." These were simple bristles whose lack of structure suggested that they might be artifacts of the fossilization process or structures from within the skin. Some of the bristles appeared to have a hollow core, which suggested a link to proto-feathers; the argument could not be settled until other fossils included enough detail for paleontologists to be certain that they represented feathers. Feathers have now joined the wishbone and other familiar avian characteristics as features with a wide distribution among ancient animals [144].

To be absolutely certain that the simple integumentary structures are truly a type of feather, we would need to know that they were produced by the same kind of cellular collar found in the follicles of modern feathers. It seems doubtful that we will ever find such detailed fossil material. In a spectacular example of the power of modern molecular techniques, however, Mary Schweitzer and colleagues elicited a response from antibodies for feather keratin applied to a fossil of the alvarezsaurid *Shuvuuia deserti* [186]. The reaction of the antibodies was restricted to the remains of what appeared to be tubular integumentary structures. As unlikely as it seems, the fossilization process apparently preserved fine details of the molecular structure of the original protein molecules after over 65 million years.

In their recent review of the distribution of integumentary structures, Mark Norell and Xing Xu [144] found reliable evidence for the occurrence of feathers in several lineages of dinosaur that all share one important characteristic. They are all rather small animals, with the longest being about 2.2 m. None of the skin impressions from larger species, such as *T. rex,* included anything like a feather. Such a distribution might be expected if insulation was an important role for feathers before the development of flight. Large animals operate at higher temperatures but lose heat more slowly than small ones [188].

On the Internet, many dinosaur websites reflect the belief that feathers were widespread among small dinosaurs. They might also have protected the young of large species, and downy nestlings of *T. rex* already appear in several museum displays. There is no direct evidence for that particular reconstruction, but *T. rex*'s pygmy relative *Dilong paradoxicus* had feathers. At 2 m from head to tail, it is still one of the larger feathered species.

Most of the fossil feathers and featherlike integumentary structures have been found on theropods but simple fibres and bristles appear in other lineages. If those structures also represent an early form of feather, their wide distribution suggests that feathers could have a very ancient ancestry that has yet to be uncovered in the fossil record.

The psittacosaur was a dog-sized dinosaur related to the much larger *Triceratops* and only very distantly related to theropods. Its tail appears to have been decorated with a sparse array of 16 cm bristles that may have been tubular, like a feather. It is not too difficult to imagine a psittacosaur with its vulnerable head thrust down a burrow while it thrashed about with a bristly tail to discourage some predator. The bristles are not as robust as those of the modern porcupine, but they might have helped.

Some sort of fur or integumentary structure has long been proposed for the flying pterosaurs, which are only distantly related to dinosaurs through their common heritage as archosaurs. There is now firm evidence that at least some of them were covered with hairlike structures. These may have been comparable to the simple bristles found on several types of dinosaurs, but there is no suggestion that they were actual feathers.

All the other feathered dinosaurs are representatives of the theropods, including some that appear to be closely related to birds (Figure 7.1). Birdlike feathers have been found in fossils assigned to oviraptors, dromaeosaurs, and *Protarchaeopteryx*. The latter fossil is fragmentary and has not been assigned to any particular lineage. As its name suggests, it might be a later relative of *Archaeopteryx*, the earliest accepted bird. The simpler bristlelike structures have been found on *Sinosauropteryx*, a compsognath that was one of the earliest feathered dinosaurs to be discovered; on *Dilong*, the tyrannosaurid mentioned above; and on *Beipiaosaurus*, a therizinosaur. None of the troodontid fossils have shown integumentary structures that could be feathers.

Long and suggestive integumentary structures have been preserved on the forearms of a fossil named *Yixianosaurus longimanus*. Unfortunately, these structures lack sufficient detail to indicate whether or not they are birdlike feathers, and the fossil itself has not been assigned to any particular lineage. Its species name is derived from its most unusual characteristic. The second to last bones in its fingers are unusually long, suggesting that it might have moved through the forest canopy like a lemur.

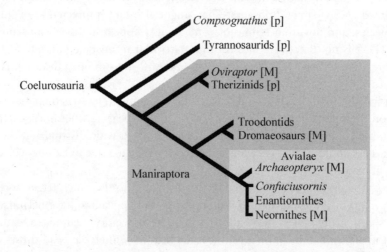

Figure 7.1   The distribution of protofeathers [p] and modern-looking feathers [M] in a genealogical relationship for advanced theropod dinosaurs proposed by Norell and Xu [144]. Protofeathers may also have been present in some ceratopsian dinosaurs.

As the feathered fossils were generating a great deal of excitement in the paleontological community, there was also excitement in the marketplace. For several years, fossils of feathered dinosaurs were bought and sold for very high prices on the international market, and a great many were excavated by non-professionals. One direct result of this interest was the disappearance of significant specimens into private collections. It would have been better if they had stayed in the ground. Only specimens available to the scientific community as a whole can be used as part of evolutionary debates.

Market interest also led to the production of fakes. Even though fakery, and accusations of fakery, have a long and colourful history in paleontology, the recent infamous *"Archaeoraptor"* took the paleontological community by surprise. Until the market developed for feathered fossils, there was little reward in the creation of sophisticated frauds, and most attempts were rather crude. *"Archaeoraptor"* was rather cleverly concocted by merging plausible specimens, and only uncovered by the kinds of forensic techniques used to investigate fine art. In spite of widespread embarrassment, no lasting damage was done and the paleontological community was put on guard – until next time.

The review by Mark Norell and Xing Xu might be described as the mainstream or conservative opinion about feathered dinosaurs [144]. Alan Feduccia and other opponents of a close relationship between birds and dinosaurs remain critical but have not produced a comprehensive explanation of their position. Gregory Paul [158] and Stephan Czerkas [42] have published descriptions of alternative theories on the link between birds and dinosaurs.

Gregory Paul's position is that some advanced theropods were flightless descendants of ancestral birds. It is an intriguing concept, referred to by early proponents as "birds came first." Just a few years ago, it would have been difficult to challenge such an opinion; today, however, between the development of advanced cladistic techniques and the successful placement of many new fossils in a cladistic context, it has become very difficult to defend. Fortunately, there is much more to Paul's book than this particular theory. Because it was thoroughly reviewed by several paleontologists before publication, it remains a well-written and valuable guide to the characteristics of theropod dinosaurs. It also includes a very large number of useful and carefully prepared illustrations, many of which appear in other publications.

The contribution of Stephan Czerkas to the discussion of feathered dinosaurs is more problematic. It appears as four papers in *Feathered dinosaurs and the origin of flight,* a publication of the Dinosaur Museum in Utah. Each paper is accompanied by an acknowledgment of professional reviewers, and in that sense they appear to be part of the professional literature. Their content, however, is not as disciplined as that of more familiar journals, and their tone is sometimes argumentative and critical of other paleontologists. They read somewhat like the technical discussions posted on the Internet, and lack a tone of detached objectivity characteristic of publications with a longer history. Most importantly, the papers are rarely cited by other paleontologists.

The papers in *Feathered dinosaurs* also illustrate a serious problem in understanding the recently discovered feathered dinosaurs – synonyms. *Archaeovolans* is apparently a synonym for both *Yanornis martini* and, possibly, *Arctosaurus; Omnivoropteryx* may be close to *Caudipteryx; Cryptovolans* is apparently a junior synonym for *Microraptor;* and *Scansoriopteryx* is probably a synonym for *Epidendrosaurus.* Czerkas' papers describe and name new fossils. Each is based on a carefully documented specimen, amply illustrated and described, and compared with other similar fossils [42]. It is not clear, however, that these fossils are worthy of a distinct name. There are no data matrices that could be used to place these fossils in a cladistic context with other specimens, and they may be too fragmentary for such an analysis. The tradition of giving a name to every newly discovered fossil, as though it were a new species, has clearly outlived its usefulness in paleontology.

Czerkas wishes to convince his readers that feathered dromaeosaurs could fly, but statements such as "dromaeosaurs were already birds and not the non-avian theropod dinosaurs as previously believed" or "The discovery of primary flight feathers in *Cryptovolans* demonstrates that these small dromaeosaurs were not the dinosaurian precursors of birds, but were in fact already birds which could fly" need far more support than that provided by the description of a few new fossils. The main technical argument against such claims is the absence of structures in the body that would support the musculature needed for flight. The scrawny *Archaeopteryx* flew, assuming it could fly, because its wings were large enough to provide lift when moved slowly. The feathered dromaeosaurs had much smaller wings and would have needed large muscles to flap them very quickly. Large muscles also need large bony attachments, for which there is no evidence in these fossils.

The interpretation of feathered dromaeosaurs as birds draws attention to a significant point about the inner-bird strategy. There is little or no sign of birdlike architecture among the reptilian theropods (or in the lanky structure of *Archaeopteryx,* for that matter). The forelimb skeleton of the dromaeosaurs shows that they had large, well-muscled arms suited to the capture of prey. There is no indication in the skeleton that the limb muscles were concentrated in a central mass and operated by elongated tendons. The large, heavy hands and claws of dromaeosaurs would have built up considerable momentum when moved quickly, and would have been very effective in disabling prey. The inner bird's concentration of muscles in a central mass and reduced weight in the extremities are characteristics of modern birds with large wings, wide hips, and a short tail.

Archaeopteryx and Its Relatives
Reptilian architecture separates the feathered dromaeosaurs from modern birds. This raises the question of what characters separate the rather reptilian *Archaeopteryx* from so-called birdlike dinosaurs. When the first specimen was described in 1861, only the impressions of feathers convinced the scientific community that

Archaeopteryx was a bird. Without plumage, it could easily have been mistaken for a relative of *Compsognathus*, a primitive theropod found in the same deposits. An astute observer like Thomas Huxley might have recognized the fossil's true character, but without feathers it might never have been brought to his attention. One specimen was misidentified as a pterosaur and gathered dust for decades until the keen eyes of John Ostrom spotted its theropod-like wrists. Fate saved the Victorians from an even more difficult puzzle. The recent flood of small feathered dinosaurs suggests that it was only a matter of chance that the fossil of *Compsognathus* lacked feathers.

Perhaps *Archaeopteryx*'s single most significant characteristic is its great age. The known fossils of feathered dromaeosaurs all date from the Upper Cretaceous, whereas *Archaeopteryx* is from Jurassic rocks some 10-25 million years older. The best that can be said about this situation is that all others are worse. Hypothetical relationships to other reptilian lineages are even more convoluted and create even greater temporal incongruities. Analyses of hypothetical phylogenies seem a particularly obtuse kind of evidence, but there are simply too few fossils of Jurassic vertebrates to clarify the situation [14]. The fossil of *Protavis* is another 25 million years older than *Archaeopteryx*, but it is based on fragments of a partial skeleton and few researchers are confident that the remains represent a bird [24]. The pieces were not handled properly after their discovery in 1983 and were not completely described until 1997. There is considerable doubt that all the fragments belong to the same animal. Although its head has several avian characteristics, *Protavis* is not likely to play an important role in the story of avian evolution until less controversial material is found.[1] With that single exception, *Archaeopteryx* remains unchallenged as the oldest bird and the earliest-known animal to fly on feathered wings.

In contrast to *Protavis*, the fossils of *Archaeopteryx* have been accepted as a bird since the first was discovered 150 years ago, and the species has been discussed and interpreted in dozens of books. It has space in every major textbook and Pat Shipman has written an excellent review of its story that should be read by everyone with an interest in birds [191].

Archaeopteryx benefited from being discovered soon after the publication of Darwin's theory of evolution and from appearing to be a convincing link between birds and reptiles. Finding a way to relate it to more modern birds has proven very difficult, however, and there is still much to learn from its fossils. Nearly all of the most birdlike features of *Archaeopteryx* are in the head. It is as though the avian head evolved first and specialized adaptations in the body followed later. In his chapter in *Mesozoic birds*, Andrzej Elzanowski points out that the head of *Archaeopteryx*, in the Jurassic, contained features that are actually more advanced than those of later birds such as *Confuciusornis*, the startlingly abundant fossil from the Cretaceous [56]. If *Archaeopteryx* was the direct ancestor of the Confucius birds, the features of

---

[1] Perhaps the most even-handed discussion of *Protavis* is contained in Lawrence Witmer's foreword to Sankar Chatterjee's book [24].

the later bird would need to have reverted to a more primitive condition. It is more likely that both species shared an earlier common ancestor that lacked the specialized features seen in *Archaeopteryx*.

With the discovery of so many feathered dinosaurs, the question arises: "How do we know that the rather reptilian *Archaeopteryx* is a bird and not just another feathered dinosaur?" Except for the feathers, the most obvious avian characteristic of *Archaeopteryx* is its brain. The volume of the brain is 1.6 mL, or about three times larger than the brain of a chicken or pigeon [2]. The range of brain volume is large in modern birds, however, exceeding 8.0 mL in parrots and crows. *Archaeopteryx*'s brain is comparable in size to brains found in Jurassic mammals, and about three times larger than the brain of coelurosaurs. It suggests that birds had an early requirement for the ability to process large amounts of sensory data. Teeth are not a feature one associates with modern birds, but we know about typical bird teeth from Cretaceous fossils of *Hesperornis* and other toothed birds.

To say that the brain of *Archaeopteryx* is more like a bird's than a dinosaur's, we need to examine its organization and structure. None of the soft tissue survives but fortunately the braincase was extremely close-fitting, as it is in modern animals, and its inner surface reveals the general shape of the brain's major components. Patricio Alonso and his team were able to make a rubber cast of the braincase of the London specimen of *Archaeopteryx* and reconstruct many of the brain's features [2].

The cerebral hemispheres and cerebellum are exceptionally large but not quite as large as those of modern birds. For instance, the cerebral hemispheres do not completely envelop the large olfactory tracts. More importantly, the optic lobes are almost as large as the cerebellum and are mounted on the sides of the cerebellum, where they are well separated from each other. In dinosaurs, the optic lobes are smaller and lie above the brain. John Maynard-Smith predicted that advanced neural connections would be necessary in an ancestral bird [125], and the large optic lobes suggest that *Archaeopteryx* was able to process the large quantities of visual information that birds collect while flying.

Casts of the inner ear of *Archaeopteryx* also reveal a specialized structure. The semicircular canals are large and arranged much like those of a modern bird, suggesting that sonic stimuli may have played an important role in the life of *Archaeopteryx*. In contrast, the large and somewhat primitive-looking olfactory lobes are not very birdlike and suggest that the sense of smell also remained important in *Archaeopteryx*. Taken as a set, however, the sensory structures suggest that *Archaeopteryx* could have had the complex neural integration necessary for aerial flight.

The remodelling of *Archaeopteryx*'s brain was already well advanced when it appeared in the Jurassic about 147 million yeas ago. Its ancestors must have begun the conversion from a more reptilian format much earlier, and made an evolutionary commitment to a unique strategy not seen in any other lineage of dinosaurs. Some aspects appear in the brains of the flying pterosaurs, but not all. Even at the end of the Cretaceous, when some theropods were experimenting with larger brains and more complex behaviour, none of *Archaeopteryx*'s supposed relatives attempted such a comprehensive redesign.

Elzanowski concludes that, taken together, the avian characteristics of the skull and the thorax are sufficient to place *Archaeopteryx* some distance from the theropod dinosaurs [56]. They are strong enough to overshadow more reptilian features in other parts of the *Archaeopteryx* skeleton. In particular, the apparent similarities of the hips and vertebrae are "extremely suspect," and the rearward tilt of the pelvis may be an example of convergence.

Only the skeleton of *Archaeopteryx*'s body suggests strong links to dinosaurs. The structure of the thoracic skeleton has features that can be interpreted as either bird-like or dinosaurlike. The glenoid in the shoulder opens to the side, as it does in birds but also in some advanced theropods, such as *Velociraptor*. The glenoid is the passage for the tendon of the pectoral muscle between the scapula and the coracoid, and it must face to the side to allow the kind of forelimb movements used by birds for flight. There is a furcula but it is the same shallow chevron seen in theropods, not the deep "U" of a more modern bird. There is a solid sternum as in a bird, but it is not keeled and it is fused to rather short coracoids. Coracoids tend to be free and rather long in later flying birds.

Some aspects of Elzanowski's review have already been overtaken by current events. At the end of 2005, the 10th specimen[2] of *Archaeopteryx* became available to the scientific community. Its existence had been known, but it remained unstudied in a private collection for many years. Some of its structural features bring *Archaeopteryx* even closer to the dinosaurs than previously suspected, especially to theropods such as *Velociraptor*. The first digit of the foot is not reversed as it is in perching birds, but is mounted in a primitive position on the side of the tibiotarsus; the second digit supports the large, extendable claw that is a trademark of *Velociraptor* and all the close relatives of *Deinonychus*. A similar claw occurs on *Rahonavis*, an apparent relative of *Archaeopteryx* from Madagascar [133].

The easily seen features of the foot are not the only peculiar features of this specimen, and not the only features that imply a much closer link between *Archaeopteryx* and the theropods. The palatine bone in the skull has four processes (tetraradiate) as in non-avian theropods, not three (triradiate) as in modern birds. Part of the lower leg is also like that of a non-avian theropod. Above the joint between the toes and the tarsometatarsus, there is an ascending process of the astragalus (as discussed in Chapter 2), not the slender "pretibial" bone found in modern neognath birds [133]. It is important to remember that theropods and primitive birds are generally very similar in structure and that such small but very specific differences in the bones are considered highly significant. This 10th *Archaeopteryx*

---

2 The urvogels, or dawn birds, whose fossils have been found in the Solnhofen limestone formations of Germany, are divided into eight skeletons of *Archaeopteryx*, a solitary feather, and a skeleton of *Wellnhoferia*. The latter species appears to have been adapted for running, but the two species are obviously closely related.

specimen clearly moves its relatives closer to the theropods, but it may force us to re-examine the position of *Archaeopteryx* in relation to other birds.

Although *Archaeopteryx* is accepted as the earliest bird, it is becoming difficult to see it as much more than another feathered dinosaur. As a bird, it appears to have some characteristics that are ill suited to flight, such as the long frondlike tail and fingers on the wing that interrupt smooth airflow. It also lacks features usually associated with flight, such as significant skeletal structures to support heavy flight muscles. One of its least avian features is the string of simple vertebrae that make up its long back and lead to rather small hipbones. There is no sign of the devices that help modern birds prevent twisting of the spine during flight.

The new *Archaeopteryx* fossil may help us understand three controversial fossils from the Cretaceous that might be its close relatives. As mentioned above, *Rahonavis* is the only bird from the Upper Cretaceous to share many important features with *Archaeopteryx*. It has the extendable claw, the long, frondlike tail, and a caudal projection on its pubic foot. On the other hand, it also has some characteristics of more advanced birds. Its vertebrae have saddle-shaped (heterocoelous) centra rather than the simple, flat faces found in *Archaeopteryx*, and its coracoid articulates with the scapula to form a flexible joint that anticipates the structure found in modern birds. The ulna has a series of small knobs to mark the attachment of the secondary feathers; these are absent from *Archaeopteryx*. *Rahonavis* also has six sacral vertebrae and five other taxonomic characters that it shares with more advanced birds.

Shenzhouraptor, or *Jeholornis*,[3] from Lower Cretaceous deposits in China appears to be a third close relative of *Archaeopteryx*. It has the long reptilian tail but other aspects of its skeleton are more modern and it may be closer to *Rahonavis*. The discoveries of *Shenzhouraptor* and *Rahonavis* suggest that birds with a body plan like that of *Archaeopteryx* could have been successful for a long time and over a broad geographic area in spite of their elementary flight capabilities. The late appearance of *Rahonavis* implies that birds of this reptilian format could compete with relatively modern birds in the Upper Cretaceous.

One of the non-avian theropods that seems to have a particularly close link to *Archaeopteryx* is the feathered *Protarchaeopteryx*. The specimen is too incomplete to assign to any particular lineage, but it has a very theropod-like set of serrated teeth. Its forelimbs are too short (70% of hind limb) to have supported flight, but other features are more birdlike. It has the flat sternum of *Archaeopteryx* and there are excavations in the sides of the vertebrae called "pleurocoels," suggesting that the air sacs invaded some bones to achieve a basic level of pneumatization. Just in front of the hip joint, there is a very birdlike feature in the form of a short process pointing down from the ilium. As in *Archaeopteryx*, its foot suggests an early stage in the development of an avian tarsometatarsus. The individual tarsal bones are clearly visible but they appear to be fused along their edge.

---

3 The near-simultaneous publication of descriptions for this fossil in two unrelated articles has led to some confusion. The name *Shenzhouraptor* appears to have precedence [102, 239].

A third potential relative of *Archaeopteryx* is the much larger *Unenlagia coma-huensis*. Its incomplete skeleton of was found in Patagonia in 1997 [145]. The fossils of *Unenlagia* show a shallow cup where the ends of its pubic bones meet, the glenoid in the shoulder faces to the side, and the scapula is like that of *Archaeopteryx*. Two structures in the hips, the extensive "apron" where the pubic bones meet and a "cranio-ventral" process on the ischium, are more typical of theropods. Some of *Unenlagia*'s features, such as its vertebrae and its leg proportions, make it look more like a dromaeosaur than a relative of *Archaeopteryx*. The relative lengths of the leg bones may be simply a matter of body size, however, and not an indicator of relationships to one group or the other. Shape and structure are more important, and *Unenlagia*'s femur is quite birdlike. The end near the body is small and the fourth trochanter, a ridge characteristic of the dromaeosaurs, is missing. Unfortunately, only 20 bones of this intriguing animal were preserved.

The Significance of Feathered Dinosaurs

Perhaps the most dramatic discovery among the feathered dromaeosaurs has been the four-winged *Microraptor*. The long plumes on its hind legs revived discussion of *Tetrapteryx*, a hypothetical type of ancestral bird that would have used all four legs for flight [9]. *Microraptor* also had a long, frondlike tail similar to that of *Archaeopteryx*. Four-winged flight is one of those ideas with superficial plausibility, but in fact there is no known action of the hind legs of birds that might generate lift, and there are no relict structures hinting at such a history in either the skeleton or the musculature of birds.

For a while, *Microraptor* was the smallest known arboreal dromaeosaur. It demonstrated that lineages of theropods other than birds were able to exploit the forest canopy and may also have followed their own evolutionary routes to alternative forms of flight. *Microraptor* lacked the specialized skeletal structures in the shoulder associated with powered flight and may well have glided from tree to tree like a modern flying squirrel. It probably represents the closest approach by the short-armed dromaeosaurs to true aerial flight.

Gliding appears in many different kinds of animals and probably offers some of the benefits of flight without specialized structures or high expenditures of energy [13]. It offers an easy route to a variety of food items that might be found on the ground. Every child and cat knows that climbing up a tree is a lot easier and faster than climbing down. Most animals must climb down backward, in a slow and clumsy manner. It would be even more difficult if your hands were covered with feathers. Only the squirrel goes up and down with equal ease because it has a special hip joint. It can rotate its hind legs to the point where the claws are reversed. There is no sign of the necessary hip structure for this feat in *Microraptor*.

In 2005, a team led by Eric Buffetaut reported the discovery of some eggs 11 mm wide and 18 mm long belonging to a tiny theropod that lived in Thailand during the Lower Cretaceous [19]. One of the eggs contained the remains of three tiny embryonic bones. Had a modern bird laid these eggs, it would not have been much bigger than a chickadee. Cretaceous animals generally laid proportionately small eggs, but the adult animal in this case must still have been considerably smaller than the crow-sized *Microraptor* and was likely another feathered dinosaur. Interestingly, the eggs have external knobs typical of the known eggs of dinosaurs, but combine it with a three-layered shell that is considered unique to birds. Perhaps the three-layered shell is another supposedly avian character that will soon lose its value.

It seems likely that many more feathered dinosaurs are going to be discovered in the near future. No matter how similar they are to *Archaeopteryx*, however, we are not going to lose sight of birds among the great mass of dinosaurs. Taxonomy is profoundly legalistic and the tail will wag the dog. "Aves" is an ancient name that has precedence over any of the Dinosauria, and if we truly cannot distinguish between *Archaeopteryx* and other feathered dinosaurs, it is the birds that will swallow the dinosaurs. *Deinonychus, Velociraptor,* and other non-avian theropods will become representatives of a sister lineage to the main avian line. The change would give some recognition to the role of *Archaeopteryx* as both an early bird and a very early, feathered dinosaur. The change would also offer some vindication to the "birds-came-first" theorists. The ideas of Gregory Paul and Stefan Czerkas would graduate from being speculations on the fringe of respectability to being prophetic insights that anticipated an important truth about the relationship between birds and dinosaurs.

Archaeopteryx on the Wing

Archaeopteryx may share many features with *Deinonychus* but its ability to become airborne on flapping wings makes it unique. Flight may have been its most out-standing behavioural feature but it could never have been accomplished without the structure that most clearly separates it from other dinosaurs. Compared with its contemporaries, *Archaeopteryx* had a very large brain, certainly comparable in size with that of any mammal of the time. Advanced coordination and control pro-vided by that brain may have facilitated flight and helped the lineage of *Archaeopteryx* gradually draw away from those of other theropods.

In spite of the feathers and wings on *Archaeopteryx*, its aerial capability has been challenged several times during the last 150 years. Feathers are expensive to produce and difficult to maintain, however. It seems highly unlikely that evolutionary pro-cesses would encourage such structures if they were not put to good use.

Sophisticated aerodynamic analyses of flight in *Archaeopteryx* began with the work of Derrick Yalden in 1971 and 1984 [235, 236]. He worked out ways to calculate the primary parameters – such as weight, wingspan, and wing area – from the rather contorted and incomplete fossils. He then used those figures to derive values for the flight parameters, such as wing loading (weight per unit of wing area), to make estimates of the bird's take-off and flight capability. In 1985, Siegfried Rietschel published a different reconstruction of the wing based on the fine detail of the

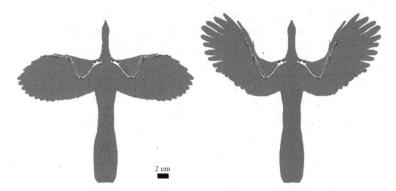

Figure 7.2 The forelimb skeleton superimposed on a recent reconstruction of *Archaeopteryx* (left) [56], showing how swinging the arms forward (pronation) would have increased the total surface area of the wing and provided greater lift (right). *Archaeopteryx* may have lacked the appropriate skeletal architecture to maintain the stiffness of its wings in such a posture, however.

feathers that could be seen in the Berlin specimen [178]. It increased the wing area by 58% and decreased the calculated wing loading by 36% (Table 7.1).

In 1984, the implications of both models were discussed at a major symposium on *Archaeopteryx* held in Eichstätt, Germany. The conference participants issued a synthesis statement describing *Archaeopteryx* as an active, cursorial predator that was facultatively arboreal. In other words, it lived on the ground but was capable of climbing trees. They also concluded that it was a glider capable of feeble flapping flight but not capable of taking off from the ground.

The Eichstätt meeting provided an opportunity for a broad spectrum of ornithologists and paleontologists to discuss a very complex problem. Its conclusions seem overly conservative, however, and may reflect the difficulties inherent in

Table 7.1

Parameters used for three models of flight in *Archaeopteryx*

Parameter	Yalden [236]	Rietschel [178]	Elzanowski [56]
Body weight (g)	250	250	276
Wing span (cm)	58	58	55
Wing area (cm²)	194	305.5	
Lifting plane (cm²)[a]	479	754	500
Wing loading (g per cm²)	0.64	0.33	0.55
Aspect ratio[b]	7.0	4.5	
Stalling speed (m per sec)	5.1-6.2	4.1-5.0	<8

[a] The lifting plane is the surface area of both wings plus the area of the body between them.
[b] The aspect ratio is calculated by dividing a circle of the circumference of the wingspan by the lifting plane (Figure 7.4).

communication between disciplines. The statements also appear to have been influenced by rather old-fashioned ideas about bird flight. Estimates of stalling speed
and the energy required for a take-off are based on the performance of modern
birds, whose body plans and weight distributions are very different from those of
Archaeopteryx. In fairness, aerodynamic models of bird flight were still rather primitive at the time of the conference, and struggled with basic problems. They were
still unable to account for the ability of the swan to become airborne. Jeremy
Rayner, who attended the meeting, was just beginning to apply the principles of
fluid dynamics to avian flight. Few ornithologists had been exposed to his ideas,
and even fewer understood their significance.

Since the Eichstätt conference, there have been significant advances in our understanding of avian flight, as well as new reconstructions of the wing and body of
Archaeopteryx (Figure 7.2) [56]. In spite of the reconstructions, the usefulness of
the wings is still challenged occasionally, while the tail has been described as more
of an encumbrance than a useful part of the flight apparatus. The structure of the
legs has also been re-examined; they are now considered to be more like those of
a walking animal, capable of running but not specifically designed for it. This has
contributed to the opinion that *Archaeopteryx* may not have been capable of sufficient ground speed for a running take-off.

In his comprehensive review of *Archaeopteryx* for *Mesozoic birds,* Andrzej Elzanowski
sees it as an animal that probably took off from convenient raised structures and
used the kinetic energy of the fall to gain air speed. It would then flap briefly to
gain a perch beyond the reach of any approaching predator. Flapping would be
slow and the flight line would be fairly straight. *Archaeopteryx*'s natural habitat does
not appear to have included any trees, but there were some shrubs tall enough to
provide a launch point and others tall enough to provide security. The shrubs do
not appear to have been so dense that *Archaeopteryx* would be required to make difficult manoeuvres in the air.

In some ways, the extended discussion of flight in *Archaeopteryx* has been neither
satisfying nor wholly convincing. There are long lists of *Archaeopteryx*'s supposed
limitations and inabilities: cannot climb trees, cannot take off from a level surface,
cannot fold the wing completely, cannot flap continuously, cannot manoeuvre in
the air, and cannot land with dignity. It must have had some capabilities because
animals with a similar body plan survived until the very end of the Cretaceous. If
its capabilities were so marginal, evolutionary forces would not have passed them
on, in modified form, to living birds.

Part of the problem is that there is a tendency to make negative comparisons
between *Archaeopteryx* and modern birds even though *Archaeopteryx* has very few of
their features [56]. The single most birdlike feature of its skeleton may be its elongated scapula. The wing bones are very thin and lack many of the structural details
in the joints that enable modern wings to work so well. The humerus, in particular,
is long and thin, and its construction seems to be oriented at a different angle from

that of modern birds. The coracoids are rectangular and not strut-like. They lack an "acronomion," which would have been the site of a ligament to prevent hyper-pronation of the humerus. The furcula is a flat, boomerang-shaped bone without obvious connection to the sternum. The sternum is a thin, flat sheet, without a keel, that seems far too fragile to function as an anchor for flight muscles [56].

Archaeopteryx may have raised its wings by contracting the deltoid muscles in its shoulder, and flapped them by contraction of muscles attached to the coracoids. Olson and Feduccia [152] suggest attachment of the pectoral muscles to the furcula, but Elzanowski feels that the furcula is too flimsy and lacks the necessary structures for active participation in flight. As we shall see in the next chapter, however, a similar furcula appears to play a very important role in the flight of another primitive bird.

Perhaps it would be useful to examine the potential aerial performance of one of *Archaeopteryx*'s close relatives among the theropods. Assuming that its reconstruction is reasonably accurate, there seems little doubt that *Microraptor* was a kind of glider that flew from tree to tree like a flying squirrel. The long plumes on its hind legs imply that it was an arboreal animal, but otherwise it shares much of its basic body plan with *Archaeopteryx*. When it leaped from a branch, it would have spread its limbs to achieve lift but probably left its wings partially folded and did not spread its fingers. If it reached forward, the feathers on its digits would have fanned out, catching the air and increasing drag. The kinetic energy of the jump would give it some forward momentum at first, but gradually drag would reduce its air speed and it would begin to fall. At that point, its spread fingers would have increased the area of the lifting surface and increased the length of its glide. If it pushed down with its hands, it would further increase lift and, in effect, achieve the first step in flapping flight.

Archaeopteryx might have flown in a very similar way, with its arms thrust further forward than those of a modern bird (Figures 7.2 and 7.3). The technique would take advantage of its seemingly loose-limbed construction. When it opened its hands, it would automatically decrease the wing loading and enable the bird to generate more lift during the downstroke. Figure 7.2 is intended only to illustrate the posture. Care has been taken to make the illustration plausible, but it is not based on the detailed analysis of original material that contributed to the reconstruction used by Elzanowski [56]. *Archaeopteryx* may not have had the appropriate structures in its wing skeleton to permit this posture [177], but the trick is widespread in modern forest-dwelling birds (Figure 7.3). They typically have rather short, round wings that generate a great deal of lift but little thrust. Such wings are said to have a "low aspect ratio." Wings with high aspect ratios are typical of seabirds and soaring birds that need additional lift to achieve high aerial efficiency (Figure 7.4).

There is one analysis of flight in *Archaeopteryx* that focuses on abilities more than on shortcomings. Jeremy Rayner sees *Archaeopteryx* as an animal capable of relatively

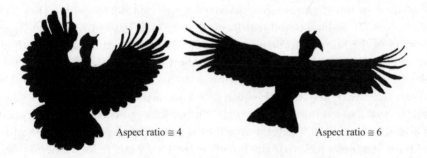

Aspect ratio ≅ 4   Aspect ratio ≅ 6

Figure 7.3 Silhouettes of a Great Hornbill in flight, demonstrating its ability to pronate its wings and increase the size of their lifting surface (based on photographs by Tim Laman [108]).

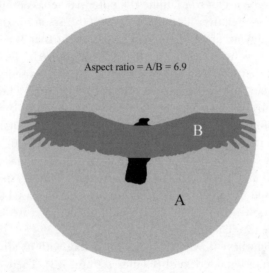

Aspect ratio = A/B = 6.9

B

A

Figure 7.4 Calculating the aspect ratio of a bird's wings. The Turkey Vulture is a typical soaring bird whose long and narrow wings give it a high aspect ratio [209].

fast cruising flight, with characteristics that gave it an evolutionary advantage over its contemporaries [177]. He bases his ideas on modern aerodynamic theory and models of flight techniques in a variety of animals.

Modern birds use two different gaits in the air. The vortex-ring gait is used for slow flight and is based on the generation of discrete vortices by each downstroke. The upstroke is merely for recovery and does not generate lift because the wing is flexed and the wrist is twisted. The wing is raised, in preparation for the downstroke, by contraction of the supracoracoideus and deltoid muscles.

The continuous-vortex gait is used for cruising flight and is based on the production of a continuous stream of vortices behind the bird, just as the name implies.

In this case, the wing is not folded but remains flat, and aerodynamic forces acting on its surface continue to generate lift. Its advantage is that there is no energetic cost of starting and stopping lift generation in each stroke.

The vortex-ring gait is used by all kinds of living birds for slow manoeuvrable flight, especially while landing. Long-winged birds simply fold the tips of the wings to increase their control. The continuous-vortex gait is typical of faster birds with relatively long, narrow wings (i.e., a higher aspect ratio). You might therefore expect the use of a continuous vortex to be more advanced than slower flight with discrete vortex rings. Rayner suggests that the opposite is the case. Continuous-vortex flight is actually less demanding aerodynamically and may be achievable with a less specialized skeletal architecture.

The structure that determined the flight style of *Archaeopteryx* must have been the simple, theropod-like pectoral girdle. There is no keel on the sternum and therefore probably no thoracic slip from the pectoral muscle. In modern birds, this branch of the pectoral muscle controls the type of wing pronation (Figures 7.2 and 7.3) that provides control during slow, vortex-ring flight. The absence of a keel, and also the absence of a pulley in the shoulder joint, implies that the supracoracoideus did not play a role in the upstroke. The downstroke of the wing could have been achieved by the deltoid muscles, but there would have been no complementary system of muscles to achieve recovery. Muscles are important only in the recovery stroke during vortex-ring flight, however. During continuous-vortex flight, recovery is achieved by the exploitation of aerodynamic forces.

The detailed architecture of the wrist may also have had a strong effect on flight in *Archaeopteryx*. Its independent semi-lunate carpal, which is characteristic of the maniraptoran theropods, was one of the structures that contributed to the conclusion that *Archaeopteryx* could not have flown. Its presence implies that *Archaeopteryx* was incapable of the complex twisting and folding that are needed for slow flight. In cruising flight, the wing is simply held flat and moved, more or less, up and down. If the tip is flexed for control, it moves in the same plane as the rest of the wing.

Taken together, the wing characteristics of *Archaeopteryx* suggest that it could cruise along, but they leave the question of take-off unresolved. Rayner sees cruising flight as a way of extending glides with a modest amount of energy expenditure [177]. Such a strategy would be consistent with the use of low platforms as envisaged by Elzanowski [56]. Most importantly, Rayner's model suggests evolutionary advantages to the flight of *Archaeopteryx* that could be stepping-stones to more elaborate aerial performance in birds with more sophisticated skeletal architecture.

In the following chapters, we will examine the contribution of flight to the evolution of more modern types of birds, and track their gradual conversion from sprawling reptilian architecture to the compact mass typical of the inner-bird strategy.

Further Reading

The discussions at the 1984 Eichstätt conference are described in Pat Shipman's book *On the wing*. Andrzej Elzanowski is very critical of the early data, however, and derides her conclusions as "sensational." That said, they still make interesting reading and are a good starting point for more detailed study.

Part of the intense debate around the origin of birds from dinosaurs focused on the problems of getting a primitive bird airborne. For an even-handed review of those arguments, Jeremy Rayner recommends:

Dingus, L., and T. Rowe. 1997. The mistaken extinction: Dinosaur evolution and the origin of birds. W.H. Freeman, New York.

Birds with a Modern Shape

8

*Birds cease to look like feathered dinosaurs
and become skilled fliers*

Because of their small size and delicate construction, bird bones have left few fossils. The great majority of ancient birds are known only from puzzling fragments. Many experimental designs that occurred during the long evolutionary history of birds have left little evidence of their passing and, without modern representatives, their significance is very difficult to appreciate. In spite of its sparseness, the fossil record suggests the development of a diverse community of birds and bird-relatives before the end of the Mesozoic. Some were spectacularly successful in their day and the ancestors of modern forms may have struggled to survive in the face of stiff competition.

The most dramatic evolutionary experimentation took place in the Cretaceous, when at least two lineages of feathered animals abandoned the reptilian structure of their ancestors and adopted aspects of the inner-bird strategy. By the end of the Cretaceous, few forms looked like feathered dinosaurs and some bore a strong resemblance to families that survive today. So far, however, only one of these animals has been identified as a member of a modern taxon with a measurable degree of certainty – that is, a thorough cladistic analysis has been able to place it within a group of living birds [30].

When the events that extinguished the dinosaurs carried away three of the four major lineages of Cretaceous birds, the survivors underwent a flurry of experimentation. They had inherited a world without significant competition from any near relatives; if they had been able to suppress the mammals, they might have come to dominate the globe completely.

For better or worse, humans and other mammals were able to gain control of life on the planet and have shaped it to their own needs. Competition with mammals was fierce and many of the birds that survived on the continental mainlands were small, inoffensive animals such as sparrows. Most of the larger avian species died out, but some thrived in unusual niches or habitats that were particularly unfriendly to mammals. The most familiar of these refugia include small islands, deserts, the high seas, and areas that have remained remote from the centres of human habitation. It is among the birds of these wilderness areas that we can find examples of the evolutionary plasticity that ensured their survival through the Mesozoic into modern times.

Mesozoic Birds

For Victorian biologists, the evolutionary history of birds leaped across great periods of time from some ancestral reptile to *Archaeopteryx,* and from toothed birds to modern forms. This rather lean story remained unchallenged until the 1980s, when Cyril Walker identified the Enantiornithes as a wholly unsuspected subclass of birds that had once lived in parallel with the modern forms. Revision of the early birds was followed by discoveries over the last few decades of a huge number of important avian, near-avian, and non-avian fossils. These new finds have helped fill out the history of birds and changed the way we look at avian evolution. Simple characteristics such as the presence or absence of teeth no longer dominate discussions of fossil birds, and paleontologists regularly call on ecologists and physicists to assist in their reconstruction of these animals.

Although fossils remain the most important source of information about the very early stages of avian evolution, it has proven very difficult to link fossil forms to living species. Avian classification, largely based on bone structure, was not particularly successful in the early 20th century. More recently, the study of avian anatomy has declined in importance as taxonomists have turned to biomolecular information to test relationships among living birds. It is not a type of analysis that can be applied to fossils. Fortunately, paleontologists have greatly increased the number of characteristics that can be detected in fossils, and thereby improved the power and effectiveness of cladistic analysis. In this way, they have been able to create convincing family trees that include the ancestors of modern birds among the dinosaurs. The same process is beginning to connect some of those ancestors to specific groups of living birds.

Cretaceous Birds

The discovery of *Archaeopteryx,* preserved in ultra-fine Jurassic limestone, was an incredible piece of good luck. Without it, the earliest stages in the origin of birds might have remained more of a puzzle than they are, and an intellectual link between birds and theropod dinosaurs would have required a great deal of imagination. In recent years, younger bird fossils have become surprisingly abundant and we are beginning to understand more about the spectacular evolutionary radiation that took place in the late Cretaceous. It is important to remember, however, that the majority of recent discoveries are coming from a small part of China. This restricted geographic distribution may be giving us a very biased view of the ancient world.

The animals represented by these newly discovered fossils have one major feature in common: they are small. Only a few are larger than a pigeon. Many have been collected from marine or aquatic sites, but they tend to have feet armed with long, curved claws that are more suitable for clinging to branches than for swimming.

Their lack of specializations for life in the water suggests that they represent terrestrial animals whose bodies fell into the water, where fossils form more readily than on land.

Confuciusornithes

The most abundant fossil of a Cretaceous bird – and perhaps the most abundant fossil of any one species of terrestrial vertebrate – is *Confuciusornis sanctus,* or the Confucius bird of China. The first example was found in November 1993 by Zhang He, an amateur collector, while he browsed the local market stalls in Jinzhou [242]. Since then, hundreds have been dug up within a single small area of China's Liaoning Province. Its quarries are incredibly productive and deserve their reputation as "the hottest spot in the world for fossil birds" [199].

The age of the Confucius bird was in doubt for some time. Liaoning Province happens to be one of the few places in the world where fossils of psittacosaurs have been found. The only other specimens of this dog-sized, parrot-beaked ceratosaur have come from Jurassic rocks in Mongolia. At first, the presence of psittacosaurs in Liaoning suggested that the Confucius bird was also from the Jurassic, raising the exciting possibility that it was a contemporary of *Archaeopteryx.* Analyses of argon isotopes have provided a more accurate estimate of age, placing the Liaoning fossils in the Lower Cretaceous, some 20 million years later than initial estimates.

The classification of the Confuciusornithes appears to be rather vague and unsettled. The group is clearly defined by an array of features, the most obvious of which is the massive humerus. The fossils of *Confuciusornis sanctus* are so abundant that they overshadow the existence of any relatives and are often the only type discussed (Figure 8.1). Some examples were originally given separate names, such as *C. chonzhous* or *C. suniae,* but these look so much like *C. sanctus* that they may be nothing more than size variants, which do not deserve the status of discrete species. Only *Changchengornis hengdaoziensis* is different enough to represent a sister genus.

There is a third, somewhat distinct type, known as *C. dui.* It is officially represented by two fossils housed in the Institute for Vertebrate Paleontology and Paleoanthropology in Beijing, and described in a 1999 paper by Dr. Hou Lianhai and his colleagues [89].[1] *C. dui* is somewhat smaller and more gracile than *C. sanctus,* but is clearly a near relative (Table 8.1). More puzzling is specimen B 065 or DNO 088 from the Wenya Museum, Jinzhou City, which may never have been described properly. It recently toured North America labelled as *C. dui* (Figure 8.2), but is clearly a different species and has very different limb proportions from its supposed relatives (Table 8.1). Most notable are the very long legs and extraordinarily elongated pygostyle (assuming that this structure is actually from the same animal). The catalogue that accompanied the tour described it as a long-legged wading bird,

---

[1] Specimens V 11521 and V 11553.

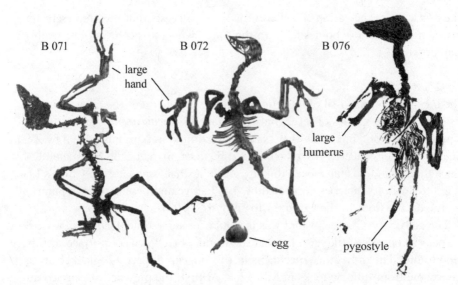

Figure 8.1 Three typical fossils of *Confuciusornis sanctus,* showing the large hands and feet and the massive humerus with the hole near the shoulder (photographed by the author while the specimens were temporarily on display in Vancouver, British Columbia). B 072 has an egg near its right foot. Fossils such as these often include clear impressions of the wing feathers [239].

Table 8.1

Limb measurements (mm) of Confucius birds compared with those of two types of modern birds

| | Confuciusornitheans | | | | |
Bone	*Confuciusornis sanctus*	*Confuciusornis dui*	B 065	*Yixianornis grabaui*	Belted Kingfisher
Humerus	53.4	42	30	49.0	49.4
Ulna	48.2	39	32	50.3	51.4
Carpometacarpus	26.5	19	–	25.6	26.9
Femur	46.5	35	33	41.0	27.9
Tibiotarsus	53.5	41	50	53.0	38.6
Tarsometatarsus	27.3	19.5	24	27.0	11.7
Pygostyle	26.2	23	135	–	4.2[a]

a The kingfisher's pygostyle is short but its central crest makes it 10.5 mm tall.

but this seems unlikely for an animal with such short toes. In life, it may have been more terrestrial than other Confucius birds, looking and living somewhat like *Archaeopteryx*. Whatever its relationship to other birds, it clearly deserves to be recovered from obscurity and described properly.

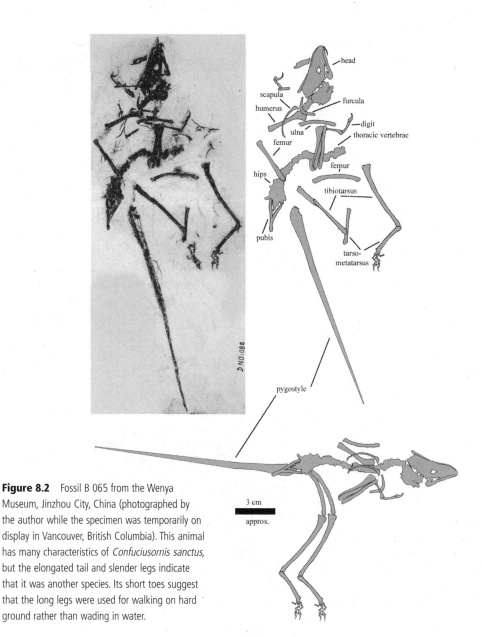

Figure 8.2 Fossil B 065 from the Wenya Museum, Jinzhou City, China (photographed by the author while the specimen was temporarily on display in Vancouver, British Columbia). This animal has many characteristics of *Confuciusornis sanctus*, but the elongated tail and slender legs indicate that it was another species. Its short toes suggest that the long legs were used for walking on hard ground rather than wading in water.

In 1986, Walter J. Bock wrote an essay that portrayed *Archaeopteryx* as an animal that climbed trees to find safe roosts and secure nest sites but probably foraged on the ground [13]. He felt that it probably climbed trees using all four limbs, like the nestlings of the modern Hoatzin or turaco, or perhaps like a tree wallaby. He saw flight as a useful way for early birds to reach foraging sites on the ground because climbing up is easy whereas climbing down poses many difficult biomechanical problems. Feathered limbs would have greatly increased those problems.

Fossils of the Confucius bird appeared a few years after Bock outlined his ideas regarding the lifestyle of *Archaeopteryx.* The younger bird is a better fit for his essay in many ways. The Confucius bird was much better equipped for clambering about branches in a forest canopy than its more terrestrial antecedent. It retained a mix of avian and reptilian characters in a unique creature that has no modern equivalent. Like *Eoraptor,* it appears to have been something of a feathered squirrel. Considering the long limbs, however, a small feathered ape might be a more appropriate analogy.

As a flying animal, the Confucius bird is exactly the kind of animal you might expect to follow *Archaeopteryx.* It certainly seems to represent the next step in the flight vortex-ring gait versus continuous-vortex gait scenario that Jeremy Rayner developed for *Archaeopteryx* [177]. It is also the earliest known species to take any significant first steps towards a modern type of skeletal architecture (Figure 8.3).

In some ways, the wing is still very reptilian, with its three independent digits and rather thick bones. If the limb was regularly used for scrambling among

Figure 8.3   A reconstruction of *Confuciusornis* in flight. The two major masses of muscle are indicated by darker shading. There are no large muscles on the breast and the muscles of the shoulder probably drove wings whose primary feathers were much longer than those in any modern bird. The skeleton and the muscles in the hindquarters are rather dinosaurlike, in spite of the abbreviated tail with its pygostyle. The femur is very long, while the tarsometatarsus and the toes are short.

branches, it may also have carried a much heavier array of muscles than is found in modern birds. Most of its length, however, is made up of asymmetrical flight feathers. In other words, the plumage or "outer bird" contributes more to the wing than the inner bird. The primary feathers create a huge lifting surface by being two or three times longer than the hand. In modern birds, the feathers are often only a little more than half the length of the hand, and approach a length equal to that of the hand only in forest birds and a few other special cases (Table 8.2). Even though the arm of the Confucius bird may have been heavy, the outer half of the wing would have been extremely light and the bird could have moved its wings rapidly without significant inertia or a build-up of momentum.

The wing represents the Confucius bird's greatest advance over *Archaeopteryx*. It is very much larger for the size of the body, and includes specialized skeletal features that reflect a greater commitment to flight. Most importantly, the hand includes a fused carpometacarpus and the semi-lunate carpal is fixed to the ulna [239]. These developments increase the bird's ability to flex and bend its wing during flight, and suggest that the Confucius bird may have been able to achieve Rayner's vortex-ring gait, at least to a limited extent. Less speed and greater manoeuvrability would have made it easier for the Confucius bird to live in complex forest environments.

Large wings would also have reduced wing loading and made it much easier for the Confucius bird to get into the air. The wings were much larger than those of

Table 8.2

Contribution of feathers to the length of the wing in several modern birds

Species	Feathers as a percentage of wingspan	Feathers as a percentage of forewing length
Common Loon	29	57
Canada Goose	34	60
Marbled Murrelet	38	62
Laysan Albatross	30	64
Horned Grebe	29	66
Prairie Falcon	44	69
Double-crested Cormorant	32	70
Black-legged Kittiwake	42	71
American Avocet	44	72
Green Heron	38	73
Turkey Vulture	48	74
Vaux's Swift	63	75
Northern Hawk Owl	44	75
Calliope Hummingbird	57	76
Common Nighthawk	51	79
Fork-tailed Storm-Petrel	54	80
Cooper's Hawk	47	81
Common Flicker	51	82

Archaeopteryx, and there is no reason to suspect that the Confucius bird could not take off from level ground. Nonetheless, it may have habitually launched itself from convenient low platforms, to save energy.

Like *Archaeopteryx,* the Confucius bird would have used continuous-vortex flight to cruise over the trees and between patches of forest. With its long, pointed wings and trailing tail plumes, it might well have looked like just another pterosaur, or perhaps a modern macaw, as it crossed canopy openings. In a world with no known threat from dangerous aerial predators, it would not have needed to achieve very high speeds or even exploit its enhanced manoeuvrability to escape. Even if a large pterosaur posed a threat, it could not follow the Confucius bird into the canopy because of the risk of damage to its fragile membranous wings.

The elongated primary feathers of the Confucius bird give its wing a rather high aspect ratio, in the neighbourhood of 5.5 or 6. In a modern bird, that would be typical of species that can achieve great speed or great efficiency. The Confucius bird's exposed fingers would have eliminated any potential for aerial efficiency by disrupting the smooth flow of air under the wing and creating drag. The fingers would also have caused problems of drag at high speeds, and other aspects of the bird's design suggest that it was not capable of extreme performance. In modern fast-flying birds, the primary feathers are relatively short and the wings are driven by massive muscles attached to large bones in the body. The wing skeleton of the Confucius bird seems too loosely jointed for a demanding aerial performance, and there are no suitable bones in the thorax for the attachment of large flight muscles.

The most unusual feature of the Confucius bird's wing is the hugely enlarged humerus. It has no modern analog, and was so large compared with other bones in the body that a large hole in the widest part may have been necessary to reduce the bone's overall weight (Figure 8.1). The shape of this bone is actually quite sophisticated and it may represent a significant improvement in flight control over the less specialized arm of *Archaeopteryx.* In his study of the Confucius bird's humerus, Rayner found that the head of the bone was rolled over, just as it is in modern birds. The shape shifted the insertion of the pectoral muscle from the midline of the humerus in a way that increased rotation when the flight muscle contracted. This rotation at the shoulder also enabled the bird to rotate the wing and provided an element of aerial control. In addition, the width of the delto-pectoral crest offered one of the few opportunities for attachment of large muscles in the Confucius bird. Its width created a broad platform that would have spread out the muscles of the shoulder and given them greater mechanical control over the movement of the humerus in the shoulder joint.

The bones that make up the shoulder joint show a mix of ancient and novel features. The Confucius bird's pectoral girdle is the result of the fusion of the scapula to the coracoid (Figure 8.4). These bones are not free as they are in most modern birds, and the pectoral girdle is one of the few rigid skeletal assemblies in the Confucius bird. The shoulder itself lacks the all-important triosseal canal, which enables modern birds to lift the wing by reversing the pull of the supracoracoideus

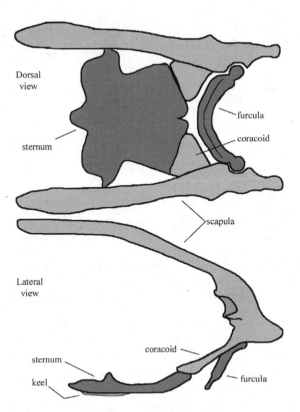

Figure 8.4 The pectoral girdle of *Confuciusornis sanctus* is more complex than that of *Archaeopteryx*. The furcula is simple but the scapulas are fused to the coracoids and the sternum may have an incipient keel.

muscle. Without a pulley, the Confucius bird may have used the deltoid muscles to raise the wing. Stubby coracoids and a short, flat sternum with a low keel suggest that its breast muscles were not particularly large. The furcula appears robust but simple. It lacks a hypocleideum, the platelike structure that often extends towards the body from where the two arms meet. A hypocleideum may be important in holding the furcula in position relative to the sternum.

Much of Confucius bird's body is reptilian and shows few inner-bird characteristics. The back is long and the centra of the vertebrae are flat, or "opisthocoelic." There are no specialized supporting structures or fused segments in the thorax. There are no uncinate processes to strengthen the rib cage, and the after part of the belly is supported by 10 pairs of floating bones called gastralia. Many dinosaurs and crocodiles have similar loose bones supporting their abdominal walls but these bones are absent from living birds. In the hips, the synsacrum is short and includes only 10 fused vertebrae. The pubic structure looks much like that of *Archaeopteryx*; it even has a small cup where the bone tips are fused. As in *Archaeopteryx* and other dromaeosaurs, the rigid pubic ring would have limited the size of the egg (Figure 8.5).

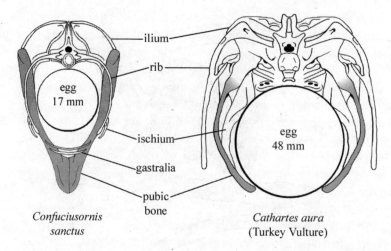

Figure 8.5   A comparison of the pelvic girdles in *Confuciusornis sanctus* and the Turkey Vulture. The size of the *Confuciusornis* egg is based on specimen B 072 from the Wenya Museum, Jinzhou City, China (Figure 8.1). The pubic bones in *Confuciusornis* appear foreshortened because they are angled strongly back towards the tail.

The skeleton of the Confucius bird's hind limb is very much like that of an advanced theropod, with two very long units (femur and tibiotarsus) and a relatively short but fused tarsometatarsus. In fossil B 065, the femur is almost as long as the back and reaches from the hip joint to the armpit, as it does in typical theropods (Figure 8.2). The feet of *Confuciusornis sanctus* are large and well clawed, but there is no sign of enlarged tendons, which are needed to operate the talons as weapons.

The Confucius bird's body skeleton is unlike anything found today. It is possible that this early bird developed a unique method of flight that has not survived in any modern animal. Perhaps its flight merely slowed the drop to the ground.

The vertebrae lack structures to enhance the stiffness of the back, and heavy legs would have weighted down the hind end. Even *Archaeopteryx* had the benefit of lift from air passing under the tail. If the Confucius bird had difficulty holding its long, flexible body horizontally during flight, it may have allowed it to droop limply, and flown more or less vertically by flapping its wings over its head so that they acted as a kind of parachute. It seems an ungainly and undignified posture but it might have been stable at low speeds, over short distances. It would have been all that was needed to drop from the canopy to the ground, with the advantage that the legs were always in position to absorb a landing or grab a passing branch (Figure 8.3).

Modern birds often hold their wings high and let their bodies hang vertically for landings at low speed. Kingfishers characteristically hang down from their wings during brief bouts of hovering. They have heavily muscled shoulders attached to a scapula that is, in turn, anchored to long dorsal spines on the thoracic vertebrae. The pectoral muscles in kingfishers are slightly smaller than those of stronger fliers.

Although the Confucius bird left no descendants, it is important in the evolutionary story because it is the earliest known animal to adopt parts of the inner-bird strategy. It abandoned the long reptilian tail and developed the lightweight, toothless head typical of living birds. Even the unusually long wing represents a proportionate decrease in the skeletal contribution and is, in a sense, a start on the concentration of the limb near the core of the body.

During the debate on the origin of flight, the shape of the claws and other features of the Confucius bird's foot attracted the attention of proponents of arboreal and terrestrial origins. The "terrestrial" side supported a running bird that would have had stubby claws, whereas the "arboreal" side supported a tree-climber with long, sharp claws.

The debate made far too much fuss over a single feature whose value we cannot really appreciate. It ignored the security that many birds find by perching only a metre or two above the ground, regardless of their claw shape. Stubby-clawed turkeys, guans, peacocks, and other pedestrian birds spend the night in trees just beyond the reach of the local type of cat, while herons regularly roost in treetops in spite of their ungainly legs and long toes. They easily grip narrow branches and even nest in the crowns of trees, moving about comfortably on the thin branches of the upper canopy. It helps that they have an amazing sense of balance. I once watched a Reef Heron attempting to fish while swinging on a thin rope looped between a boat and a wharf.

Except for a few birds that are highly specialized for walking, most birds (and bats) have a sophisticated tendon locking mechanism in the foot that is far more important for perching than curved claws (see Chapter 2). Curved claws with sharp tips help, but they are most useful in small birds that forage in the crevices of tree bark and need an enhanced grip to cling to a vertical branch or trunk.

The Confucius bird is the most primitive avian to develop a pygostyle. Instead of a long, tapering string of vertebrae, there are just five free caudal vertebrae and then a well-formed pygostyle, which is much longer than that of a modern bird but shares many of the same structural features. The joints are still visible. There is a crest or lamina along the midline, with ridges along each side so that its cross-section is similar to modern versions. In living birds, the pygostyle plays an important role in the tail's ability to change shape and function as a control surface. Since no fossil of a Confucius bird sports a modern fan of tail feathers, the role of the large pygostyle is a mystery. Perhaps a relative with a larger and more useful tail on its pygostyle is waiting to be discovered.

In some Confucius bird fossils, the tail includes only two very long plumes but they are not likely to have been important in flight unless they acted like the tail of a kite. It seems more likely that the long plumes were a sexual display for the males, but there is little fossil evidence to support such an idea. In China, the Wenya Museum has a fossil (B 072) with an egg lying below a short-tailed specimen, where it could have been dropped as the bird died. A sample of one could have been an accident, however, and it is possible that the long tail feathers signal maturity in both sexes, or even the readiness for reproduction in females alone.

Sexual dimorphism is often related to breeding strategy in modern birds, and large differences are typical of species where the female does all the incubating. It is also typical of forest birds but much less frequent or less dramatic in species living in more open environments. Both sexes of cranes, storks, shorebirds, gulls, auks, loons, albatrosses, and penguins tend to be similar and share incubation and nestling care. Male and female hawks often look alike but the female is typically significantly larger and performs most of the incubation duties.

A Confucius bird specimen with fossilized medullary bone could settle the question of sexual dimorphism. Medullary bone appears only in female birds because it contributes calcium to eggshells (see Chapter 2). Bird fossils are so rare that destructive examination is prohibited, but there are so many fossils of the Confucius bird that some have even been used as architectural decoration at the Wenya Museum. Surely a few long bones could be sectioned without significant loss of future opportunities for scientific study.

The cursorial/arboreal debate at the end of the 20th century left us with a wide variety of reconstructions of the Confucius bird. A Google search for images on the Internet will turn up several dozen. The "mad hen" style, with strangely unorganized flight feathers and deformed hands, appeared on the cover of *Scientific American* in 1998. This view descends from Roger Tory Peterson and Arthur Singer, whose early reconstructions portrayed *Archaeopteryx* in mid-scream and gave it an improbable, but artistically satisfying, array of head plumes. The pose is dramatic but has no more scientific foundation than the screams and roars that dramatize the soundtracks of animated dinosaurs on television.

In another popular reconstruction, a Confucius bird appears as a kind of primitive songbird sitting on a carefully constructed nest of twigs. It is suggestively coloured like a jay but perched vertically like a becard or some other unfamiliar tropical species. Such pictures certainly make the Confucius bird look more like a bird and less like a feathered dinosaur, but the images often carry the analogy too far. Most of the arboreal birds in the surviving basal lineages, such as woodpeckers, owls, and kingfishers, do not build elaborate nests but incubate their eggs in burrows or tree cavities. We know nothing of the nature of a Confucius bird's nest, but it seems unlikely that the Confucius bird built a free-standing nest attached to a forked branch. The solitary fossil egg also argues against an open nest. It was slightly smaller than the adult bird's head, and such small eggs are typical of birds that lay rather large clutches and care for their nestlings over an extended period. If this was true of the Confucius bird, tree cavities might have offered greater security than a more exposed location.

Many reconstructions imply an aggressive flight style for the Confucius bird by enhancing its sleekness or its supposed similarity to a modern type. Several images on the Internet look like green tropicbirds, a seabird that happens to have long tail plumes. Most of these reconstructions should be used with caution, keeping in mind both the drag created by the fingers and the absence of large muscles in the abdomen.

Sapeornis

In 2000, paleontologists found three specimens of a very large bird in Lower Cretaceous deposits at Liaoning. They named it *Sapeornis chaoyangensis* to salute the role of the Society for Avian Paleontology and Evolution (SAPE) in promoting the study of early birds and drawing international attention to the Chinese fossils. The fossils share some characteristics with the Confucius bird, such as a pygostyle and a large humerus with an opening through its wider portion, but other features appear to be more primitive. For instance, the fossils include teeth and a coracoid bone that is short, like that of *Archaeopteryx*. The bones of pectoral girdle are not fused as in the Confucius bird, however. There has been no sign of a sternum in any of the three specimens but this is not uncommon in fossils of early birds. There are sternal plates on only one of the 10 fossils of *Archaeopteryx*. Zhonge Zhou and Fucheng Zhang place *Sapeornis* close to the base of the avian family tree, between the reptile-like *Archaeopteryx* and the more birdlike *Confuciusornis* [240]. In many ways, however, it appears to resemble the Confucius bird very closely, and may have shared some aspects of its behaviour.

One of the outstanding features of *Sapeornis* is its peculiar furcula (wishbone). It is quite a large and robust bone with wide, flat arms like those of both *Archaeopteryx* and *Confuciusornis*. *Sapeornis'* furcula is a somewhat deeper U-shape, and there is a large and robust hypocleideum where its arms meet. The hypocleideum is not particularly long, about twice the width of the furcula arms, but it is the first such structure to appear in avian evolution and a feature of most later fossils. Although many modern birds do not have such a structure, they are all descended from ancestors that had one.

Enantiornithes – Ball-Shouldered Birds

Besides the Confucius birds, the Cretaceous rocks of the Liaoning area have produced examples of two other avian subclasses. Some are Enantiornithes, or ball-shouldered birds; others are early versions of the Neornithes, or modern birds. The Confucius birds have been found only in China, whereas the other groups have a much wider distribution.

In 1981, when Cyril Walker realized that *Ambiortus, Gobipteryx, Alexornis,* and some other avian fossils formed a wholly unsuspected lineage discrete from modern birds, he called them Enantiornithes, meaning "opposite birds." Members of this new subclass shared certain characteristics of the shoulder and the tarsometatarsus that were the exact opposites of the constructions in the Neornithes. By the end of the 20th century, the new subclass was being hailed as a great evolutionary experiment whose members were probably more abundant and widespread than the ancestors of modern birds. They have been reported from all continents except

Antarctica and India, which was an isolated island-continent in the Cretaceous. As yet, none have been found on the mainland of Africa, although their fossils have been dug up in Madagascar. They are so widespread that Kurochkin and Walker have speculated that the success of the ball-shouldered birds prevented the expansion of the modern, socket-shouldered types [111]. The recent discovery of *Apsaravis*, a very early modern bird, suggests that we need to be very cautious about such claims. It seems likely that the advanced nature of *Apsaravis* will force ornithologists to reinterpret the relationships between the two subclasses, and may require the reconsideration of many fossils. Unfortunately, many of fossils are so fragmentary that they will never be convincingly reclassified.

The ball-shouldered birds provide early examples of the inner-bird strategy. They had a much more compact body than the Confucius birds and their general appearance was probably indistinguishable from that of some modern birds. The fossil record indicates that early types of ball-shouldered birds had teeth but later ones did not. Early members of the Neornithes, such as *Hesperornis* and *Ichthyornis*, also had teeth. The presence or absence of teeth may be misleading, however, and many different groups of dinosaurs experimented with toothlessness. It may be more significant that the ball-shouldered birds from early in the fossil record are small and arboreal, like the Confucius birds, while larger forms adapted for running or swimming appear later. There is good evidence that ball-shouldered birds developed a variety of feeding techniques and exploited both animal and vegetable resources [88].

Some features of the ball-shouldered birds suggest that they were a relatively primitive type of bird. As a rule, their back was longer than in living birds, and the vertebrae of their synsacrum were not fused. This suggests that their spinal columns were very flexible and not particularly well adapted for the stresses generated by energetic flight styles. Without specialized structures in the hips or fusion of vertebrae into the synsacrum, it is unlikely that such loosely built birds would have been able to flap their wings with great vigour. Potential predators in the group would not have been able to make powerful strikes with their feet to disable their prey. Consequently, they probably did not hunt in the air.

Although ball-shouldered birds were widespread and successful, they appear to have disappeared with the dinosaurs. Why they should have disappeared while the Neornithes survived is a great mystery. A tenuous clue may lie in the distribution of their fossils. Ball-shouldered birds appear to have been most successful in the northern continents. If the Mesozoic was terminated by the earth's collision with an asteroid, areas near the earth's equator might have suffered the worst effects. In contrast, the Neornithes appear to have been concentrated in Gondwana (now Antarctica, Australia, New Zealand, and South America). By being closer to the South Pole, they may have been protected from the worst effects of the catastrophe by distance and time.

Although ball-shouldered birds have been extinct for 65 million years, they, as we shall see below, have found an imitator in one of the oddest of modern birds, the South American Hoatzin. Before we can discuss this unusual situation, we need to examine the early history of the subclass of birds that is still with us.

Neornithes – Socket-Shouldered Birds
The thin walls of bird bones make them exceptionally rare as fossils; in addition, preservation processes usually crush and distort the material. Interpretations of fossil birds are therefore often controversial in spite of painstaking examination. The fossils of *Archaeopteryx* are merely flattened but paleontologists are still making important discoveries and offering new interpretations after nearly 150 years. Other fossils of early birds are more distorted and much less complete. The numerous fossils of the Confucius bird are an important exception, and recently another important bird fossil has been found immaculately preserved.

APSARAVIS *Apsaravis ukhaana* is an incredibly rare example of a modern bird that was already fully committed to the inner-bird strategy some 83 million years ago, in the Campanian Period of the Upper Cretaceous. It is preserved in three dimensions so that the bones are in an articulated position, just as they would have been in the living animal. Except for a few breaks and inevitable losses, the fossil of *Apsaravis* is almost as easy to examine as a modern skeletal preparation. Only the skull is fragmentary, but there are ossicles from the eye ring and a piece that appears to be the quadrate bone. One of the jaw fragments shows that there were no teeth in the dentary bone.

Mark Norell and Julia Clarke found that *Apsaravis* shares 27 derived features with other modern birds and conclude that it is close to the crown clade in the Class Aves [29, 142]. Most of *Apsaravis*'s backbone is present, and it is not very different from that of other early examples of modern birds. There are 12 neck vertebrae with typically heterocoelic (saddle-shaped) centra, 7 dorsal vertebrae with flatter "opisthocoelic" centra, and 10 sacral vertebrae fused into the synsacrum. The spine ends with 5 free caudal vertebrae and a terminal pygostyle. Bones in the legs and wings show a level of reduction and fusion much closer to the condition in modern birds than to *Archaeopteryx*. The structure and arrangement of the hip-bones is also essentially modern. Most significantly, the tips of the pubic bones are not fused together and there is no sign of the pubic cup found in *Archaeopteryx* and *Confuciusornis*.

The absence of the pubic cup is much more significant than the incidental loss of a characteristic shared with theropod dinosaurs. For the first time in any terrestrial vertebrate, the oviduct is not forced to pass through a solid ring created by the pubic bones. The tips are unconnected and there is no skeletal structure in the abdomen to limit the size of the egg. From *Apsaravis* on, birds were free to create a truly huge egg. This characteristic, more than any other, separates them from the dinosaurs.

In a fossil of such a modern-looking bird, it is not surprising to find evidence that *Apsaravis* was a capable flier, perhaps as competent as the much later *Ichthyornis*.

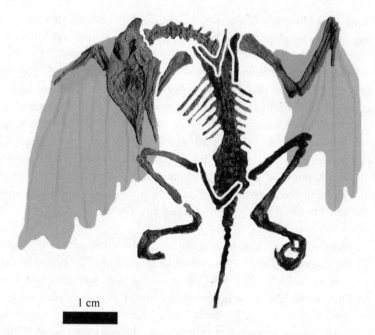

1 cm

Figure 8.6 *Liaoxiornis delicatus,* a modern type of bird from the Cretaceous, with a well-developed furcula.

It is one of the oldest birds to have a wing skeleton with all the structures necessary for the automatic opening of the primary feathers as the wing is extended.
This is an important adaptation in the development of effective flight, indicating
that this bird was fully able to flex and bend its wing for slow flight. There is no
doubt that it could make full use of both vortex-ring and continuous-vortex flight.

The fossil also shows a central keel that extends onto the forward part of the
sternum. It is only a low ridge but it may represent the beginnings of the structure
that supports strong flight muscles in later forms. Low keels occur in the Confucius
birds and other early birds, but they are restricted to the posterior part of the sternal
plate. Birds more advanced than *Confuciusornis* often had a very large Y-shaped furcula. Its elongated hypocleideum extended over the sternum and provided a tall
keel for the attachment of large flight muscles. No furcula was preserved in *Apsaravis*
but it was likely similar to that found in *Liaoxiornis,* another Cretaceous representative of modern birds (Figure 8.6), or *Sinornis,* one of the ball-shouldered birds.

Although *Apsaravis* is a modern bird, it has many characters that were previously
thought to be primitive. Some of these had been erroneously described as being
unique to ball-shouldered birds. For instance, the deep trench that runs down the
inner side of the coracoid appears in "non-avialan theropods" and the Confucius
bird, as well as in ball-shouldered birds, but it is supposed to be absent from modern birds. As we shall see, however, there is a similar trench in the coracoid of the
Hoatzin. The arrangement and position of openings for nerves within the coracoid
trench of *Apsaravis* is another primitive feature that was supposedly unique to

ball-shouldered birds. Additional similarities of the feet and legs in *Apsaravis* and ball-shouldered birds suggest that the two lineages may not be so far apart.

In all, only 12 of the 30 characters originally used to define the ball-shouldered birds are unique to that group. As a result, we can no longer be certain of the identity of many fragmentary fossils. The single fossil of *Apsaravis* may eventually lead to fundamental changes in the interpretation of birdlife in the Upper Cretaceous. It is no longer certain that ball-shouldered birds were either more abundant or more diverse than modern birds at that time. The presence of *Apsaravis* among continental fossils also shows that ball-shouldered birds were not able to exclude modern birds from upland habitats or restrict them to marginal marine and aquatic areas.

YIXIANORNIS In March 2006, Julia Clarke, Zhonghe Zhou, and Fucheng Zhang described another remarkably complete fossil that throws further light on the early development of modern birds [31]. Like so many other important bird fossils, *Yixianornis grabaui* comes from Lower Cretaceous deposits in China that are so fine that they preserve the remains of both bones and feathers. *Yixianornis* is even more complete than *Apsaravis* and may displace it as a benchmark for the assessment of future discoveries.

Yixianornis is more than just another interesting toothed bird. Other fossils may hint at the evolutionary events leading to modern birds but the mixture of primitive and advanced characteristics in *Yixianornis* implies a specific sequence for those events. It is also the first fossil to suggest that it might be possible to link the eventual survival of modern, or "socket-shouldered," birds (Neornithes) and the extinction of "opposite" or "ball-shouldered" birds (Enantiornithes) to the advantages offered by structural differences.

The most thoroughly modern features of *Yixianornis* appear in the pectoral girdle. Its wing skeleton is essentially that of a modern bird, although the fusion of the digits is not complete and a terminal claw is retained on some digits. The proportional lengths of the wing bones are very similar to those of a Belted Kingfisher (Table 8.1). The shoulder bones are almost indistinguishable from those of a modern forest bird, so there was no shortage of space for the attachment of large flight muscles. The sternum has a well-developed central keel whose leading edge projects forward and down between the long and strut-like coracoids. The blades of the scapulas are flat and extend back across the rib cage as they do in most modern birds. The arms of the furcula are very long and slender but have no hypocleideum where they meet. They are also curved and could have carried parts of the flight muscles in front of the shoulder joint to increase the power of the downstroke.

The vertebral column is similar to that of many modern birds. The neck is long, with 12 vertebrae; the thoracic region is rather short, with only 10 vertebrae in front of the fused hipbones. At the rear of the body, 9 vertebrae appear to be fused with the hipbones to form a synsacrum, but at least 5 of the caudal vertebrae remain free. The exact number is uncertain because some of those small bones may have been lost during fossilization. At the tip of the spinal column, at least

4 vertebrae have fused to form a very modern-looking pygostyle, with the distinctive ploughshare shape found in many modern birds.

There are indications that the structure of the spinal column had already responded to the stresses generated by flight. The articulating faces of the thoracic vertebrae are tall and narrow and would have resisted vertical bending.

All the thoracic vertebrae and possibly some of the vertebrae fused to the hipbones have ribs. As in modern species, the thoracic ribs usually have slender uncinate processes that extend back, over two of their neighbours. The ribs appear to be rather simple and there is no indication that they had the modern complement of upper and lower components. Instead, *Yixianornis* has an array of transverse gastralia, the elongated, free-floating abdominal bones seen in many dinosaurs. They also appear in *Archaeopteryx, Confuciusornis,* and other early birds. None of *Yixianornis*'s gastralia is connected to a rib in the fossil, but a slight swelling at the ends of some suggests that there might have been such an attachment in the living bird.

The gastralia suggest that *Yixianornis* is more primitive than *Apsaravis,* but they are not the only primitive feature of this new fossil. Although its hipbones are fused together, they remain rather small and include some primitive characteristics reminiscent of earlier birds. It also has rather dinosaurlike legs. The femur is long, and the proportional lengths of the leg bones are more like those of *Confuciusornis dui* or of specimen B 065 than the bones of a modern Belted Kingfisher (Table 8.1). Most importantly, the tips of its pubic bones are fused to form a solid ring and its ischia are longer than in *Archaeopteryx;* they are not flared or widened to form the roof of a broad abdominal vault. Consequently, *Yixianornis* could not lay a very large egg.

The tail skeleton of *Yixianornis* appears very much like that of a modern bird, and its modern type of pygostyle may be one of this bird's most significant features. The pygostyle sits at the end of a very short series of free caudal vertebrae. Muscles connecting the pygostyle and the caudal vertebrae to the somewhat elongated hipbones would also have been short and attached to the elongated ischia at a very steep angle. Such an arrangement may given *Yixianornis* fine control over the movement of its tail and tail feathers. Small contractions of the short muscles would have created disproportionately large movements of the tail fan, and they could have held the fan in position against the moving air with little effort. Control would have been much more difficult and energetically expensive in birds with longer, more reptilian tails and smaller hips. These birds would have been able to move the tail, but holding it in position against a moving flow of air would have required the contraction of long muscles for considerable periods of time, thereby consuming a great deal of energy. As a result, *Yixianornis,* with its short tail skeleton and broad fan of tail feathers, may be the earliest known example of a bird with aerial capabilities similar to those of a modern species.

In living birds, the tail feathers are mounted in a pair of fatty bulbs that sit on either side of a crest that rises from the midline of the pygostyle. When muscles squeeze the bulbs, their shape is distorted and the tail feathers spread out as a fan.

Forest birds and other aerobatic species change the shape of one bulb at a time to twist the tail and enhance their aerial manoeuvrability.

Confuciusornis, Sapeornis, and the Enantiornithes also had pygostyles and are classified with birds more advanced than *Archaeopteryx* as members of the Pygostylia. The primitive pygostyle found in these groups is little more than a simple rod of fused vertebrae, however. It lacks the enlarged central crest of modern species. Such simple pygostyles are not restricted to birds and have been found in *Nomingia,* a relative of the flightless but birdlike dinosaur *Chirostenotes.* The primitive pygostyle may be linked to other characteristics of the tail. We do not have tail feathers from the fossils of *Sapeornis* but *Confuciusornis* is famous for its tail, with just two long central plumes. Clarke and colleagues point out that all known examples of tail feathers in the ball-shouldered birds are also limited to a pair of central plumes [31]. Only modern birds have both a proper tail fan and a specialized pygostyle.

The absence of a proper tail fan may have been a much more significant difference between the Enantiornithes and the Neornithes than the ball-shoulder/socket-shoulder discrepancy. The lack of a tail fan would have limited the aerial manoeuvrability of the Confucius birds, and ball-shouldered birds may have had limited opportunities and ability to exploit specialized ecological niches. Without enhanced aerial manoeuvrability, they could not evolve into the equivalent of flycatchers, hawks, kites, or other types of aerobatic specialists. It may even have been difficult for them to produce an effective soaring bird.

It would be ironic if the tail fan was the key feature that gave the socket-shouldered birds a great edge over the other groups, because no sooner had modern birds begun to diversify than the elaborate tail fan began to disappear in some groups of modern birds. Some basal lineages with small pygostyles, such as tinamous and the extinct Lithornithiformes, may never have used a tail fan as an aerial rudder, while the large pygostyles of the Galloanseriformes are used only for the display of special feathers. Cranes, storks, ibises, and other groups that trail their legs have small, rather useless tails and may have evolved from birds that found a rudder unnecessary to their flight style. Large tail fans have also disappeared or become simple splitter plates in birds that fly at very high speeds or with great efficiency. Some of those groups show a lot of structural diversity. Fast, short-tailed birds often have slower, long-tailed relatives. Short-tailed auks are closely related to longer-tailed terns, while albatrosses are related to long-tailed storm-petrels and the peculiar Bulwer's Petrel, with its heart-shaped tail.

GANSUS   In June 2006, a team led by Hai-Lu You described five new specimens of *Gansus yumenensis* [238]. This bird from the Lower Cretaceous had been known since 1984 from the fossil of a single webbed foot [86], but the new material includes well-preserved bones from all other parts of the animal except the head. It appears to have been a swimming bird that fits into the avian evolutionary tree between *Hesperornis* and *Ichthyornis.* Unlike *Hesperornis, Gansus* was a strong flier. Its skeleton looks surprisingly modern, with well-developed wings and a sturdy keel mounted forward on the sternum. Its hipbones are fairly large and the short tail ends with a small pygostyle.

Aquatic and marine lifestyles may have offered several attractive opportunities to primitive birds [106], and *Gansus* is the earliest known example of a flying species that seems to have been adept at foot-propelled diving. In life, it had webbed toes and its knee had a stubby cnemial crest for the attachment of the strong swimming muscles needed by divers. Other bones suggest, however, that it was not specialized for particularly deep dives, and it may not have pursued fish. Its femur is quite long and narrow and would have been a poor design for absorbing the stress of vigorous underwater swimming (Table 8.3). Its other leg bones are considerably longer than those of a modern diving duck and not nearly as robust or specialized as those of a loon or grebe. Although its hipbones are large, they are not as narrow as those seen in fossils of *Hesperornis* or in a modern loon. They are somewhat similar to those of a merganser. As in most diving birds, *Gansus*'s thoracic vertebrae have hypapophyses that extend inward between the lungs.

The skeletons of the wing and the pectoral girdle are those of a moderately strong flier (Table 8.3). The furcula and other bones are simple, with no sign of the specializations seen in modern aerial specialists. The humerus is similar in length to the radius and ulna, a little longer than in most modern birds but not massively developed like the humerus of the Confucius bird. The bones of the hand are fused but a little shorter for the length of the wing than the hand of a modern bird (Table 8.3). The hand is long enough to suggest, however, that it supported a modern type of wingtip with fairly short primary feathers.

The structure of the hips and tail suggests that *Gansus* was very manoeuvrable in the air. The hipbones extend far behind the hip joint, capturing five of the caudal vertebrae, creating a rigid central base for the body. The hipbones also extend past four or five of the independent tail vertebrae. Consequently, any muscles between their trailing edge and the pygostyle would have been short and quick to move the tail fan when the bird wished to change direction.

Table 8.3

Limb proportions in *Gansus* compared with two similarly sized modern diving birds, the Red-breasted Merganser and the Horned Grebe

Limb element	*Gansus*[a]		Merganser		Grebe	
	Length (mm)	Percent of limb	Length (mm)	Percent of limb	Length (mm)	Percent of limb
Humerus	48.4	36.7	86.4	34.5	79.0	37.1
Ulna	46.8	35.5	73.1	29.2	71.9	33.8
Manus (hand)	36.6	27.8	90.9	36.3	61.8	29.1
Femur	31.0	23.4	46.3	26.8	33.6	21.7
Tibiotarsus	65.0	49.1	80.6	46.7	75.8	48.9
Tarsometatarsus	36.3	27.4	45.6	24.1	45.6	29.4

a [238, electronic supplement].

The hipbones also include a pronounced primitive feature. The tips of the pubic bones are fused as they are in dinosaurs and so many other primitive birds, but the arms of those bones are exceptionally long and thin. *Gansus* is the earliest flying bird capable of laying a very large egg.

OTHER CRETACEOUS BIRDS   Two other Cretaceous birds appear to be more primitive than *Apsaravis* but still belong within the Neornithes. *Vorona* is known from only a single well-preserved leg found in Madagascar, and is perhaps most significant as an early fossil of a modern bird from an area outside of Gondwana. *Patagopteryx* was a flightless resident of the South American section of Gondwana. It was a land bird of about 2 kg that probably bore a superficial resemblance to a large rail, perhaps somewhat like the modern Takahe of New Zealand. Unlike *Vorona*, it is known from many fossils. There is enough material from this one species to define a large number of characteristics for use as a baseline in comparison to less well-represented birds.

The most familiar examples of modern birds in the Cretaceous are the two types discovered by O.C. Marsh in the 19th century. One is the toothed flightless diver *Hesperornis* (along with the more gracile *Baptornis*); the other is the toothed flying bird *Ichthyornis*. The divers are large and spectacular but so specialized for life in the water that they have little to offer our understanding of flying species. We have already seen how their superficial similarity to loons had an important effect on the construction of evolutionary trees in the 19th century. *Ichthyornis* should also have played an influential role, but soon after its discovery, many of its most important components were embedded in plaster for display purposes and so were effectively removed from further consideration. Additional fossils of both types were found throughout the 20th century.

Like the *Patagopteryx, Ichthyornis* is one of only a handful of modern birds from Cretaceous fossils represented by more than one or two bones. With the addition of Confucius and ball-shouldered birds to Class Aves, it became imperative to undertake a modern anatomical study of the *Ichthyornis* remains. Some 20th-century scientists felt that *Ichthyornis* would tell us nothing new because it was a chimera, constructed from the remains of several different kinds of birds; others were dividing *Ichthyornis* into as many as seven discrete species.

Recently, its fossils were removed from the plaster plaques and subjected to a comprehensive cladistic analysis by the same Julia Clarke who worked on *Apsaravis* and *Yixianornis*. She showed not only that most of the bones represented one or the other of the two species of *Ichthyornis* that Marsh had originally described but also that the rest included important remains of three other kinds of birds (Figure 8.7) [28].

Ichthyornis is often portrayed as though it was a kind of tern, a charadriiform bird. Clarke found, however, that it is also clearly separated from that group by significant structural characteristics. Those characteristics imply that any similarity is superficial and just another example of convergence. It is useful to understand

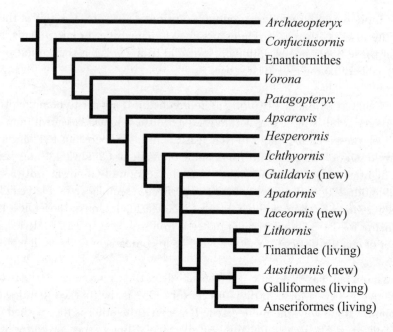

Figure 8.7   The family tree for birds developed by Julia Clarke after her analysis of the *Ichthyornis* fossils [28]. Fossils from other species that had been included in the display plaques of the Peabody Museum of Natural History are marked "new." *Austinornis* is attached to the galliform birds on the strength of only a single character. Clarke used *Dromaeosaurus* as the outgroup (not shown).

the level of convergence between *Ichthyornis* and charadriiform birds because the graculavids, another group of Cretaceous shoreline birds, also converges on them. That group's fossils are very abundant but they have never been successfully linked to a living species. If *Ichthyornis* converged on charadriiform terns in response to life on the shoreline, perhaps the graculavids moved in the same direction for the same reason.

New Fossils and New Phylogenies for Fossil Birds
For over a century before 1981, the evolutionary story of birds jumped 80 million years or so, from *Archaeopteryx* to *Hesperornis* and *Ichthyornis,* with only a few poorly understood fossils in between. Two concepts dominated thought about the early evolution of birds. The first concept was that birds could be divided into two major lineages, the extinct Sauriurae (lizard-tailed birds) and the surviving Ornithurae (bird-tailed birds). In the second, birds were seen as a single monophyletic lineage running from the ancestor of *Archaeopteryx* to modern forms.

The event that sparked new discussion was Cyril Walker's recognition of the Enantiornithes in 1981 [221]. At that time, many comparative morphologists rejected a theropod origin for birds, and they placed the Enantiornithes (and later the Confucius birds) among the Sauriurae, with *Archaeopteryx. Hesperornis* and *Ichthyornis* appeared with the modern birds in the Ornithurae. The concept led to a contorted family tree that required independent development, in each lineage, of an alula, a pygostyle, a keeled breastbone, and the other specialized features needed for advanced flight.

Luis Chiappe is perhaps the most prominent spokesman for the second interpretation of avian history. In the closing chapter of *Mesozoic birds,* he reviews the history of research since Cyril Walker's discovery and offers a new phylogeny based on a cladistic analysis. It arrays 17 taxa according to 169 characters: 17% from the skull, 82% from the postcranial skeleton, and 1% from the plumage [25]. He uses the mallard to represent modern types in the crown clade, and anchors his tree with an outgroup based on three different kinds of theropods – allosaurs, velociraptors, and troodontids. All 169 characters were shared by at least two taxa. Unique (autapomorphic) characters may be useful in increasing confidence in the definition of certain groups but they do not help the analysis distinguish between groups and he decided against their use.

Chiappe's phylogeny shows birds as a single lineage, with each of the major groups leaving the main stem in sequence (Figure 8.8). It lays out the whole early history of the birds in a simple and logically satisfying way. The theropods are at the base and the peculiar alvarezsaurid, *Mononykus,* which has so many birdlike features, appears as a sister group, immediately outside the birds. *Archaeopteryx* and the slightly more advanced *Rahonavis* appear as the earliest birds. All the rest, from the Confucius birds to the mallard, have a pygostyle at the end of their tail, and Chiappe gives them the name "Pygostylia." Needless to say, all the major features associated with flight evolve only once.

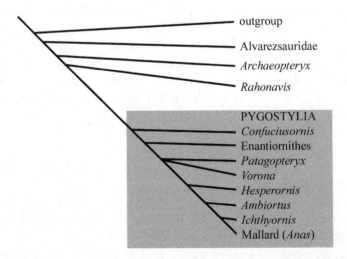

Figure 8.8 An abbreviated version of L.M. Chiappe's phylogeny for early birds, showing the groups included in the Pygostylia [25].

In spite of the large data set, the members of the Pygostylia share only three advanced or derived characters (synapomorphies): (1) the relation of the nostril to the rostrum of the skull, (2) the presence of a prominent acromion[2] on the shoulder blade, and (3) the relative size and shape of the hipbone in front of and behind the hip joint. Of these, only the acromion is directly related to flight because it forms the outer edge of the triosseal gap, through which the tendon from the supracoracoideus muscle passes to the humerus. This position makes the acromion a vital part of the pulley system that allows a muscle on the breast to raise the wing. Some of the remaining 166 characters may eventually prove to be synapomorphies but at this itme, they are simply not preserved in one fossil or another. The rest have been lost or changed during the long march from *Archaeopteryx* to modern birds.

Modern Families of Birds in the Age of Dinosaurs

Based on gene sequences and other evidence from the molecular clock, it seems likely that some of the major divisions within the modern birds date back far into the Cretaceous. Assigning Mesozoic fossils to living taxa has always been very controversial. Ironically, the recent flurry of new avian fossils has weakened evidence for modern birds in the Mesozoic even as new biomolecular data strengthen it. The problem lies in both the quality of the fossils and they way they have been analyzed.

According to Julia Clarke, only 13 modern birds known from Mesozoic fossils are represented by more than a single bone, and are therefore based on enough characters for a reliable analysis [28]. Even conclusions about the significance of such fossils is based on tenuous evidence. Her own assignment of the Cretaceous *Austinornis* to galliform birds is based on a single character.

In another recent review of Mesozoic fossils, Sylvia Hope tentatively assigned 39 fossils of modern birds from the late Mesozoic to living orders: Galliformes (6), Anseriformes (6), Charadriiformes (19), Procellariiformes (2), Gaviiformes (2), Pelecaniformes (3), and Psittaciformes (1). Ten others appeared to represent modern groups of birds but could not be assigned to a living group. Of that list, only the anseriform *Presbyornis,* the Antarctic *Polarornis,* which appears to belong to the Gaviiformes, and two of the unassigned fossils were represented by more than two or three different bones [86]. Unfortunately, both charadriiform birds and pelecaniform birds are only weakly defined by skeletal characters. Neither may be a natural assemblage. Consequently, the placement of fossils in either of these groups may signify nothing more than widespread convergence among early groups of marine and coastal birds.

Hope dismisses all fossils of supposedly modern birds from the Lower Cretaceous because their features are ambiguous or the fossils are of such low quality that they preclude a confident decision. She selects *Apatornis,* from the Upper Cretaceous, as

---

[2] The acromion is a distinct prominence on the forward edge of the scapula, near the articulation with the humerus. It forms the outer edge of the triosseal canal.

the earliest undoubted example of a modern bird,[3] and she finds some significant similarities between it and geese or flamingos. Unfortunately, *Apatornis* was originally described from a worn and broken synsacrum that had later been poorly repaired. The relationship between that fragment and better-preserved material that was found later is uncertain. The best that can be said is that all the material shares characters with modern anseriform birds, which implies a very early origin for that group.

Anseriforms are neognath birds. An unchallenged relationship to the ancient *Apatornis* would imply a split between neognaths and paleognaths over 84 million years ago. Acceptance of such an early date would create one of the few occasions when the fossil record agrees with biomolecular analyses.

An early origin for the anseriform birds implies an equally early origin for their sister group, the galliforms. *Palintropus,* from the Campanian stage of 74-75 million years ago, provides some physical evidence. It is known from five coracoids and a scapula that seem very similar to those of the Quercymegapodidae, a more recent group of fossil galliforms from Quercy, France. In turn, those fossils bear a strong resemblance to living megapodes, sometimes called bush turkeys, that are found in Australasia.

Another likely representative of the Galloanserinae is called *Presbyornis.* It is known from a very large number of fossils and appears to have been a long-legged duck. It was widespread in the Cretaceous and fossils have been found in Antarctica [140], North America, and perhaps Mongolia. It is particularly noteworthy for surviving the great extinction events at the end of the Cretaceous, to appear in Paleocene and some later deposits [53, 58, 116].

Presbyornis is not the only fossil anseriform found in Antarctica. Early in 2005, a team led by Julia Clarke completed a cladistic analysis of *Vegavis iaai,* a fossil from Upper Cretaceous deposits on Vega Island in Western Antarctica [30]. Unlike most other Cretaceous bird fossils, it does not consist of a single isolated bone but includes several vertebrae, wing and leg bones, and a complete pelvis. The fossil shows 20 unambiguous synapomorphies with modern anseriform birds. Because the dating is reliable and the cladistic analysis is based on a large number of characters, *Vegavis* becomes the first, and so far the only, Cretaceous fossil that can be placed within a group of living birds with a quantifiable degree of confidence.

As mentioned above in the discussion of *Ichthyornis,* the Graculavidae are an extinct group of shorebird-like animals whose fossils occur in very large numbers. They appear to have enjoyed a long period of evolution in the Cretaceous, which contributed to their variability and makes them difficult to interpret. They simply share too many characters with other groups for confident assignment to any of the modern orders. Their uncertain affinities have generated considerable controversy. Alan Feduccia argues that once the dinosaurs were extinct, the graculavids

---

[3] At the time of her analysis, neither *Apsaravis* nor *Yixianornis* had been discovered.

were the source of an explosive radiation that gave us all living modern birds [60]. His theory rejects both the identity of most neornithean fossils from the Cretaceous and the interpretation of biomolecular evidence. It also denies the importance of Gondwana as a refugium for animals that managed to survive the great extinction at the end of the Cretaceous.

The fossil evidence for a refugium in Gondwana continues to grow. Both the loonlike *Polarornis* and the anseriform *Vegavis* have been found in Antarctica, and it is likely that fossils of other neornithean birds are waiting to be found in such places or are hidden under the ice fields of Antarctica. Other parts of Gondwana were also important. W.E. Boles has reported the world's oldest known songbird from Eocene deposits in Australia, while recent excavation on the Chatham Islands off New Zealand suggest that the islands were a late refugium for dinosaurs.

The *Polarornis,* found in Antarctica by Sankar Chatterjee, has special significance for evolutionary discussions. Loons appear to belong among recently evolved clades with the penguins and albatrosses. If the biomolecular phylogeny of Sibley and Ahlquist [192] reflects the true evolutionary history of birds, all other lineages – from forest birds, to shorebirds, to cranes – must also have existed in the Cretaceous. Fossil evidence for such diversity has yet to be found.

The main problem with interpretation of Cretaceous fossils is that few of them have been subjected to the kind of cladistic analysis that Julia Clarke used on *Ichthyornis* or *Vegavis* [28, 30]. Without such an analysis, fossils such as the loonlike *Polarornis* cannot be placed in an evolutionary context. Other fossils are less complete but suggest that we will eventually have evidence that other important groups existed in the Mesozoic. An albatross-like furcula from Mongolia and the North American *Lonchodytes* could place the petrels in the Cretaceous. There are also recognizable cormorants from those locations.

Unfortunately, the problem of relationships is not limited to fossil birds. As we have seen in earlier chapters, ornithologists are still struggling to apply cladistic techniques to living birds: "How can we know to which group a fossil under study belongs if the characters considered diagnostic for the major clades within Neoaves have yet to be formulated, much less tested?" [54]. We will not be able to link fossils to living birds with any confidence until there is a phylogeny for Class Aves based on a comprehensive cladistic analysis constructed from a large number of characters.

The Ball-Shouldered *Enantiornis* and the Enigmatic, Socket-Shouldered Hoatzin

The general similarity from one skeleton to another among forest birds has contributed to an amazing situation in which evolutionary convergence has apparently reached back in time to create a modern bird with both basic features and structural details that are almost identical to those found in a long-extinct and completely unrelated group. There are no modern animals that truly mimic dinosaurs, and there are no modern equivalents to *Archaeopteryx* among living birds. Even the more advanced Confucius bird has no modern imitator. We appear to have a modern equivalent to a ball-shouldered bird, however. The subclass of ball-shouldered birds disappeared with the dinosaurs at the end of the Cretaceous, but their fossil skeletons echo strongly in the modern South American Hoatzin.

The name "Hoatzin" has been attached to the word "enigma" so often that the phrase has become an ornithological cliché. The Hoatzin (*Opisthocomus hoazin*) looks much like a chicken but does not live on the ground. It spends its life among the branches of dense shrubs that overhang the water. As you might expect in any large, edible bird, most of its feathers are rather drab and discrete. It has a long, broad tail and the wings are unusually big and round, larger for its size than the wings of a parrot or toucan. They generate a slow, almost ponderous flight. Like many larger forest birds, the Hoatzin makes a few deep flaps and then glides until it needs to flap again to stay airborne. It looks vulnerable in the air and you might expect a little discretion, but the Hoatzin advertises its movements with vivid chestnut feathers that flash in the sun as the wings open. It attracts even more attention by repeating loud calls as it flies. It seems an easy target for any passing hawk or eagle; this might be one reason why you rarely see a Hoatzin in the air.

Even among the branches of lakeside trees, Hoatzins are often surprisingly visible. They are very clumsy and make the branches rustle and bounce up and down as they climb about. Again, the birds emit loud calls every now and then. Perhaps their defence mechanism is rather subtle – they are known make an unappetizing meal. They supposedly smell like fresh cow dung, but I have never been able to get close enough to find out. The smell might discourage predators, but snakes have no sense of smell and jaguars regularly eat rank, decomposing meat. Perhaps the big cats turn up their noses at a bird whose body is little more than a sac of fermenting leaves: a quarter of the bird's weight may consist of partially digested plant material, and there certainly isn't much nutritive value in the rest of a Hoatzin. The flight muscles are long and stringy, and indigestible feathers make up a disproportionate amount of the bird's remaining volume.

A Hoatzin eats large quantities of leaves that are both low in nutritional value and rich in the exotic toxins with which plants discourage browsers. To digest its leaf clippings, it has a hugely enlarged crop that creates the space necessary for fermentation. Just like the rumen of a cow, this sac carries a whole community of specialized microbes that are responsible for chemical reduction and detoxification of the food. Cowlike behaviour seems to follow cowlike anatomy, and the Hoatzin spends much of its life dozing while the microbes carry out their activities. The weight of the fermenting "cud" may be one reason that the Hoatzin rarely takes to the air.

Making space for the crop in the highly specialized skeleton of a bird has required some unique structural adaptations. Although the Hoatzin does not lead a very active life, its skeleton consists of surprisingly large and robust bones. The production of bone and eggshell requires a great deal of calcium. Many birds find it hard to come by and are forced to make do with a very modest supply. Small birds often eat insects, which are usually a particularly poor source of calcium, and many birds take it up directly by eating land snails or the discarded shells of their own eggs. On the other hand, plants can be an excellent source of calcium, and a vegetarian diet may explain the Hoatzin's ability to grow some spectacularly large bones. Older adults lay down so much new bone within sheets of connective tissue that parts the skeleton often become rigidly fused and disappear as discrete structures.

In keeping with the bird's weight and rather basic and laborious flight style, parts of the pectoral girdle are particularly large and many of its bones are welded together. Only the scapula seems relatively short compared with those of other birds, but it is also thick and stiff. Instead of lying over the ribs, to support to the dorsal vertebrae, it is angled upward. In adults, sheets of bone grow over the shoulder joint and make it rigid. The fusion is more superficial than real, and the internal structures remain distinct. For instance, the ball-and-socket joint between the scapula and the coracoid bone remains an articulating surface in adult birds in spite of its immobility.

The pectoral girdle is further stiffened by fusion of the furcula to the sternum. The Hoatzin's furcula is huge and consists of three roughly equal arms – two forward and one back. The forward arms lie alongside the coracoid bones, which are unusually long and quite strongly built. Together they create an oblong frame in the space between the shoulders and the breastbone that supports the enlarged crop. The rearward arm of the furcula is an extension of the hypocleideum. It travels down the face of the sternum towards a small keel that is just a thick ridge on the after part of the sternum. This enlarged and extended hypocleideum functions as part of the keel in the sense that it is the site for the attachment of flight muscles (Figure 8.9). As the bird matures, sheets of new bone obscure the joint between the extended furcula and the breastbone.

The crest of the keel on the breastbone is thickened and covered by a pad of connective tissue and thickened skin. This pad creates a comfortable cushion on which the Hoatzin rests during the long hours it sits around waiting for microbes to ferment its lunch. Theropod dinosaurs appear to have rested on a similar pad at the end of their pubic bones. Their bipedal legs may have made it difficult to get up from a completely prone position, but resting comfortably propped up on their belly left their eyes in a good position to spot approaching threats.

In the Hoatzin, the coracoids become fused to the front edge of the sternum shortly after hatching. They look typical but have a rather unusual internal construction. The coracoids of many large birds have a flattened inner surface. A simple "D" takes up less space and is only slightly less resistant to bending than an "O" of similar size. In a Hoatzin, the cross-section of the coracoid only looks like a simple "D." The flat inner surface is created by a sheet of new bone that forms a roof over a deep groove lying on the inner side of the coracoid. Without taking up more space, the groove creates a very strong box of bone that has four vertical walls to resist bending (Figure 8.10). The roof that extends over the groove provides extra support, effectively creating a super-strong double-box or figure-eight beam.

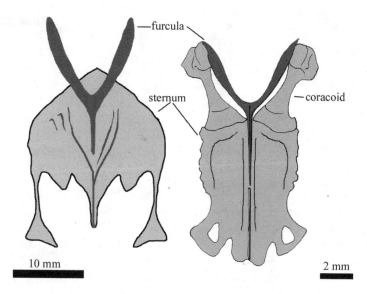

Figure 8.9 Reconstruction of the sternum and furcula from a member of the Enantiornithes (ball-shouldered birds) (*Cathayornis*) (left) [239], and the same bones from a late embryo of the Hoatzin (right) [155].

In a sedentary bird with exceptionally long coracoids like the Hoatzin, both strength and space saving are likely to be advantageous when small increases in weight are not significant considerations. The roof over the groove in the coracoid is not complete. It ends in a ragged edge, near the joint with the sternum, and the original construction is visible on the inside, where it meets the edge of the sternum.

In 1891, W.K. Parker prevailed upon a friend to provide him with some carefully pickled Hoatzin embryos, from eggs that had been close to hatching. Careful dissection enabled him to draw some of the important skeletal features of the Hoatzin before they could be hidden by adult development (Figure 1.1). The results appear stunningly similar to the reconstruction of the fossil *Enantiornis*, the stereotypical enantiornithean bird, known only from Cretaceous fossils. The most dramatic similarity to *Enantiornis* is the nearly keel-less face of the sternum in the Hoatzin embryo. The extended tail of the furcula reaches back to a low ridge on the back part of the sternal plate. Of course, the significance of such a similarity was not available to Parker. It would be another hundred years before Cyril Walker discovered the ball-shouldered birds.

I am itching to suggest that the Hoatzin might be a surviving member of the ball-shouldered birds but, sadly, it is not so. The similarity between the Hoatzin and *Enantiornis* is just another example of convergent evolution. Just one living member of the ball-shouldered group for comparison with modern birds would probably make the convergence obvious. It would also offer much-needed insight into the various evolutionary options available to birds during their long history. Perhaps the most important question, after 65 million years, is: "How can we

The simple, tubular
coracoid found
in most birds

In-folded coracoid
found in *Enantiornis*
(extinct)

Roofed and in-folded
coracoid found in the
Hoatzin

Figure 8.10 Diagrammatic cross-sections showing the types of construction used in the coracoid bone. Most birds have a simple tubular form. The Enantiornithes (ball-shouldered birds) had a deep groove down the inside of the bone. Hoatzins also have a deep groove, but growth of new bone in later life forms a roof over it to create a more complex structure.

know that it is merely convergence?" The practitioners of comparative anatomy and cladistics have not been so resoundingly successful in their attempts to classify the living birds that we can be wholly confident in their ability to distinguish between a living bird and an ancient fossil without resorting to DNA.

The gross structural similarities between the Hoatzin and *Enantiornis* suggest that the latter also ate leaves. We cannot tell whether it "chewed" its food like a Hoatzin but it may have digested it in an enlarged crop. Accommodating a crop makes unusual demands on the structure of the pectoral girdle. As we have seen above, the unusually elongated coracoids of the Hoatzin are not the simple tubular bones found in most birds, but complex "U" or box-beams.

The combination of long coracoids and a short sternum has left both the Hoatzin and *Enantiornis* with little room for ribs. In both groups, the ribs are rather broad and short, looking more like protective armour than part of the respiratory apparatus. They have the typical flexible two-headed connection with the vertebrae, a joint near their midpoint, and, ultimately, a connection to the sternum. As a group, however, they are only large enough to encase the heart and upper surface of the liver. Almost all modern birds have uncinate processes that extend rearward from the middle of their ribs. These have not been found in the fossils of *Enantiornis*, but if they were like the Hoatzin's, they could have been lost or misinterpreted. The uncinate processes in the Hoatzin are unusual. They are not the long extensions that cross over neighbouring ribs but flat biscuits of bone that sit between the ribs like a designer's afterthought. Parker's drawing of the embryonic Hoatzin shows them as oblongs of bone that are much wider than the adjacent ribs. In the adult Hoatzin, there is often so much secondary development of bone that the actual boundaries of the ribs and uncinate processes are obscure.

In spite of the structural similarities between the Hoatzin and *Enantiornis*, there are two pieces of skeletal evidence that place them in unrelated groups. First, the tarsometatarsi fuse from the top in all living birds and begin development as an M-shaped fusion of long tarsal bones. *Enantiornis* is indeed the opposite: its tarso-metatarsus fuses from the bottom and looks more like a "W." Parker's 1891 illustrations from the late embryo of the Hoatzin show just the basic bony elements and suggest that the Hoatzin's development is the same as that of any other neornithean bird. Second, the pectoral girdle separates the Hoatzin from *Enantiornis*. The Hoatzin's shoulder is clearly that of a modern bird. Beneath the layers of secondary bony deposits that help make the pectoral girdle rigid, the adult Hoatzin has the familiar round boss on the scapula that fits into the pit at the top of the coracoid. *Enantiornis* is a typical ball-shouldered bird, with a cup on the scapula to hold the round boss on the coracoid.

Although the Hoatzin and *Enantiornis* have solved specific biomechanical challenges in different ways, their overall structural similarity suggests that the two birds shared some basic features in their lifestyles. It seems likely that *Enantiornis* was also a herbivorous bird that led a relatively inactive life. It may have been so sedentary that it never needed fully developed homeothermy. We do not know why *Enantiornis* and all its relatives became extinct with the dinosaurs, but the limited success of the Hoatzin offers a clue. Its survival has been a matter of luck. The Hoatzin is so specialized that it survives only as a solitary species in a single, very narrow ecological niche.

The primitive-looking fingers on the wings of juvenile Hoatzins have always attracted attention. Amazingly, the other structural peculiarities of the Hoatzin are also all reminiscent of features in primitive birds. The Y-shaped furcula with the long hypocleideum that functions as a keel appears in many early birds. The grooved coracoids appear in the early ball-shouldered birds. Even the rigid fusion of the shoulder joint echoes the fusion of the scapulas and coracoids in the Confucius birds. The Hoatzin cannot be an enantiornithean but it may have retained a remarkable suite of primitive features. Perhaps its lifestyle was widespread during the Cretaceous, and many early neornithean birds fermented leaf clippings in enlarged crops.

A Possible Role for the Furcula in the Early History of Flight

Most of the muscles in a typical bird are concentrated in two large masses, one to drive the hind legs and one to power the wings. These masses are either attached to or closely associated with the two large bony structures in the body: the wide hipbones that roof the abdominal vault or the huge sternum with its massive central keel. It is reasonably easy to understand how the small hipbones of a dinosaur expanded to swallow the tail because the events involved existing structures and their remnants are visible in the final product. The history of the sternum and pectoral girdle is much less obvious. There is no sign of a massive keel in primitive birds, but it is so dominant in modern birds that it is difficult to imagine flight without it.

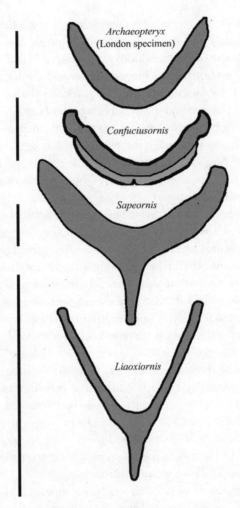

Figure 8.11   The furcula in four fossil birds, showing the increase in the length of the hypocleideum and the arms (the bar to the left of each furcula indicates approximately 1 cm).

One possibility is that the furcula played a key role in the early evolution of flight and later contributed a central keel to the small and fragile sternum of early birds. As mentioned in Chapter 7, Storrs Olson and Alan Feduccia have put forward the idea that the flight muscles of *Archaeopteryx* were attached to its simple barlike furcula [152]. Andrzej Elzanowski argues that the furcula was mounted too high on the shoulder and only the coracoids are large enough and have sufficient elaboration to carry flight muscles [56]. Later fossils suggest that both bones had roles to play (Figure 8.11).

In *Confuciusornis*, the furcula is still rather simple, although it does have a peculiar flange or groove along its lower edge. On the other hand, its coracoids are fused to its scapulas to create an elaborate rigid framework. The coracoids in *Confuciusornis* are still rather stubby, and flight muscles attached to them would be much shorter than those in more modern birds. Some workers have reported a slight keel on the sternum, but the sternum appears too flimsy to have played an important role.

Sapeornis is the earliest fossil bird to exhibit a significant midline structure that could have supported flight muscles [240]. Its furcula has a robust but short hypocleideum that extends towards the breast. The furcula is also more strongly curved than in other birds, but its arms and the coracoids are elongated and suggestive of the strut-like structure found in modern birds. All more advanced birds have a furcula that can be described as a "Y," "U," or "V."

In the fossil of *Liaoxiornis* (Figure 8.6), we can see that the long arms of the furcula are matched by a long hypocleideum where they meet. The coracoids are also elongated and the pectoral girdle could have supported the long muscles needed for strong flight. The elongated hypocleideum may also have made an important contribution by carrying some muscle slips further towards the breast, closer to their point of attachment in modern birds.

The elongated hypocleideum in early birds such as *Liaoxiornis* may have been the precursor for the type of keel found in the modern Hoatzin or the extinct *Enantiornis*. In both birds, the hypocleideum fuses to the breastbone (Figure 8.11). Besides the Hoatzin discussed above, there may be one other example of this scenario in living birds. The pelican lacks a keel on the back part of its sternum, and such keel as there is fills the space between a hugely inflated furcula and the forward part of the sternum (Figure 8.12). Embryological studies would be required

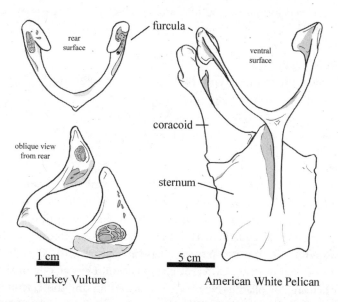

Turkey Vulture   American White Pelican

Figure 8.12 The furcula in two modern birds. On the left is the elaborate recurved furcula of a Turkey Vulture. On the right is the greatly elongated and inflated furcula of an American White Pelican. In the pelican, the furcula is fused to a short, wide sternum with a greatly reduced keel.

to determine the source of these bones. Most other living birds have a keel very similar to that seen in the fossil of *Yixianornis*. Perhaps the development of this keel provided another important advantage over more primitive birds.

While the hypocleideum has declined in importance, the arms of the furcula in a living bird usually play an important role in flight and they exhibit a wide range of shapes and sizes. There are exceptions: in some groups, the arms are reduced to a thin strap or loop of ossified ligament between the clavicles; in scattered cases, they disappear altogether. Significantly, the species that have such a reduced furcula are forest birds, which make the most use of slow, vortex-ring flight. None are long-distance migrants or any other form of aerial specialist.

In aerial specialists, the furcula retains or regains some of its importance and is often a large and complex bone. In terrestrial soaring birds, the arms may be as robust as the coracoids (Figure 8.12). The furcula is often closely associated with the sternum and may be fused to its leading edge. It is also large and important in birds that achieve high speeds, where it is usually placed some distance forward of the sternum. Typically, it is deeply curved and sweeps forward from the shoulder in an arc before returning to the body. In some species, there is a short hypocleideum with a ligament attached to the leading edge of the sternum. In others, the furcula is a smooth recurved tube with a wide space between its base and the sternum. The deep curve of these bones carries a segment of the pectoral muscle forward, increasing its length and power. Consequently, a strongly recurved furcula is typical of fast aerial predators, such as falcons; wing-propelled diving birds, such as auks; and fast-flying birds with a high wing loading, such as loons or grebes. It is also strongly recurved in marine soaring birds, such as the albatross, which need fine control of their wings for efficient flight.

Post-Cretaceous Fossils

Unlike earlier fossil birds, most types that appeared in the early Tertiary are quite similar to modern forms. Except for a few flightless giants, none would look out of place in the modern world. Not all can be assigned to a modern order, but all are members of the same subclass, Neornithes, or modern birds.

Lithornithiformes

Some of the best-known and best-understood avian fossils from the early Tertiary belong to the Lithornithiformes. We tend to think of bird fossils as having a short history in human terms, but Richard Owen published a description of a *Lithornis vulturinus* from the London Clays in 1841. He saw them as relatives of the New World vultures, and one of his genera is called *Paracathartes*. In 1896, however, Frank Beddard proposed a very different classification. He was one of the first to see "no good reason" for placing New World vultures among the Falconiformes but thought that there were enough similarities between the lithornithid birds and accipiter hawks to place the fossil group among the Falconiformes.

It was not until 1988 that Peter Houde demonstrated that the lithornithids should be looked upon as an extinct order of flying paleognath birds, somewhat similar to the modern tinamous [92]. Considering that tinamous, galliform birds, and accipiters share many morphological similarities and have sometimes been lumped together taxonomically, his idea does not require a great deal of imagination. The skeleton of the lithornithids implies that they were likely better, or at least more frequent, fliers than tinamous. They had a keeled sternum, a robust furcula, and a large, curved humerus. It had been the curved humerus that reminded early workers of New World vultures. In the lithornithids, the shape suggests that they made fairly deep strokes during flight, but perhaps only for take-off. The attachment points for the muscles are not appropriate for the vigorous flapping that tinamous and galliform birds use to get airborne.

The small, unspecialized furcula of the lithornithids has nothing in common with the deeply recurved furcula of vultures or accipiter hawks, but it is, with a little imagination, comparable to the rather simple furcula found in rails and cranes. It was not fused to the sternum. Houde suggests that lithornithids probably flew by combining periods of wing beats with periods of gliding. Many animals that undertake an extended activity such as flight use short rests to recharge their muscles and increase their overall endurance. It is one of the habits that contributes to bounding flight in many birds that are too small to glide or soar effectively.

The tail of the lithornithids is typical of living paleognaths and other living birds without strong flight capabilities, and it may not have played much of a role in their flight. The pygostyle is particularly tiny and suggests that the tail fan was too small to be an effective rudder. Perhaps lithornithids trailed their feet behind, like herons or egrets, and flew rather slowly in long, straight lines.

If lithornithids shared their flight style with small herons and egrets, they may have lived in a world with few aerial predators. In North America, Bald Eagles are notoriously lazy predators and rarely pursue birds unless made desperate by hunger. Winter roosting areas along the coast are usually littered with the remains of gumboot chitons, abalone, and dead fish, but only a few bird wings. Some birds offer exceptional foraging opportunities to eagles by gathering in large numbers [105]; others, such as cormorants, are large, slow targets. Under those circumstances, some of the eagles seem to become specialists in hunting other birds.

There is some evidence that such eagles are effective enough as predators of other birds to limit the distribution and abundance of some slow-flying birds, such as herons. Bald Eagles and Great Blue Herons coexist along the coast of British Columbia, but not peacefully. For most of the 20th century, Bald Eagles were rather rare in central North America. They were shot as vermin and the effects of DDT decimated their breeding populations after the Second World War. Even in British Columbia, where salmon and other fish provided abundant food, there were very few breeders along the heavily urbanized southern coast. The spread of cities also affected herons. As suburbs consumed suitable sites, Great Blue Herons began to concentrate in a few very large sites. Further north, where there were more eagles and fewer people, the herons nested furtively in ones and twos. As the eagle population responded to lower levels of DDT in the environment, it began to expand and more eagles began to breed along the south coast of British Columbia. They soon discovered rich hunting opportunities at the large heron colonies, and began to harass them regularly. If the eagle population continues to expand, it seems likely that the breeding success at large heron colonies will decline and the remnant of the southern population may be forced to adopt the secretive and solitary strategy of their northern relatives.

Bald Eagles may also have had a more direct effect on a smaller heron, one closer in size to the extinct lithornithids. Several years ago, the Cattle Egret spread across the United States, and for a short while it appeared regularly along the Pacific coast of Canada. The area supports small populations of Green Herons and Black-crowned Night Herons, but the egrets never got established. A small white bird that flew slowly was too easy a target for Bald Eagles. Of the 10 or so egrets that I saw, eagles killed two while I watched. The eagle was simply faster and more manoeuvrable, and had no problem casually plucking the egret out of the air as it passed by. It did not even bother to dive at the egret for a hard strike. If lithornithids were slow fliers like the egret and also flew in long, straight lines, they would have been easy prey for early predators. There is no evidence that predation caused their eventual disappearance, but it may have made a significant contribution. Other wetland birds with similar flight capabilities, such as rails and bitterns, have learned to depend on stealth and secrecy for survival.

Peter Houde reconstructed some of the habits of the lithornithids from the position of muscle attachments and articulating surfaces in the leg joints. Like rails, they had fairly long legs that were not specialized for running. It seems likely that they walked with their ankles close together and their legs somewhat parallel. The elegant walk of modern cranes and other birds that move their feet through ground vegetation is based on a similar pattern. When cranes want to run, they straighten their legs to change both their gait and their stance.

In life, lithornithids were smaller than most tinamous and much smaller than any living ratite. They had long beaks and very short tails, and may have looked more like a rail than a crane because of their size. Like sandpipers, they probably fed by probing for invertebrates and appear to have been able to open the tip of their beak without moving the whole jaw. This trick is called rhynchokinesis, and it appears in birds such as the Dunlin, which probes for food with a sensitive beak. The tip of the beak in lithornithids shows many small fovea, or openings for nerves. These probably served the sensory receptors along the margin of the beak, which most birds use to detect food by smell or taste.

The discovery of lithornithid fossils in western North America and in Western Europe suggests that the modern paleognaths, including the flightless ratites, had flying ancestors. The peculiar southern distribution of the surviving ratites is rather puzzling and may not be entirely consistent with theories based on the break-up of Gondwana. Flight capabilities would need to have been widespread among their early ancestors to offer an alternative explanation, however. At this time, no lithornithids have been found in Asia, but the neognath fossil *Apsaravis* suggests that the distinctions between the paleognath and neognath lineages developed much earlier than previously thought. We have yet to find fossil evidence supporting a gradual dispersal of ratites to more northern continents early in the Cretaceous.

Post-Mesozoic Survival and Experimentation

Whatever the nature of the event that closed the Cretaceous period, it created the opportunity for a great deal of experimentation and diversification among the birds. Many lineages revisited the strategies that had made dinosaurs successful and produced a series of large, flightless browsers and fierce bipedal predators. If the heavyweight vegetarian *Diatryma* or the carnivorous phorusrhacids had survived into modern times, birdwatching would be one of the more thrilling adrenaline sports[4] [3]. The giants played a major role in the Eocene ecosystems and left many spectacularly ferocious-looking fossils, but their similarity to dinosaurs was superficial. None achieved the great size of their ancestors and none abandoned the inner-bird architecture or regained a usable hand.

Flightless birds are particularly vulnerable to the depredations of mammals, especially during the nestling period. Today, most survive only in special circumstances, such as isolated oceanic islands. There are a few modern types, however, that have recast themselves as neo-dinosaurs and managed to survive on continental mainlands. The ostrich, emu, and rhea live on open plains and the bad-tempered cassowary prowls dense forests. Not only is the cassowary overly sensitive to disturbance but its inner toe is armed with a fierce dagger, reminiscent of the extendable claw on the foot of ancient theropods. Other examples of flightless giants survived until humans arrived to change the local ecology. The sad stories of the dodos, elephant birds, moas, Great Auks, and other bizarre birds are well-known examples of the impact of humans as predators.

The survival of modern flightless birds may be related to their ability to mimic the most successful strategies of their reptilian ancestors. Birds that have found themselves on small islands without mammalian predators have frequently experimented with a pedestrian existence. Flight is an energetically expensive form of locomotion and not particularly valuable in relatively small spaces or on small islands. The rails, a group not noted for particularly strong flight in the first place, have produced the greatest number of examples, but there are, or have been, flightless pigeons, owls, and ibises as well as flightless marine birds such as auks and cormorants. Many of these experiments failed as soon as humans or other

---

[4] Apparently, relatives of the cranes made several independent attempts at gigantism. Herculano Alvarengo and Elizabeth Höfling have recently reviewed the Phorusrhacidae in relation to some of the other giant flightless birds [3; available on the Internet at http://www.scielo.br.paz.htm].

mammals arrived on the scene, but penguins are proof that flightlessness is not always a failing strategy.

There are many fossils of an intriguing group of flightless seabirds from the North Pacific that offer an example of a post-Cretaceous experiment. The birds have no modern equivalent, although they share some features with the penguins and may be one of their closer relatives. Several species of the fossil *Plotopterus* have been found along the west coast of the United States and in Japan. The great Californian paleontologist Hildegarde Howard was the first to recognize their fragmentary fossils as the remains of a giant flightless, wing-propelled diving bird. At first it was thought to be a type of pelecaniform bird, something like a giant, flightless gannet, but recent studies suggest that it was more closely related to the penguins, with which it shared aspects of its lifestyle [130]. If it nested or roosted on coastal beaches, it may have lost out in a competition for space with newly evolving seals.

Another group of giant flying pelecaniform birds has been found all over the world among marine fossils of the early Tertiary. Recent cladistic analyses suggest that they may actually be an anseriform bird, even though it is difficult to imagine a relative of the ducks converging on the albatross. Their name, Pseudodontornithes, refers to peculiar serrations along the edge of the mandibles that bear a superficial resemblance to teeth. They are neither as heavy nor as strong as real teeth, but they probably helped grip slippery prey such as squid. As in modern pelicans, the lower jaw was hinged midway along its length so that it could bow open to accept exceptionally large prey. These birds probably looked like a version of the modern gull or fulmar, but their specialized soaring wings spanned as much as 6 m.

It seems unlikely that pseudodontorns needed to nest on beaches like *Plotopterus*, but their reproductive strategy may have had other weaknesses. The young of a modern marine giant, such as the albatross, may spend over a year in the nest even though it hatches from an exceptionally large egg. Safety during a long period of helplessness as a nestling can be achieved only on isolated islands that are free from both mammalian and avian predators. No modern pelecaniform or anseriform bird produces a particularly large egg compared with its body size, and the young of the extinct giants may have been helpless for an inordinately long time. Perhaps the rapidly diversifying world of the early Tertiary produced too many nest predators for such a strategy to be viable.

Conclusion

Compared with the controversies and complexities that surround the origin of the first birds, major features of their subsequent evolution appear surprisingly straightforward. Perhaps adoption of the inner-bird format is the only practical way to solve the biomechanical problems of flight in a bipedal vertebrate. It would be incredibly exciting to find the fossil of an animal that clearly followed some other path, but at the moment we have only the route from *Archaeopteryx* to moderns birds and some unhelpful examples that evolved from quadrupedal ancestors, such as pterodactyls and bats.

Archaeopteryx may be the most likely candidate for ancestor of the birds but structurally it was only a little more specialized than other theropods. It showed no tendency towards centralization even though that is such a fundamental characteristic of the inner bird and a feature of all three of the other major branches in the avian family tree. It is least obvious in the Confucius bird, with its elongated body, heavy legs, and fingered wings, but even it had a lightly built head and a shortened tail ending in a pygostyle. The ball-shouldered enantiornithean birds and the socket-shouldered neornithean birds appear to have followed parallel approaches towards the inner-bird format, and are distinguishable only by differences in the details of skeletal construction. In the latter, the head loses its teeth, the fingers of the wings lose their independence, the breast develops a keel from the furcula or sternum, and eventually wide hips surround the base of a tail that ends in a pygostyle.

The pygostyle of *Yixianornis,* a modern bird from the Lower Cretaceous, was associated with a tail fan and looks modern. It probably steered the bird in flight. In other groups of early birds, the function of the pygostyle remains a puzzle. In the Confucius birds, it seems to have been little more than a series of fused terminal vertebrae. There was no tail fan to act as a rudder, and the pygostyle may have lacked a locomotory function. Similarly, the ball-shouldered enantiornithean birds appear to have had no use for such a bone. The known examples of their tails are limited to two central feathers that would have been of little use in the air. Only in the modern birds does the pygostyle form a critical part of a tail that functions as a rudder.

Evolutionary history within the modern birds remains much more of a puzzle. Many of the best fossils appear to represent ancestral forms that might link one modern family to another, but the evidence is often equivocal and weakened by biomolecular studies suggesting that the living representatives of these candidate groups are unrelated. Unfortunately, neither morphology nor biomolecular chemistry has produced a generally accepted phylogeny. Without some agreement on the general shape of avian phylogeny, there can be no meaningful evolutionary story for the group. We can, however, find evidence for some of the significant events in their story.

In the last two chapters, we will examine how birds have exploited flight and increased egg size to become one of the most successful groups of vertebrates in history.

Further Reading

Surprisingly little has been written about the Confucius birds or ball-shouldered birds. Perhaps not enough time has passed since their discovery for ornithologists to digest their significance. The best sources are the books by Gregory Paul and by Lowell Dingus and

Timothy Rowe, although my own interest in the subject stems from early work by Alan Feduccia. *The age of birds* (1980) and *The origin and evolution of birds* (1996) discuss many Cretaceous and Tertiary fossil groups, but you have to beware of his interpretations of evolutionary processes. He talks about evolutionary mosaics and is one of the last holdouts against the theropod origin of birds. He is also dead set against the idea that modern groups of birds repeatedly radiated northward from refugia in Gondwana.

Paul, G.S. 2002. Dinosaurs of the air. Johns Hopkins University Press, Baltimore and London.

Dingus, L., and T. Rowe. 1998. The mistaken extinction: Dinosaur evolution and the origin of birds. W.H. Freeman, New York.

Feduccia, A. 1980. The age of birds. Harvard University Press, Cambridge, MA.

Feduccia, A. 1996. The origin and evolution of birds. Yale University Press, New Haven, CT.

9

A robust design helps birds compete with mammals

If ornithologists agree on any aspect of avian evolution, it is probably the importance of flight. Variations in flight capability and technique have played a central role in the evolutionary success of birds by giving them access to a wide range of resources and habitats. Although the great majority of birds spend more time walking or swimming than flying, and only a few aerial specialists catch prey on the wing, flight offers birds a strategic advantage over more pedestrian animals. They can often escape earthbound predators by flying only a few feet, up into a shrub, and they can locate resources scattered across the landscape more quickly than any walking animal. Some migrate thousands of miles between insect-rich summers in the north and fruit-rich winters in the south; others fly across oceans to find localized concentrations of squid.

All modern birds except hummingbirds use two basic methods of flight: vortex-ring and continuous-vortex. Continuous-vortex flight requires little more than that a straight wing move up and down to provide the bird with a relatively inexpensive way to cruise from site to site. Vortex-ring flight is based on the bird's ability to bend and contort the wing. It is used to lift the bird directly off the ground or to achieve complex manoeuvres at slow speed. Hummingbirds have achieved a special form of flight all their own. The remaining modern birds have combined the two basic methods to achieve six common flight strategies. Most use only one or two strategies, which are closely related to their foraging techniques or other important behaviours:

High-speed escape. Pheasants, grouse, and other galliform birds are noted for their ability to explode off the ground. Rapid wing beats drive them at high speeds, but the muscles soon tire and the birds are not very manoeuvrable in the air. Once their muscles are exhausted, they glide back to the ground. The paleognathous tinamou uses a similar tactic but is not capable of such an explosive launch.

Forest flight. The great majority of birds, roughly 85% of all species, live in forested habitats or are descended from ancestors that recently lived in such habitats. Almost all these birds share a very similar body plan and flight technique. They are small, have a low-density body, and fly at low speeds on short, rounded wings. They typically have rather long tails. Even the largest forms, such as ravens and hornbills, share these structural characteristics. Most forest birds are not capable of flying long distances but some species make intercontinental migrations. Typically, the migrants have longer wings than their less adventurous relatives.

Aerial pursuit. Some forest birds have become specialists at aerial pursuit. Most are very small species that pursue insects in forest gaps or over the canopy. This strategy has been adopted by larger species, such as the falcons, which prey on other birds. Specialists in aerial pursuit tend to have moderately long, pointed wings for speed and long tails that enhance their manoeuvrability during an attack.

Thermal soaring. Habitats beyond the edge of the forest offer many opportunities to birds but the resources tend to be widely dispersed. Birds that move into open country need to be able to search across very large areas. To reduce the energetic costs of lengthy flights, they have learned to exploit rising thermals and other forms of updraft that may carry them great distances with little muscular effort. There appears to be a lower limit to the size of bird that can exploit updrafts. Crows are able to exploit the occasional strong thermal, but the larger ravens use them regularly. Most soaring birds are exceptionally large compared with typical forest birds. They include groups such as cranes, hawks, eagles, and vultures. Teratorns, soaring vulture-like birds from the Pleistocene, were the largest flying birds in history.

Marine soaring. The sea offers particularly rich resources but, as in open terrestrial habitats, birds must be able to search across huge distances. Updrafts are not reliable over the sea but winds tend to be strong and steady. Marine soaring birds use the control and efficiency provided by long wings to exploit the smaller-scale events associated with waves. Because the wings enable extremely fine control of movement, the tail plays little role in flight and is usually very short.

High-energy flight. In many environments, there are unusual resources that offer extremely large amounts of energy, but a bird is likely to need a highly specialized flight style in order to exploit them successfully. The most familiar example is the sugar-rich nectar that powers the hovering flight of the hummingbird. Hovering gives this tiny bird ready access to flowers but consumes huge amounts of energy and has required extensive modifications to the bird's skeleton. No other bird has such a gigantic keel to carry its flight muscles.

There are also concentrated energy sources in the ocean. Many marine organisms use calorie-rich oils for buoyancy but they are patchily distributed and often far from shore. Auks have evolved a special flight tactic to help them exploit these animals. They fly from site to site at very high speeds so that they can fit foraging bouts into the feeding requirements of their young. The energetic cost to such a large bird is huge, but it is apparently easily met with their oily prey. Some other aquatic birds, such as ducks, loons, and grebes, have a similar but less demanding flight style. They also exploit less calorie-rich foods. All these birds have rather narrow, pointed wings that are specially reinforced to withstand the stresses of very rapid wing beats.

In terms of the evolutionary history of birds, all modern birds are far more advanced than *Archaeopteryx* in that they are clearly able to use both vortex-ring flight and the less energetically expensive continuous-vortex flight. If other flight styles evolved, they did not make it past the great extinctions at the end of the Mesozoic. In this chapter, we will focus on the flight of forest birds and other terrestrial types. We will save the unusual flight skills of marine birds for the final chapter.

Forest Birds Look Alike

If you ask a child to sketch half a dozen different mammals, chances are the first picture would look something like a basic dog or cow – four legs, a toothy head, and a simple tail. The other pictures might portray a person, a porpoise, a bat, or even a kangaroo, depending on the child's imagination and interests. All are perfectly good mammals whose specialized bodies are easily recognizable from rough sketches. If you asked for a half dozen birds, it would be more of a challenge, and you might to get a series that varies greatly in colour but not much in shape. There are no four-legged birds and – except for long legs and neck on an ostrich, longer wings on an albatross, or more fat on a penguin – there are not many outstanding features that can make the major groups easily recognizable. For most people, even the most specialized species look much like a "typical" bird.

This tendency of many different kinds of birds to look alike has had several important effects on ornithology. It is one of the factors that contributed to the inability of ornithologists to build a convincing family tree. Basically, information from the living species of birds has not been very useful in the discussion of avian evolution or the development of related subjects such as historical zoogeography. The degree of similarity also encouraged anatomical specialists to lump some groups together and split others without compelling evidence one way or the other. There are still ornithologists who are convinced that owls are closely related to hawks or that grebes are close to loons; others are equally convinced that these groups are unrelated. Taxonomists have compounded the problem by depending on only 41 features for classification, making it almost impossible to establish ranks for taxa represented by only one or two species. Should the Hoatzin be the sole representative of a family or an order? Alfred Romer, one of the 20th century's foremost anatomists, expressed his frustration with the limited structural variation among the birds by suggesting that the whole group hardly merited separation as a distinct class [179].

The plumage tends to emphasize birds' general similarity in shape by hiding structural differences and distracting us with the variety of colours and shapes in individual feathers. Without strong clues offered by feet or beaks, most naturalists would find it very tricky to sort a mixed group of museum skins as woodpeckers, kingfishers, bee-eaters, cuckoos, collies, parrots, pigeons, and songbirds. Even their skeletons look alike and are difficult to identify without expert knowledge of the fine details of individual bones.

The similarity among birds is a result of the series of biological and historical factors examined in earlier chapters. First, birds are a natural group whose basic skeletal architecture is the product of their descent from a shared common ancestor that was bipedal. Second, the ancestors of all living birds passed through a filter at the end of the Mesozoic that eliminated all their near relatives. All feathered animals except those with a highly centralized body mass, short tail, and well-developed three-unit leg were excluded. Even many advanced birds that fit these criteria, including all of the Enantiornithes, or ball-shouldered birds, failed to survive. Third, the avian genome remained small after the mass extinction at the end of the Mesozoic and has not been colonized by the huge numbers of viruses that appear to have produced so much variation in mammals over the last few million years.

Life on the Forest Floor and in the Canopy

Birds are descended from a long line of pedestrian animals whose success was based on their ability to find food on the ground or other substrates. Although many of these animals failed to survive the events at the end of the Mesozoic, their extinction cannot have been caused by their combination of foraging strategy and body design. This combination still offers many opportunities, especially among animals with the additional capability of flight, and is the lifestyle used by about 85% of all birds.

The abundance of dinosaurs in the ancient world and the combination of birds and mammals today illustrate the huge number of potential niches for animals that find food on foot. We should not be surprised to find that the ability to fly has not reduced the value of the foot among modern birds. Searching for food on foot is the only technique available to flightless, or nearly flightless, birds such as the paleognaths and galliforms, but it is also a very widespread technique in groups with better flight capabilities. Forest birds, in particular, include thousands of examples of pedestrians that prowl about on the forest floor looking for food. There are ground hornbills in the coracids; anis and roadrunners in the cuckoos; and pitas, antbirds, thrushes, and many others among the songbirds. There is even an extinct flightless owl that can only have hunted on the ground. Other forest birds have carried the pedestrian habit into the third dimension and forage while clinging to vertical surfaces on the trunks and branches of trees. Beyond the forest edge, many very large birds have also succeeded as pedestrians and stalk their prey through open grasslands and marshes. The feeding techniques of secretary birds, jacanas, sandpipers, rails, coots, cranes, and storks are not very far removed from that of the domestic chicken.

In effect, a version of the pedestrian habit has even been adopted by many waterbirds. Ducks, geese, and swans forage on the ground, but the "ground" that they explore happens to be the bottom of a water body. The more active species dive beneath the surface but most remain floating on top and simply stretch their long necks downward to reach their food. Only a few forest birds, such as flycatchers and swallows, have evolved into active predators. The ducks have also produced an active predator in the form of the merganser, which pursues fish in the water column.

If the Metaves truly represent a slightly more primitive group of birds, some of them may have avoided competition with the more advanced Coronaves by abandoning the pedestrian foraging technique. Many Metaves are unusual specialists that survive on the continental landmasses because their behaviour has no real equivalent among the Coronaves. Hummingbirds are unique as hovering nectivores; swifts are the only aerial insectivores that hunt at very high altitudes; frogmouths, nighthawks, and owlet nightjars are the only avian insectivores that hunt at night. Several of the Coronaves eat leaves, but the metavian Hoatzin's use of large-scale fermentation has no real equivalent in other birds. The Metaves that retained pedestrian foraging typically survive on isolated islands such as New Caledonia or Madagascar, where competition is not so intense. On the continental mainlands, the unique ability of the dove to produce a kind of "milk" for its young may have given it an edge over less specialized Coronaves.

Among the few kinds of birds that actively pursue vertebrate prey, metavian grebes might appear to be similar enough to coronavian loons for there to be competition between them. If so, grebes are arguably the more successful design. They boast more species (22 as opposed to 5) and are more globally distributed. Perhaps grebes owe their success to the exploitation of small lakes. Loons appear to require large water bodies, which are rare in many parts of the globe, and competition between these two groups may be more apparent than real.

The surviving Metaves represent groups that have avoided competition by finding habitats with few other occupants or evolving the specialized structures needed for unusual lifestyles. These options are not available to the great majority of birds. Their survival often depends on finding new ways to exploit the niches in a crowded forest environment. Fortunately, the forests are incredibly diverse and even small changes in behaviour, plumage colour, bill length, or song can give a bird a competitive advantage over its relatives. It is the study of such small changes that give one bird the advantage over another that attracts the attention of ornithologists, and many have spent their careers examining the complexities of the speciation process in forest birds.

Flight among the Trees
Of all birds, the small forest species are among the most habitually pedestrian and make the most limited use of flight. Many rarely fly more than the short distances between one branch and another or from the ground to the nearest tree limb. Nonetheless, they are also the most abundant kind of bird. Not only do they include about 85% of the world's species but they also make up about half of the higher taxa. Birds are also the most obvious kind of animal in the forest, and they greatly outnumber small mammals. Except for some parts of the tropics, they also outnumber reptiles and amphibians.

The abundance and diversity of small forest birds suggest that the canopy has been their home for a very long time. In fact, it may well have been the evolutionary nursery for Class Aves. Birds are probably the descendants of the first vertebrates to move into the canopy when trees became the dominant form of plant life on the planet, and they have been co-evolving with trees for the last 200 million years or more. The forest canopy would have provided a valuable refuge from earthbound predators and a rich smorgasbord of foods, including swarms of insects and spiders among equally edible flowers and fruit. At first, those earliest of birds would have been able to enjoy this novel habitat without competition or the risk of predation. By the time effective predators were able to move into the canopy, birds may already have learned to flit from branch to branch and established themselves as a dominant life form in the forests. They are one of the most abundant kinds of animals in the trees but remain one of the least accessible types of prey.

As success bred competition among the residents of the forest canopy, birds learned to be useful to the trees. At first, birds may have done no more than help trees by reducing plagues of insect pests, but later the trees came to exploit birds for special tasks. Birds were just as useful as insects in moving pollen from flower to flower; unlike insects, however, they were large enough to move whole seeds or fruit from place to place. Today, the seeds of a great many forest plants travel to new opportunities either on or within birds. Bird feces also represent a valuable redistribution of mineral nutrients that are useful to trees but often rare in forest soils. Small birds broadcast tiny fecal pellets throughout the forest; large species, such as storks or eagles, deposit significant amounts of nutrients at regular roosts. Such predators also carry whole fish or other prey into the forest and leave behind scattered bones and remains that are rich in nitrogen, calcium, and phosphorus.

Life in the forest helped shape the wings of forest birds and had a profound effect on their flight style. Among the trees, flight is an apparently uncomplicated matter of using wing muscles to generate enough lift to overcome gravity and enough thrust to create forward movement. Forests do not offer room for running starts, and the trees inhibit the updrafts or stiff breezes that are useful to birds in more open habitats. On the other hand, there are many opportunities to jump off elevated structures or gain air speed during a fall. Tests on sparrows show that their short, wide wings are the ideal shape for generating lift and that they can generate about six times more lift than they actually need for take-off [18]. Such wings keep most forest birds close to home, however. They create too much drag for either speed or effective gliding, and they are of the wrong shape to take advantage of the wind or other air movements. Migratory species among the forest birds have changed their wing shape to reduce some of these problems. The primary feathers have become longer to increase speed and energy-efficiency in cruising flight. Nonetheless, seasonal migrations are still major challenges that claim the lives of millions of small birds every year.

Even small forest birds can convert thrust into lift when they have an emergency requirement for speed. We have all seen films of birds twisting and turning to avoid the attack of a falcon or other aerial predator, and birds will roll over on their side as they dive across the path of an oncoming car. These rollovers are an attempt to increase speed as well as change direction. During the few seconds when a bird is on its side, the flapping wings cannot be generating lift but the energy from the wing's vortices must go somewhere. If it is not going into lift, it can only go into thrust. When it goes into thrust, it increases air speed.

What the wings of forest birds lack as generators of speed they make up in flexibility. The wider the wing, the greater the lift it generates for a specific amount of effort. Forest birds can greatly increase the effective lifting surface by swinging the outer part of the wing forward until the tips meet (Figure 7.4). At the same time, they will fan the tail to create even more surface. In effect, the wings become a circular disc that flexes along its midline. The same posture is very useful when the bird wants to slow down for a gentle landing, and various versions of it can be seen in much larger birds with stiffer wings. In combination with a long, variable tail, the flexible wings of a forest bird produce the manoeuvrable flight that is exactly what is needed in a crowded and complex environment. Unfortunately for our understanding of avian flight, changes in the wing's shape and the extreme flexibility of the feathers impairs the aerodynamicist's ability to calculate the forces that occur during flight.

Flight is a matter of counteracting weight with lift and drag with thrust, but when it comes to calculating values for these forces, the best that engineers can do is determine the average effects. Even in the most complex machines, they can calculate the instantaneous forces acting on individual parts during each millisecond. In birds, however, the flexibility of the wings and feathers and their tendency to change shape during a stroke make such precision impossible. Consequently, mathematical models of avian flight portray it as a kind of equilibrium in which all the forces are in a state of balance [81]. Some modellers, such as U.M. Norberg [139], have based their analyses on the forces acting on the blade of the wing and have tried to work around the limitations of comparing a rigid theoretical model with the more variable real-life wing. Jeremy Rayner and his students have avoided issues of wing flexibility by developing models based on the effect of the bird on the medium it is moving through. They look at a bird as though it were just another body moving through a fluid medium, and use the principles of fluid mechanics to assess the kinetic energy of vortices in its wake [172, 173, 176]. Their findings have led to the concepts of vortex-ring flight and continuous-vortex flight. In the final chapter, we will see what one of these vortex models can tell us about the aerial performance of a bird that depends on exceptionally high-speed flight.

For forest birds, the most important aspect of flight appears to be the ease with which the wings can lift the body off the ground and into the canopy. The consequent emphasis on lift, as opposed to thrust, has led them towards intense forms of weight reduction and made the small forest birds some of the most extreme examples of the inner-bird strategy. Their most frequent approach has been to reduce their overall density by greatly increasing their volume. They bury minimal bodies within a great halo of feathers. Although this lightweight cloud of feathers reduces total density, it greatly increases the effect of air resistance or drag and would significantly increase the energetic cost of fast flight if the bird needed to achieve it. Lower density can be very useful in a fall. Air resistance quickly reduces the drop speed, and the thick layer of trapped air lessens the impact of collision with solid objects or the ground. Birds such as Wood Duck and Bufflehead depend on this phenomenon to protect their young when they drop from the nest. Some must leap 10 or 20 m when they are just small balls of down, long before they are capable of flight.

Birds further reduce their weight by making temporary reductions in the size of unnecessary structures in their soft tissues. For instance, the organs in the reproductive system do not grow until the bird reaches sexual maturity, and then shrink temporarily after the breeding season. There is even evidence that the part of the brain associated with song decreases in size when it is not in use.

Birds decrease the weight of their skeleton more permanently by simplification and fusion. Fusion is particularly widespread among forest birds, where it reduces weight by lowering the number of relatively heavy articulating joints. Joints such as knuckles contain relatively heavy fluids and pads of connective tissue. Many of the bird's fused joints occur in the head, where they are replaced by flexion areas of exceptionally thin bone. The best known of these flexion areas connects the upper beak to the braincase. Figure 2.3 shows that not all types of forest birds make extensive use of fusion; parrots require a strong skull because they need to be able to peel tough-skinned fruits, and have retained independent bones. Most other birds ingest very small prey items in one piece.

Variation in average weight from group to group may offer clues about important events in the history of birds. There is a theory among zoogeographers that modern birds survived on the continent of Gondwana (now Antarctica, Australia, New Zealand, and South America) when events at the end of the Cretaceous exterminated the large dinosaurs and many other life forms. The survivors prospered and radiated out of their refuge during the early parts of the Tertiary. The Metaves may represent one of the earlier radiations that was overtaken by the later expansion of the Coronaves, but the strongest evidence for this event is subtle variation in the genetic coding that identifies members of those two subclasses [59]. The exceptionally fine fossils from Messel in Germany, Quercy in France, and the Green River formations of North America may offer physical evidence for a similar kind of radiation event. It was just as important and may have left a modern echo in the unusual distribution of weights in living forest birds.

All over the world, most of the smaller types of forest birds (<50 g) belong to the songbirds in the Order Passeriformes. The other groups of tiny forest birds, such as hummingbirds and swifts, are unusual aerial specialists that rarely interact with the more terrestrially oriented songbirds. Songbirds, however, do have ecological interactions with some of the other groups of forest birds, especially the woodpeckers (Order Piciformes) and kingfishers (Order Coraciiformes). Surprisingly, the latter groups are somewhat larger, on average, than songbirds. For years it was assumed that these two groups could not produce tiny species because of some hypothetical evolutionary or architectural constraint.

Among the Oligocene fossils of Europe and North America, however, there are examples of small birds related to woodpeckers or kingfishers. One of the most spectacular is a hummingbird-sized wood-hoopoe from the Messel shales of Germany [127], but there are others [119, 136, 128]. Significantly, there are no songbirds among these early fossils, even though songbirds now contribute about half the species in any modern mixed community. Songbirds appear quite suddenly in the Miocene; suspiciously, the tiny hoopoes and other small nonpasserine birds disappear from the record. It seems plausible that the songbirds replaced the smaller members of these other orders by competitive exclusion, as wave after wave arrived in the North from Gondwana.

The ancient success of the Miocene songbirds may be reflected in modern distribution of average body weight among forest birds. Since the Miocene, small songbirds have held on to most of the niches for small birds all over the world, and they thoroughly dominate the classes under 40 g (Figure 9.1). Modern woodpeckers and kingfishers tend to be relatively large and some are much larger than any songbird. Perhaps songbirds are unusually skilled at being small but may not be particularly good at being large. The group includes only a few species over 100 g, and even fewer examples as large as the raven (6 kg).

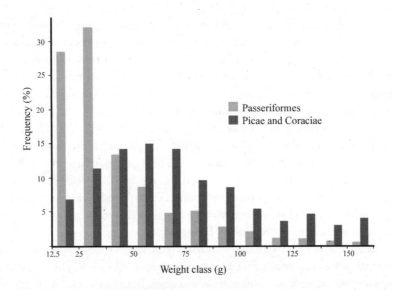

Figure 9.1 The global distribution of weights among species in three of the most numerous groups of forest birds. All the very small members of the Picae and Coraciae (dark grey) disappeared in the Oligocene, when they were replaced by tiny Passeriformes (light grey). The weights are from J.B. Dunning [52].

Aerial Specialists in the Forest

Not all forest birds forage while walking on the ground or clinging to branches; as suggested above, competition with songbirds for a share of the forest's resources may be a strong incentive to find a less crowded niche. Consequently, several groups of forest birds have experimented with special foraging techniques and become aerial specialists capable of foraging on the wing. One of the most popular is a form of fly catching that exploits the opportunities created by gaps among the trees. Many of these birds use the elementary lurking technique favoured by crocodiles and other primitive animals. They sit on some vantage point and assault passing prey when the opportunity arises. The technique takes advantage of the lift produced by the moving wing of a typical forest bird, and few of these species chase their prey more than a few metres. Bee-eaters, fly-catching relatives of the kingfisher, are an exception. They are capable of longer flights in the pursuit of food and are one of only a few small nonpasserine birds that undertake lengthy migrations.

Two groups of songbirds have become true aerial specialists and have learned to make extended pursuits of insects flying in the air above the trees. Swallows (a family of songbirds related to warblers) and wood-swallows (a family of songbirds related to crows) hunt insects at low elevations during daylight in forest gaps, over lakes, and above the canopy. It is a foraging technique that requires speed, and these birds have developed long, pointed wings and streamlined profiles. They look a little like swifts but have achieved a successful design largely by modifying their feathers. Their wing skeletons are not much different from those of their more pedestrian relatives (Figure 1.10).

Nightjars and their relatives have long wings designed for pursuing insects in the air, but these groups specialize in species that swarm at night. Like bats, they are usually most active in twilight hours but may hunt from dusk to dawn. These nocturnal hunters have exceptionally lightweight skeletons and very large mouths but are otherwise unexceptional. As in the swallows, their wing characteristics are largely the product of extended primary feathers.

Secretive, nocturnal animals have always been viewed with suspicion, but these birds' activities are accompanied by haunting calls and various alarming noises. As a result, they have accumulated suitably evocative names: whip-poor-will, poorwill, potoo, nightjar, nighthawk, goatsucker, and frogmouth. Ornithologists usually consider them to be somewhat poorly understood relatives of owls, but recent genetic studies that place them among the Metaves suggest that they are an artificial group consisting of several distantly related lineages [59]. A related species, the Oilbird, is notable for the ability to function in total darkness. Like bats, it has developed an elementary form of echolocation for navigating in the dark. It is a fruit eater and its sonar is not sophisticated enough to track flying insects, but it can avoid daylight predators by roosting in the absolute darkness of deep caves.

The swifts are yet another group of aerial insectivores, recently assigned to the Metaves. Unlike the swallows or the nightjars, which have specially adapted plumage, swifts have skeletons that are modified for high-speed aerial pursuit. Traditionally, they have been seen as close relatives of the hummingbirds (also in the Metaves), with which they share some specialized wing structures (Chapter 2). Both are usually placed in the Order Apodiformes, a name that is supposed to refer to their tiny feet. The name may just as likely have originated with the failure of early collectors to keep feet attached to the skins of specimens. At one time, the order included the birds-of-paradise, partly because both hummingbirds and birds-of-paradise boasted spectacular metallic feathers and partly because the skins of birds-of-paradise brought home from Magellan's expedition by Antonio Pigafetta in 1522 lacked their feet.

Energetically Expensive Flight in Small Birds

According to the group's position in the Sibley and Ahlquist biomolecular phylogeny, swifts are among the most ancient of the aerial specialists [192]. At first glance, a swift looks somewhat like a swallow. Both are active in daylight, although swifts usually pursue insects at altitudes far beyond the reach of swallows. Because they fly so high, they are rarely noticed by ground observers. Birders most frequently see swifts at dawn and dusk or immediately before the arrival of intense summer storms, when insects are abundant at lower altitudes. Many species of swifts forage in the updrafts created by major mountain ranges such as the Andes and Himalayas. As in other birds that fly at very high altitudes, oxygen carried by the myoglobin in the muscles helps the bird function in the thin air 4 km above sea level. The swift also depends on its hemoglobin, and its red blood cells are much larger than those of other birds.

Although the swallow-like appearance of swifts may confuse novice birders, the rapid, vibrating beat of their stiff, crescent-shaped wings is quite distinctive. The wing is also very narrow and noticeably curved, because the primary feathers at the tip are exceptionally long while the secondary feathers, closer to the body, are exceptionally short – much shorter in relative terms than the secondary feathers of a swallow. This elongated wing helps swifts achieve extremely high air speeds. They have been tracked on radar at 110 km per hour, and may be able to reach speeds as high as 169 km per hour. To achieve such speeds, swifts must be dense enough to overcome drag. Unlike their relatives among the forest birds, they are compact masses of muscle covered by a thin, dense mat of small feathers. When you hold one in your hand, it feels like some kind of bullet that is much heavier for its size than another small bird.

Although hummingbirds are supposed to be closely related to swifts, they are best known for time spent hovering in front of flowers while they collect nectar. They do not appear to be specialized for the aerial pursuit of insects. Nonetheless, they regularly feed that way to add much-needed protein and a supply of minerals to their sugar-rich diet [237]. Hovering is a highly specialized form of flight that consumes a great deal of energy, and these tiny birds are dependent on the exceptionally high energy density of their food. Although it is a successful flight style, imitated by a handful of equally tiny insects and bats, it appears to be beyond the capability of other birds, particularly the sunbirds, flower-piercers, and other small nectar collectors among the tropical songbirds. The fact that hummingbirds are confined to the New World suggests that hovering flight may be a very recent innovation among birds.[1]

Although hummingbirds and swifts have different flight styles, both depend on very rapid wing beats, with the small hummingbirds achieving 70-80 Hz.[2] Even at lower frequencies, the unaided human eye sees the moving wing as no more than a blur, and for a while ornithologists were convinced that swifts could achieve such a high frequency of wing beats only if the limbs moved alternately. It took super-high-speed photography to show that the wings move synchronously, as in other birds [184]. Both swifts and hummingbirds have a short, stubby humerus that is deeply sculpted with channels for the passage of large tendons. Muscle fibres could not twitch fast enough to drive a longer humerus at the same rate, and the thickness of this bone, compared with the other bones in these birds, is indicative of the stresses generated by this exceptional flight style. The radius and ulna of the apodiform forelimb are also rather thick but are about the same length, proportionately, as bones in a songbird. The bones of the apodiform manus (hand) are greatly elongated and strongly flattened. A swift's hand is thrice as long, and that of a hummingbird almost four times as long, as that of a songbird (Figure 1.10).[3]

Small size protects most of these birds from the effects of momentum in the moving wing. Hummingbirds are generally smaller than swifts. The smallest hummingbird (*Mellisuga helenae*) weighs only 2 g; the largest (*Patagona gigas*) weighs about 21 g. A very small swift, such as *Collocalia troglodytes,* weighs 5.4 g. Momentum generated by the movement of large wing bones could potentially be a significant problem for giant swifts such as *Streptoprocne semicollaris* or *Hirundapus celebensis,* which weigh more than 175 g. Like all swifts, however, the giants also use a very shallow flight stroke. Their wings vibrate through very short arcs so that the momentum generated is minimal.

The size of the largest swifts overlap the size range of the smallest auks and provide us with an opportunity to contrast the flight styles and anatomy of two unrelated fast-flying groups. Both beat long, narrow wings very quickly to achieve high

---

[1] A single fossil of a hummingbird has been found in the Messel oil shales of Germany, suggesting that the group once had a wider distribution. Perhaps an early expansion of the group was cut short by subsequent glaciations, because it is hard to believe that hummingbirds would not continue to thrive in the forests of Africa and Asia if they had been able to get there. The fossil may also represent a previously unknown but convergent group within Metaves. The full significance of the division of birds into Metaves and Coronaves has yet to be explored [128].

[2] In the days before stroboscopes and high-speed cinematography, the speed of a hummingbird's wing was estimated by comparing the pitch of its "hum" to something easier to measure, such as the pitch of a vibrating violin string.

[3] The Apodiformes have sometimes been given the name "Macrochires," a reference in Greek to their unusually large hand bones.

air speeds. Most birds fly at less than 50 km per hour and have a wing beat rate of less than 8 Hz, but a fast-flying auk, such as the Marbled Murrelet, beats its wings at 12.8 Hz to achieve speeds as high as 154 km per hour. Even that rate is much lower than the wing beat of a swift. Smaller auks may be able to beat their wings even more quickly, and they certainly use rapid rates to climb away from attacking falcons.

The most important difference in flight between swifts and auks may be the depth of the stroke. Murrelets swing their wings through an arc between 120° and 140°, while the wings of the swift seem to hardly move at all. They vibrate very quickly through an arc of less than 20° and perhaps as little as 10°. These differences in flight technique are reflected in the wing skeletons. Compared with the greatly elongated hand of a swift, the manus of an auk is tiny and has little bony tissue in the outer wing. At the other end of the limb, the swift has a stubby, robust humerus, whereas the auk's humerus is the longest and heaviest bone in the wing. It is also flattened and slightly arched.

In a large bird such as an auk, the use of relatively heavy bone to elongate the hand would generate a great deal of momentum as the bird moved its wingtip. It would need to expend extra energy at both the top and bottom of the stroke when the wing changed direction, and the effect would be amplified if the bird used deep strokes. Within the size ranges of hummingbirds and swifts, however, there is no significant difference between the weight of a bone and the weight of a feather. It is important only that the wing stand up to the stresses of the flight style. The combination of a large hand and stubby humerus may be beyond the evolutionary capabilities of other aerial specialists, all of which have elongated their wings by lengthening their primary feathers.

Rapid wing beats cause the muscles to generate a great deal of excess heat. Swifts can depend on their small body size to get rid of excess heat quickly, but size is a two-edged sword, and smallness makes it difficult to retain heat during periods of inactivity. Like hummingbirds, swifts become torpid at low temperatures, when their insect food is unavailable. Temperature control in the fast-flying auks is a bit of a mystery. They are exceptionally well insulated but they are also members of the only fast-flying group of birds that lacks pneumatized bones. Other fast birds, such as falcons, can transfer heat from their flight muscles into the air moving through the hollow bones of their pectoral girdle. We do not know what method auks use to radiate excess heat.

Both auks and swifts have a short, stubby tail, which reflects their choice of speed over manoeuvrability. Aerodynamically, it probably functions as a "splitter plate," a design feature that appears in many other fast-flying birds such as loons and also in highly efficient fliers such as albatrosses and shearwaters [124]. In all these species, a streamlined body helps reduce drag by maintaining laminar airflow above and below the bird. The splitter plate effectively extends the bird's length so that turbulence created by the movement of its body through the air occurs after it has passed. The bristles that extend from the trailing edge of the tail in many swifts are useful props that hold the birds against the walls of their roost, but they may also have an important aerodynamic role. They may channel airflow and enhance the effectiveness of the splitter plate. A sharp, pointed tail would release a single large vortex, whereas a tail ending in a row of spines might reduce energy expenditure by releasing a series of small vortices, working in a way similar to the slotted wingtip of a vulture.

The Wing Skeleton and Aerial Specialization

Just as feathers are responsible for the long, sleek wings that distinguish swallows from less aerobatic relatives, comparable variations in feather shape and arrangement distinguish the birds of open country from residents of the forest. The wing skeletons of most birds are very similar, with bones that vary in length according to the body size but are generally similar in shape. The stubby humerus and the elongated manus of the swifts and hummingbirds, or the long, flattened humerus of auks, are dramatic but exceptional skeletal modifications. The most obvious differences between major taxa are in the sizes and locations of muscle attachments related to variations in flight style.

One indicator of the relative similarity in wing skeletons is a parameter called the brachial index. It is simply the ratio of the humerus to the ulna. Unlike other flight parameters, such as wingspan or wing loading, which can be measured accurately only on a fresh or living wing, the brachial index can be determined from bones and fossils in museum collections and even from accurate drawings or photographs. Recently, Robert Nudds, Gareth Dyke, and Jeremy Rayner found that each family and order of birds had its own characteristic value for the brachial index (Figure 9.2) [146]. Values of the index varied little between species within a family but to a much greater extent between higher taxa. Not surprisingly, flight styles and life history characteristics, such as foraging technique, also tend to vary between orders but are more consistent within families.

The brachial index is highest in birds with fast, powerful flight, such as ducks (1.12), grebes (1.15), and loons (1.24). It is exceptionally high in auks (1.32) and penguins (1.37), which use wing propulsion underwater, and is highest in the Mancallidae (2.34), an extinct group of flightless wing-propelled diving birds whose fossils have been found along the Pacific coast of North America. Considering their vigorous flight style, the index seems surprisingly low in the swifts (0.69); with their unusually long hand, however, their brachial index may not have the same significance as it does in other birds.

The high values among heavy-bodied waterbirds suggest that the brachial index is closely related to the load placed on the wing skeleton during flight. A low-density aerial specialist should place only a moderate amount of stress on the wing skeleton and might therefore have a low brachial index. The spectacularly aerial frigatebird is just such a low-density bird; its brachial index is a mere 0.78. In the heavier and more densely muscled tube-nosed seabirds or petrels, the average brachial index is 1.03. Within this group, however, the diving-petrels use their wings for underwater

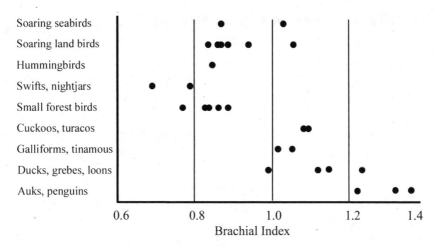

Figure 9.2   Brachial indices for various groups of living birds [146, supplement].

propulsion. This form of swimming places a large load on the wing skeleton and diving-petrels have a predictably auk-like brachial index of 1.23.

The brachial index is somewhat variable among low-density terrestrial birds but usually lies close to 1.0. Most values range from 0.8 to 1.1. The galliform birds and tinamous, which expose their wing skeleton to the stress of sudden take-offs, have a brachial index close to 1.0. The index is a little lower (about 0.9) among the groups of large soaring birds. There are no obvious differences between the brachial indices for groups assigned to the Metaves and those in the Coronaves. The brachial indices for swifts, hummingbirds, and nightjars are all exceptionally low, but indices for other groups in the Metaves have values similar to comparable birds in the Coronaves. Generally, the index appears to be more related to the bird's flight style than to its classification.

Brachial indices for some familiar fossil species suggest that a specialized wing architecture is a fairly recent development in birds. The index is greater than 1.0 in *Archaeopteryx* but close to 1.0 in many of the early fossils. Values much less than 1.0 do not appear until after the Cretaceous except in the unusual wings of the Confucius birds (0.90-0.92, Table 8.1). The index is slightly higher than 1.0 in some ball-shouldered birds, such as *Otogornis* (1.18), but near 1.0 in others, such as *Cathayornis* (0.99) [242]. Two recently discovered modern birds from the Lower Cretaceous also have a brachial index close to 1.0. In *Yixianornis*, it is 1.03 [31], and in *Gansus* it is 0.97 [238, supplement]. Among post-Cretaceous fossil groups, the brachial index is more variable. It is sometimes quite high, and Nudds and colleagues [146] report the average for *Anneavis* and three of the Sandcoleiformes as 1.10, which is similar to values for the modern colies, to which they may be related. Two species of *Primobucco*, an extinct coraciiform, have indices of 0.81 and 0.94, close to the values for small living kingfishers.

Varied Shape in the Birds of Open Country

Flight beyond the Forest

The variation in the brachial index is almost negligible compared with the great range of wing sizes and shapes among birds. Nearly all that variety is concentrated within the small number of groups that live in open habitats. The birds of open country tend to be very much larger than typical forest birds. They can fly faster and over greater ranges because they are capable of carrying a great deal more fuel [209]. Several groups of terrestrial birds and most seabirds exploit this ability to fly great distances in search of scarce patches of food that are often scattered over very wide areas. Taken together, these birds account for only about 15% of the species in Class Aves, but they have such distinctive structures and varied behaviours that they account for half of the taxonomic orders.

The transition from closed forest habitats to open spaces implies a shift from flight styles that carry birds upward very quickly to flight styles that move them long distances from site to site. In other words, there is a shift from designs that specialize in the generation of lift to designs that generate thrust, either in modest amounts over longer periods or in larger amounts for quick, short trips. These innovative flight styles require specialized wing shapes, but the narrow range of brachial indices (Figure 9.2) suggests that the terrestrial species have been able to make the transition by modifying feathers more frequently than bones. The higher brachial indices among some of the seabirds suggest that modifications for that environment have taken a different route and may involve both feathers and bones. We will examine the structure of seabirds in the next chapter.

Successful modification of the feathers to accommodate advanced flight styles has had an unfortunate consequence for the study of avian evolution. It is difficult enough to find fossilized bird bones, but feathers leave fossil evidence only under the most unusual circumstances. Feathered dinosaurs presumably had large and robust plumes but, after studying dinosaurs carefully since 1832, it is only in the last two decades that we have found unambiguous evidence that any dinosaurs had feathers. Even when clear feather impressions accompany well-preserved skeletons of primitive birds, as in *Archaeopteryx* and many fossils of the Confucius birds, it can be very difficult to reconstruct a flight technique. Feather impressions in fossils of the earliest representatives of modern birds suggest that they used unspecialized flight styles and were forest birds.

The specialized function of feathers in flight has become apparent only since the invention of very high-speed photography. Small birds move their wings so quickly that the unaided eye cannot detect subtle changes in wing shape, and the dry, dusty wings of museum specimens provide little evidence of their use in the living animal. Only high-speed photography gives us the opportunity to appreciate the full capability of the feathered wing and the profound importance to the bird of keeping its feathers soft and flexible through constant care. As mentioned earlier, forest birds generate extra lift or soften a landing by temporarily changing the

Figure 9.3 A chickadee rolls its wings into a tube at the beginning of the recovery stoke (L. Lougheed).

shape of the entire wing. In doing so, they put their flight feathers through amazing contortions. On the downstroke, the primary feathers may arch back until the tips are almost parallel to the direction of the stroke. At the beginning of the recovery stroke, the primary feathers may lag behind and bend so far that the wings briefly form a hollow tube around the belly (Figure 9.3). Among small birds, only the swifts, with their giant bony hands, fly on stiff wings.

Variation in wing shape and wingtip design in larger birds was one of the first characteristics to attract the attention of early ornithologists. Many of these birds beat their wings rather slowly and they frequently soar with their wings outstretched in ways that make the structure easy to observe without special equipment. Leonardo da Vinci and other early students of flight made many sketches of gulls in an effort to uncover the secret of avian flight.

There are basically two types of wings among the larger terrestrial birds. One is narrow and tapers to a point. It enables birds to travel long distances rather quickly but is energetically expensive. Such wings occur in fast, manoeuvrable species that are either predators, such as falcons, or potential victims of such predators, such as sandpipers. Both groups also include long-distance migrants. In flight, the pointed wing generates a series of large vortices whose departure from the tip at the end of the downstroke represents a significant loss of energy. As a result, when one of these birds is flying slowly, progress seems to consist of a series of pulses, but the wing beats are often so rapid that there is no noticeable hesitation between strokes and the vortices are continuous.

The other type of terrestrial wing is rather broad and rectangular, with a tip that consists of a series of independent feathers. Such tips are said to be "slotted." This kind of wing is typical of very large birds that soar in broad, smooth circles but do not demonstrate much manoeuvrability in the air. Each feather at the tip develops its own small vortex that travels along the feather's edge into the notch with the next feather. When it reaches the point of overlap, the vortex is released, carrying a small amount of energy away from the bird. The total energy of these small vortices is less than that of a single large vortex, and the arrangement of the primary feathers staggers the notches so that the vortices are freed piecemeal and not all at once [172]. Consequently, the energy changes are small and not concentrated at any one moment in time, and the bird's flight appears exceptionally smooth. The wings are typically so large that the wing loading is small and the bird can save energy by exploiting updrafts and other currents once it has achieved take-off.

Large predators, such as eagles, and scavengers, such as vultures, are among the most familiar soaring birds. Most are members of the Falconiformes,[4] a group that demonstrates how one basic body plan can be adapted with modest change to meet the needs of several different lifestyles. Eagles and buzzards (buteo hawks) are typical of the open-country birds that use soaring flight to search for prey. They typically kill their prey on the ground because they lack the manoeuvrability to make a kill on the wing. Their wings are long and broad with slotted tips, and they fly with their tails spread to increase the lifting surface. In some of the soaring buteo hawks, the tail fan creates an unbroken trailing edge with the wing; from below, it is often difficult to see where one ends and the other begins. Although eagles and hawks will often search for food on the wing, they frequently save energy by lurking at some vantage point, from where they can drop on passing prey.

Accipiter hawks and falcons are active predators that pursue and kill other birds in the air. Their flight technique must provide more speed and manoeuvrability than that of their prey, and they exhibit a whole suite of specializations associated with aerial hunting. In many ways, the accipiters that hunt in the forest are larger versions of the birds they eat. They have the short, round wings and very long tails typical of the highly manoeuvrable forest birds. Falcons hunt in the open air and merely outperform other fast birds. Their narrow, pointed wings are powered by massive flight muscles, and they simply knock their victims out of the air. A kite is a more ethereal and very lightly built version of the falcon. It may use its long, pointed wings to generate speed but, like a swallow, it can use its very long, forked tail to perform aerobatic manoeuvres. Kites are able to take erratically flying insects on the wing or drop out of the air on small, zig-zagging rodents.

Old World vultures, related to eagles, and New World vultures, related to storks, are iconic terrestrial soaring birds. No animal is more noted for its ability to search vast areas for a meal or for its inability to kill its own prey. A basic analysis of energetic requirements recently led Graeme Ruxton and David Houston to conclude that only large, soaring animals could earn a living as scavengers and that this may also have been the lifestyle of the large, soaring pterodactyls. It is a lifestyle beyond

---

4   New World vultures were long considered to be members of the Falconiformes. Recently, however, their similarity to Old World vultures was recognized as convergence, and the American birds have been placed among the storks (Ciconiiformes).

the locomotory capability of large mammals [183]. A mammal inevitably resorts to killing because it will run out of energy before it covers an area vast enough to include a ready meal.

Avian Aerodynamics

Because the details of flight style and technique did not play much of a role in taxonomic analyses, it was not until the mid-20th century that field ornithologists began collecting measurements specifically for aerodynamic studies. Some of the impetus for the collection of these kinds of data came from ecologists who wanted to investigate a link between evolutionary pressures and animal behaviour. The theories of David Lack had helped shift the focus of ecology from community structure to cost/benefit analyses of specific activities, and flight seemed to be one activity that was particularly demanding and energetically expensive. At first, ecologists could apply only outmoded aerodynamic concepts, but gradually the ideas of C.J. Pennycuick and Jeremy Rayner began to make themselves felt. Investigators began to examine the finer points of flight energetics, and it soon became clear that the proportion of the daily energy budget consumed by flight varied greatly according to the flight style of the bird.

Ecologists tend to focus on two elementary measurements of the bird. The simplest is body weight, a parameter that is easy to measure but often varies greatly with time of day, life-history stage, and season. For the last hundred years and more, ornithologists have collected weights because they were one of the few useful indicators of size in birds and, unlike other measurements, they relatively easy to take without mistakes. For ecologists, short-term changes in weight became the easiest way to measure the cost of particular activities, especially the seasonal effort adult birds made to feed their young.

For more sophisticated studies of flight, it is necessary to understand the shape of the wing. Length, width, area, total wingspan, and wing cord have all been used. Although these are all relatively simple to measure with a ruler, they are also surprisingly easy to get wrong. For instance, wing cord is a measurement taken by bird-banders of the distance between the tip of the longest primary feather and a notch in the wrist (between the carpometacarpus and the ulna) that can be felt easily when the wing is folded. It sounds like a simple measurement but it varies with the wear on the tip of the primary, the degree to which the feathers are straightened along the ruler, and the angle of the bend at the wrist. The major national bird-banding organizations have published carefully worded protocols for this measurement but it still varies from country to country and often from station to station. Fortunately, it is a value used only by bird-banders and has no significance for the analysis of flight.

Wing area is particularly important for understanding flight, but replicable measurements are very difficult to achieve. The degree of straightening has a large effect on the final value. Most scientists try to adhere to a standard protocol set out by C.J. Pennycuick [160]. Wing area is used to calculate wing loading, a good index of the effort that it takes for a bird to get off the ground. It is no more than the body weight of the bird divided by the total area of the wings. Wing area is also used to calculate a more sophisticated parameter, the aspect ratio, a value that describes the specific shape of the wing. For the purposes of aerodynamic study, aspect ratio is the ratio of the lifting surface to a circular disc whose diameter is the wingspan. The lifting surface is the total area of both wings plus that part of the body lying between them (Figure 7.2). A long, narrow wing has a high aspect ratio; a short, round wing has a low aspect ratio.

It is relatively easy to determine the aspect ratio of a rigid aircraft wing, but the flexibility of a bird's wing makes accurate measurement difficult in spite of Penny-cuick's protocol. It is often difficult to tell whether values taken by one worker can be compared with those offered by another. Accurate measurement is not the only problem with interpreting an aspect ratio. As mentioned above, small forest birds flex their wings into a variety of shapes, including a disc. That flexibility of their wings greatly reduces the usefulness of the theoretical aspect ratio as a flight pa-rameter. The aspect ratio of a disc-shaped wing would be close to 1.0, but in most forest birds, the aspect ratio lies between 5.0 and 6.0. Because it is the ratio of two areas, it has no units of measurement.

Many of the values for aspect ratio in Table 9.1 are from a large collection pub-lished by Jeremy Rayner in 1985 [175]. The rest I have measured using Pennycuick's protocol in order to get additional values for under-represented groups. The table uses the taxonomic rank names and sequence from the biomolecular phylogeny of Sibley and Ahlquist [192], but it is arranged according to increasing values of the biomolecular parameter $\Delta T_{50}H$, for reasons that will soon become clear.

One of the most obvious features of the data in Table 9.1 is that the weights, wing loadings, and aspect ratios appear to fall into three large groups, and the groups bear a strong resemblance to those set out by Aristotle over two millennia ago: land birds, waterbirds, and shoreline birds. If we plot average weights for each group against average aspect ratios, we can use a statistical process called discrimi-nant analysis to draw lines that isolate the three groups objectively (Figure 9.4):

- birds with short, wide wings (low aspect ratio) and low weights, typically associ-ated with forested habitats (e.g., sparrows, wrens, and other songbirds)
- birds of open country with moderate to high aspect ratios and low body densi-ties, such as birds that use thermal soaring (e.g., cranes, vultures, pelicans, eagles)
- birds with long, narrow wings (high aspect ratios), pointed wingtips, and high body densities, which exhibit very efficient or very fast flight and typically live on the water (e.g., albatrosses, shearwaters, boobies, loons, grebes, and auks).

There are no flying birds that combine a low aspect ratio (less than 7.0) with a large body weight (greater than 1 kg). The galliform birds come very close to this category and some individual species may fall into it, but otherwise it includes only flightless types, such as penguins.

Table 9.1

Average wing parameter values for groups of flying birds in the biomolecular phylogeny of Sibley and Ahlquist

		Flight parameters			
		Body weight		Wing parameters	
Biomolecular classification	Analytical parameter $\Delta T_{50}H$ (°C)	Mean ± SD (N)[a]	Number of species in sample	Wing loading mean ± SD (N per m²)	Aspect ratio of the wing mean ± SD	Number of species in sample
Taxon						
Gallimorphae	21.6	8.24 ± 0.6	230	106.7 ± 9.5	6.6 ± 0.2	18
Anserimorphae	16.1	19.1 ± 3.5	165	129.3 ± 6.3	9.9 ± 0.4	33
Picae	26.3	1.0 ± 1.8	301	29.2 ± 6.7	5.1 ± 0.5	4
Coraciae	25.0	1.8 ± 4.8	316	35.2 ± 3.0	5.9 ± 0.7	3
Coliae	24.5	0.5 ± 0.1	6	34.53	4.46	1
Cuculimorphae	23.7	1.3 ± 1.3	96	23.9	7.0	2
Psittacimorphae	23.1	2.0 ± 2.4	248	38.7 ± 8.1	6.7 ± 0.6	3
Strigimorphae-Apodimorphae	22.5	1.3 ± 0.03	495	25.3 ± 7.4	7.8 ± 1.6	12
Passeriformes	21.6	1.4 ± 0.4	3,450	21.7 ± 9.8	6.4 ± 1.1	39
Columbiformes	20.8	2.5 ± 2.7	236	47.3 ± 10.7	7.1 ± 1.1	4
Gruiformes	20.1	14.1 ± 23.3	150	95.8 ± 33.7	7.6 ± 0.5	6
Charadrii	18.7	3.0 ± 1.6	319	61.6 ± 48.4	9.7 ± 1.4	34
Falconides	16.4	11.9 ± 19.7	322	53.4 ± 19.9	7.3 ± 0.9	32
Podicipedidae	14.9	5.8 ± 4.9	18	132.9 ± 33.5	9.7 ± 1.4	4
Phaethontidae	14.0	5.7 ± 2.1	3	56.7	10.9	1
Sulida	13.3	16.0 ± 6.1	51	85.2 ± 22.9	10.5 ± 2.4	7
Ardeidae	12.4	11.2 ± 9.5	19	40.0 ± 8.5	7.7 ± 0.4	6
Scopidae	11.9	4.23	1	–	–	0
Phoenicopteridae	11.5	22.9 ± 4.5	5	87.0	7.7	1
Threskiornithidae	11.1	11.7 ± 4.9	19	59.3 ± 4.2	6.8 ± 1.1	3
Ciconiidae-Pelecanidae	10.9	42.6 ± 27.8	28	64.9 ± 18.7	7.5 ± 1.6	10
Fregatidae	10.7	12.4 ± 3.8	4	40.0	14.7	1
Gaviidae	10.4	32.4 ± 16.8	3	186.5 ± 7.8	12.4 ± 3.1	3
Procellariidae	10.0	11.4 ± 18.6	74	58.9 ± 34.2	10.7 ± 2.3	74

[a] Body weights are given in Newtons (N). 1 N = 100 g × the acceleration due to gravity.
Sources: Biomolecular data [192]; body and wing measurements [52, 174, and the author].

Figure 9.4 emphasizes the similarity among the small forest birds and the great variety in the other groups. The small types of forest birds form a very cohesive group in the lower left quadrant of the graph, while the larger birds occupy the rest of the space. In several cases, groups appear to fall on the wrong side of the line. These are groups whose members have a wide variety of body sizes and flight styles. For instance, the charadriiform birds seem out of place in the upper ranges of the terrestrial forest group because they include long-winged gulls and terns. The heavy-bodied auks and many species of small sandpipers with short wings affect the group's

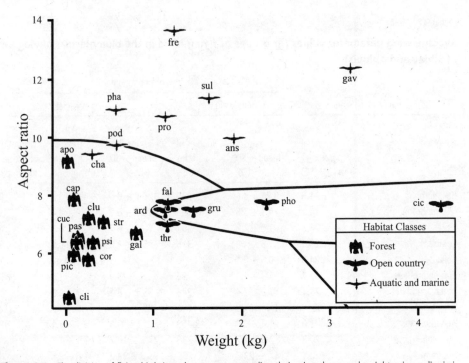

Figure 9.4   The division of flying birds into three groups according their wing shape and weight using a discriminant function analysis. Legend: **ans** = Anseriformes (ducks, geese), **apo** = Apodiformes (hummingbirds, swifts), **ard** = Ardeiformes (herons), **cap** = Caprimulgiformes (nightjars), **cha** = Charadriiformes (sandpipers, gulls, auks), **cic** = Ciconiidae-Pelecanidae (storks, pelicans, New World vultures), **cli** = Coliiformes (mousebirds), **clu** = Columbiformes (pigeons), **cuc** = Cuculiformes (cuckoos), **fal** = Falconiformes (hawks), **fre** = Fregatidae (frigatebirds), **gal** = Galliformes (grouse, peafowl), **gav** = Gaviiformes (loons), **pas** = Passeriformes (songbirds), **pha** = Phaethontidae (tropicbirds), **pho** = Phoenicopteriformes (flamingos), **pic** = Piciformes (woodpeckers), **pod** = Podicipediformes (grebes), **pro** = Procellariiformes (petrels), **psi** = Psittaciformes (parrots), **str** = Strigiformes (owls), **sul** = Sulida (boobies, cormorants), **thr** = Threskiornithiformes (ibis).

average score, however. Similarly, the Order Falconiformes includes many large soaring birds as well as smaller species, such as short-winged forest falcons and accipiter hawks. The groups of large birds that show considerable structural variety also contain more species and have wider distributions than more conservative types. For example, frigatebirds and tropicbirds, which show little anatomical variation among their few species, are confined to the tropics and subtropics, while all five species of very similar loons are confined to northern areas. Both groups also tend to have rather extreme flight styles and occur on the outer edges of the figure.

Figure 9.5 compares the ranges of weight and aspect ratio for individual species in the two most successful groups of seabirds: the Charadriiformes and the Procellariiformes. Where these groups overlap in weight, the members tend to have very similar aspect ratios. The Procellariiformes contain some very heavy birds with high aspect ratios (albatrosses and giant petrels) that have no equivalent type among the charadriiforms. The only charadriiform birds with large weights are the auks,

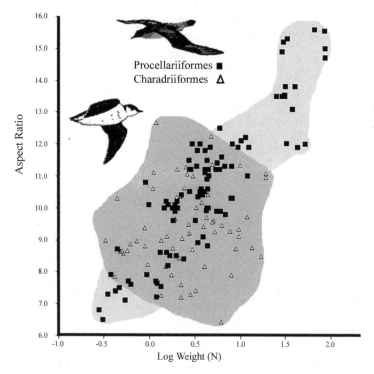

Figure 9.5 The range of variation in weight and aspect ratio for two highly successful groups of seabirds, the Charadriiformes (dark grey) and the Procellariiformes (light grey). Values for weight have been converted to a log function because weight increases exponentially with aspect ratio.

which have lower aspect ratios. Both orders include some very small species. Interestingly, the small members of the Charadriiformes, such as the smaller sandpipers, have higher aspect ratios than the small storm-petrels in the Procellariiformes.

Flight and an Evolutionary Story for the Birds

After all the television specials with their spectacularly animated reconstructions of dinosaurs and the publicity given to the discovery of unusual new fossils, it must come as a shock to students to learn that we still know almost nothing about the evolution of birds. When Gareth Dyke and Marcel van Tuinen [54] undertook their recent attempt to reconcile molecular clocks with the fossil record, they concluded that the resolution of relationships within Neoaves "may not yet be close at hand." The lack of an accepted evolutionary story is a major stumbling block to our understanding of birds, and there is a long way to go. Both the cladistic analyses of Joel Cracraft [35] and the gene-based phylogeny with which Matthew Fain and Peter Houde [59] introduced the Metaves end with a large unresolved cluster of unsorted orders and families. The only phylogeny to portray relationships among all the various groups of birds is the heavily criticized biomolecular tapestry of Charles Sibley and Jon Ahlquist [192] (Figure 6.2).

Even without an acceptable phylogeny, however, it may be useful to explore the idea that variations in flight capability are linked to the evolutionary success of birds. It is one of the few ideas about the evolution of birds that is widely accepted. Brian Maurer took the first step shortly after Sibley and Ahlquist published their biomolecular phylogeny, when he noticed that its arrangement was correlated with a general increase in the average size of species in the various orders [123]. In other words, the biomolecular tapestry agrees with Cope's Rule, which says that as lineages get older, they tend to include larger and larger animals [85].

If gradual increases in weight represent part of the evolutionary signal from the biomolecular phylogeny, perhaps changes in characteristics related to flight follow suit. Such an idea is very easy to test because the phylogeny (Figure 6.2) is constructed from the single parameter ($\Delta T_{50}H$) that Sibley and Ahlquist calculated from changes in the melting points of mixed samples of DNA [192] (see Chapter 5). Consequently, any quantifiable characteristic that can be measured for a sample of individual species in a clade can be examined in the context of the whole clade's value of $\Delta T_{50}H$ and related to comparable values determined for other clades. Because the scores of $\Delta T_{50}H$ form a simple linear progression, other characteristics can be arrayed along a similar gradient and examined with elementary statistics.

Figures 9.6, 9.7, and 9.8 compare the values of average body weight, average wing loading, and average aspect ratio of each of the major clades of "Neoaves" in the Sibley and Ahlquist biomolecular phylogeny. I have left out the flightless ratites and the galloanseriform birds on the grounds that they branched from the main group very early and appear to have followed their own evolutionary paths. In each case, the correlation with $\Delta T_{50}H$ is statistically significant. The result for weight was expected after Brian Maurer's observation (above), and wing loading follows suit because it is a function of body weight. Aspect ratio is strictly a function of wing shape, and is functionally independent of body weight. Its correlation with $\Delta T_{50}H$ provides further support for an evolutionary signal in the Sibley and Ahlquist phylogeny. The gradual increase in the three flight characteristics across the phylogeny implies steady progress from a base of small birds with lower aspect ratios to larger birds with higher aspect ratios among the later branches. There are no exceptions.

Aerodynamically, the relationship between phylogeny and flight parameters supports an evolutionary trend from simple flapping flight that overcomes lift to specialized flight styles that move large birds very long distances. Some of these birds travel on long, narrow wings that are inherently unstable but have very long glide paths. To control such wings, these groups have had to develop very advanced neuromuscular systems, and their position among the crown clades is not unexpected.

Unfortunately, a correlation is merely a comparison. It tells us only that two trends are moving in the same direction; it is not evidence for a cause-and-effect relationship. To demonstrate that the changes in flight parameters are linked to branches in the phylogeny, it is necessary to apply a test that examines the values at each evolutionary step. This process is called "independent contrast" or "ancestral states analysis." It has become the standard for comparing phylogenies with

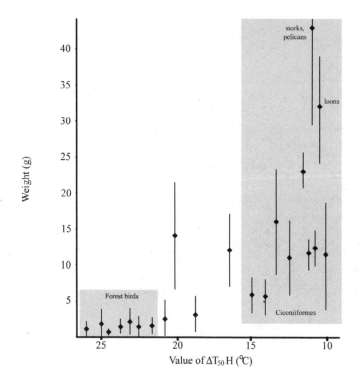

Figure 9.6
Increases in average weight among clades of flying birds in the sequence presented in Sibley and Ahlquist's biomolecular phylogeny [192].

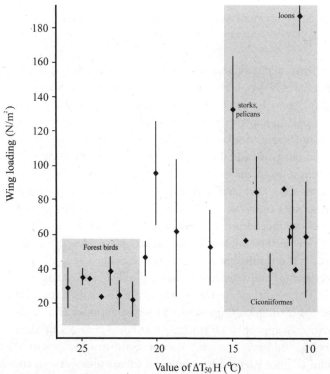

Figure 9.7
Increases in average wing loading among clades of flying birds in the sequence presented in Sibley and Ahlquist's biomolecular phylogeny [192].

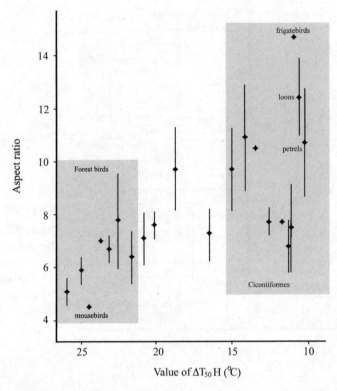

Figure 9.8   Increases in average aspect ratio among clades of flying birds in the sequence presented in Sibley and Ahlquist's biomolecular phylogeny [192].

evolutionary changes in physical characteristics because it can be applied to any shape of dendrogram, not just the peculiar linear conformation of the Sibley and Ahlquist biomolecular phylogeny. In this case, it also enables us to test any other avian phylogeny against variations in the flight parameters.

The analysis is both simpler and more complex than it sounds. It takes the branches of the dendrogram in pairs and calculates a value for the physical characteristic (flight parameter) as it might have appeared in some hypothetical ancestor. The size of the statistical error in that calculation tells us how confident we can be that the values for living groups are consistent with the ancestry portrayed by the phylogeny. Computers can produce large numbers of such values very quickly, but unfortunately the actual calculations are hidden in the machine. This makes it difficult to unravel problems if the results are not what you expect.

A colleague, Falk Huettmann, and I ran the values in Table 9.1 through a program called *Compare 4.0*. The results proved both intriguing and frustrating. First, we looked at body weight and wing loading. Branches near the upper end of the Sibley and Ahlquist phylogeny appear to be strongly related to changes in these weight-based parameters but the lower branches do not. Surprisingly, tests of a very different dendrogram (Figure 9.9) produced almost identical results. In this

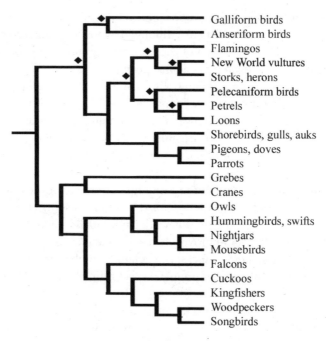

Figure 9.9 A dendrogram derived from a phylogeny of birds posted on a "Vertebrate Notes" website that has since been closed [228]. Ancestral states analysis indicates that the branches marked by a diamond are consistent with variation in the average weight of those clades.

second phylogeny, the classification of large birds is somewhat similar to that in the first, but the branches for smaller birds are very different. Perhaps the changes in weight among large birds includes a strong evolutionary signal that is either weak or missing among small birds.

As mentioned earlier in this chapter, the fossil record suggests the existence of an important interaction between songbirds and other groups of small forest birds during the Miocene. Perhaps the evolutionary signal was erased or corrupted at that time. Those events still seem to cause a significant echo in the modern distribution of bird weights, in spite of the amount of time that has passed (Figure 9.1). If the weight data for small birds is no longer linear, it may fail to meet the assumptions of the ancestral states analysis and greatly reduce the statistical power of the tests [204].

The problem may also be due to variations in the size of the differences in neighbouring values of $\Delta T_{50}H$. They are very large between clades at the beginning of the phylogeny, where there has been time for discrete genetic variations to accumulate in each branch. In higher branches that supposedly represent recently evolved clades, there has been less time for the accumulation of differences in the DNA; consequently, the measured values of $\Delta T_{50}H$, which indicate the separation of clades, are much closer together (Table 9.1). The Order Passeriformes is separated from the Order Coraciiformes and the Order Piciformes by 2.9°C and 3.4°C, respectively, even though all three groups look very similar and traditional taxonomists have had a great deal of difficulty finding diagnostic characteristics to separate them. On the other hand, each of the last seven groups in the phylogeny – flamingos, ibises, storks, frigatebirds, loons, penguins, and petrels – is separated from its sister clades by no more than 0.4°C. These are very small differences, considering the great structural differences that distinguish those groups. Perhaps the analysis found that the large differences in $\Delta T_{50}H$ among the basal clades were simply inconsistent with small and irregular variations in weight.

An entirely different type of result was obtained for tests of the aspect ratios. Variations in that parameter appear to be completely consistent with all the branches in both dendrograms, even though the dendrograms are very different. There does not seem to be a reasonable explanation for this result. In some clades, the average aspect ratios have very large standard deviations; in others, the sample size is very small. Perhaps the results will improve when someone increases the sample size by undertaking the tedious chore of measuring a larger number of live bird wings. Such an event seems unlikely, even though aspect ratio is an important parameter in most flight models.

The brachial index might be a useful alternative as an indicator of wing shape. It is certainly much easier to measure with precision and has the advantage of being available from fossil material. Unfortunately, its values appear to be more related to stress on the wing skeleton than to wing shape, and they show a strong correlation with wing loading but not with aspect ratio.

Although flight characteristics have played a key role in bird evolution, it is unlikely that variation in a single characteristic or feature is going to track the evolution of such an ancient and varied group. Birds are adapted for life in a variety of complex environments, and other aspects of their biology need to be considered. Recently, Kenneth Dial has offered a model with a more integrated approach [49]. He compares a trend towards increased importance of the forelimbs relative to the hind limbs with variations in nestling development and parental care, nest type and design, body size, and flight style. It seems a plausible approach, but his conclusions tend to reflect the interests of paleontologists and he emphasizes the link between birds and dinosaurs. By focusing on the difficulties in getting the ancestral bird airborne, he oversimplifies the ecological and behavioural parameters separating modern groups of birds from each other.

Dial's paper provides an interesting contrast to the approach that I have taken with the flight parameters, because he correlates the trends in each of the parameters to a modified version of the avian phylogenetic tree laid out by Max Fürbringer in 1888. Loons are near the base and the songbirds appear at the top. This arrangement begins with birds that have strong legs for walking or swimming and follows a trend for increased use of their forelimbs and increasingly manoeuvrable flight. Dial proposes concurrent trends towards increased intensity of parental care, increased complexity of nest structure, and decreasing size (a contradiction of Cope's Rule). It appears very satisfactory on the surface, but in each case, the description of the parameter seems rather unsophisticated from an ecological or behavioural point of view. The underlying message in Dial's model is that several different trends should be considered in an integrated way. This is cogent advice, but each parameter needs to be considered in a broad and meaningful context.

The three-dimensional manoeuvrability exhibited by forest birds may be an advanced technique but that does not make all other forms of flight primitive or as similar to each other as they appear on the surface. There is a great difference between the flight of a galliform bird and that of a grebe. The projectile-like flight of galliform birds may well be primitive. It is more or less limited to a brief burst of power followed by a moderately controlled glide. Grebes, on the other hand, exhibit an advanced form of flight that is a continuous performance and demonstrates the exploitation of sophisticated aerodynamics (discussed in the next chapter). For birds that live in open environments, speed appears to be the adaptation of choice regardless of body size. A few others have chosen efficiency and energy saving, but only a handful of unusual species, such as kites, are specially adapted for manoeuvrability.

Similarly, the elaborate nest of a songbird may be evidence of highly evolved reproductive behaviour but it is no more specialized than the year-round territoriality of ducks such as the Barrow's Goldeneye or the elaborate colonial behaviour of murres and other seabirds. Nor is the parental care of an albatross or the puffin either less or more advanced than that of the songbird. Most groups of birds appear to balance the cost of egg production against the cost of parental care. The galliform birds and ducks are at one end of a spectrum, with numerous eggs and offhand care, while the albatross is at the other end, with a single large egg and an exceptionally long period of care. Songbirds are in the middle, with many eggs divided among several clutches and a relatively brief period of very intense care.

Perhaps the most important aspect of Dial's analysis is that he calls attention to factors other than flight that have affected the overall success of birds. Flight may have given birds an edge over flightless dinosaurs but avian evolution was probably driven by intense competition among the birds themselves, especially in the forest. Some 8,000 species have found niches in the canopy, where they depend on elaborate displays, complex songs, specialized nest construction, and other behaviours to ensure their continued success. Nearly all these birds share a single type of slow, manoeuvrable flight.

Outside the forest, variations in flight style may have been the single most important factor in the success of birds. Their range and speed in open country provide ample evidence of their aerodynamic skill and specialization.

In the next chapter, we will examine some birds that appear to have abandoned the evolutionary trends that characterized the earlier types of birds. Seabirds are robustly muscled inhabitants of a hostile environment. They appear to have little in common with the scrawny gnomes that live in the sheltering forest. In seabirds, the airy cloud of feathers has been exchanged for a compact, close-fitting body suit. This toughened plumage enables some to fly very long distances at very high speeds or with great efficiency. It enables others to dive deep beneath the sea in search of food. Although many seabirds range far out to sea, when it is time to nest, many return to the forests of their ancestors.

Further Reading

There are very few books that explain the flight of birds according to modern aerodynamic principles. One of the best and easiest to understand is by Henk Tennekes:

Tennekes, H. 1998. The simple science of flight: From insects to jumbo jets. MIT Press, Cambridge, MA.

Perhaps the best and most lavishly illustrated introductions to the different groups of living birds can be found in the 16 massive volumes of the *Handbook of the birds of the world*, edited by Josep del Hoyo, Andrew Elliott, and Jordi Sargatal (Lynx Edicions, Barcelona). The first was published in 1992 and the last is planned for 2011.

Birds at Sea

10

Sophisticated aerodynamics take birds onto oceans

Dinosaurs were once true masters of the land. They lived in the deserts, forests, and grasslands of every continent. As far as we know, however, they never made it into the sea.[1] All the giant reptiles that patrolled Mesozoic oceans were more closely related to lizards, snakes, or crocodiles. This failure to exploit one of the world's largest and richest habitats is all the more surprising because so many different kinds of the dinosaurs' closest relatives now make it their home. Life in the oceans requires the expenditure of a great deal of energy to overcome the physiological and behavioural challenges posed by the sea, but many birds appear to have found the necessary resources one way or another.

Salt concentrations present one of the biggest challenges to any vertebrate that tries to return to the sea. Seawater is saltier than blood and there is a risk of accumulating so much salt that the osmotic balance is disrupted and normal physiological processes are impaired. Animals that live at sea must find a strategy that reduces the salt load or pay the energetic and anatomical costs associated with salt extraction. In birds, this means that they must either stay within flying distance of fresh water or be able to carry the weight of unusually enlarged kidneys and accommodate the extra physiological cost of operating them. Second only to the salt load is the problem of staying warm. The sea is extremely cold compared with most terrestrial environments, and there is no shelter from its effects. Marine birds must be able to rely on their plumage and their ability to generate sufficient body heat under all sorts of conditions.

Challenges of a Life at Sea

Salinity

In spite of the challenges posed by salinity, species that spend at least part of their

---

[1] *Quianosuchus,* an archosaur of the Middle Triassic, suggests that the ancestors of dinosaurs were not inherently unable to live in marine environments.

lives at sea can be found in all groups of waterbirds.[2] Gulls are the most familiar among groups that avoid the energetic costs of salt control by rarely straying far out to sea. They live over the continental shelves, where they can easily fly to sources of fresh water for a drink and a bath. Large flocks of gulls frequently gather on beaches where coastal streams enter the sea. Loons, grebes, ducks, and pelicans spend much of their lives on salt water and are normally able to divert energy to the process of salt extraction. When they wish to reproduce, however, they migrate to freshwater lakes, where they can exploit seasonally abundant foods and their young can develop without the extra cost of maintaining osmotic balance.

Frigatebirds also tend to be confined to the continental shelves. Like swimming, however, drinking is not one of their activities described in the literature and it is difficult to imagine where the birds that nest on dry tropical islands such as the Galapagos would find fresh water. Perhaps they depend on their prey for water and the continental shelves simply provide the best hunting opportunities for their life-style. Only boobies, petrels, penguins, and auks spend their entire lives on salt water and come ashore only to breed.

The truly marine birds represent only 3% of all bird species. Only they are able to balance the additional physiological and developmental costs of elaborate salt-control systems. Without those adaptations, they would not be able to reach the rich food resources far from shore. It is very difficult to study live seabirds in a laboratory, however, and we have only a rough idea of the amount of energy that they devote to osmotic balance. Research on the avian kidney has usually stopped short of auks and petrels, and we must base our understanding on studies of part-time marine species such as gulls and sea ducks. All we can say at this point is that life at sea has touched every aspect of the existence of the true marine birds, and made them very special animals.

The lack of fresh water makes oceans as hostile as any terrestrial desert but seabirds get some help from their prey. Although crustaceans are as salty as the sea around them, fish spend their entire lives excreting excess salt for their own physi-ological purposes and actually contain fresh water. They are also rather oily, and their body's lipids break down in the bird's digestive system to produce yet more water. In spite of such help, however, seabirds would still tend to accumulate salt if it were not for kidneys almost as large as the liver and large salt glands that sit in bony pockets over the eyes (Figure 1.2). The size of those troughs in the skull is dramatic evidence for the high cost of maintaining osmotic balance in a challeng-ing environment. The salt glands developed from specialized tear glands and are the largest and heaviest special structure carried on the head of any bird. Although they are much smaller than the kidneys, they are more important for the excretion of sodium chloride. Unlike the kidneys, they can be turned on and off to save en-ergy. They operate only when needed.

---

2 None of the Gruiformes is marine, but American Coots (*Fulica americana*) will spend part of the winter in brackish estuaries or in the salty lakes of the American Southwest.

Cold

I am just old enough to have travelled on one of the last winter crossings of the North Atlantic in a Cunard liner. I still remember the ponderous roll and pitch of the *Ascania* as huge walls of green water slapped at the heavy clouds. I wish I could say that I also remember the birds that passed while I was hanging over the rail, but at least I experienced something of their daily existence at first hand. The winter trials faced by oceanic birds are not one of those exotic natural wonders brought into your living room by television. At best, you may see a few emotive frames of albatrosses sailing over ocean rollers in an Antarctic summer, but professional camera crews are not enthusiastic about exposing either themselves or their expensive equipment to the dramatic climatic events of a winter at sea. There are not even very many first-hand accounts. Most scientists prefer the warm and pleasant summer weather, and insurance companies ensure that polar nature cruises for tourists are active only under gentle conditions.

There is some old but very emotive footage in gritty black and white that captures the reality of winter at sea. It appears every now and then on television when the History Channel airs a documentary about the search for the German battleship *Bismarck* or the rigours of the North Atlantic and Murmansk convoys during the Second World War. In one spectacular clip, an ice-covered heavy cruiser buries its nose in green (grey) water. As it disappears under a mass of spray and foam, storm-petrels and fulmars can be seen silhouetted against the foam and sailing by, unconcerned, in the gale.

The key to life at sea in both winter and summer is possession of the appropriate protection. Conditions at sea are much more extreme than on land – temperatures are lower, winds are stronger, and food items are usually hidden by metres of icy water instead of a few fallen leaves. Even brief contact with cold water can be a fatal shock for a bird, and it is a special problem for species in the polar areas, where the water is often freezing and fully 40°C below the bird's body temperature. Even the food in such regions can contribute to physiological problems. The prey may be exceptionally nutritious but it is also extremely cold and sucks up valuable body heat during digestion. If a bird can keep warm in such an environment, all other problems seem relatively easy to solve. It is not surprising, then, to find that most seabirds are exceptionally well insulated. They are typically covered in a dense, watertight plumage that is quite unlike that of any terrestrial bird.

The watertight plumage of marine birds is analogous to the "dry suit" used by human divers in very cold water. The feathers trap a layer of air against the body because the strong surface tension of water prevents it from squeezing between the plumules. The effect is enhanced by the natural hydrofuge[3] effect of the keratin in

---

[3] "Hydrofuge" is a French word that is creeping into English to describe a water-resistant surface on clothing, building products, and makeup. The surface molecules do not become wet and the water droplets are shed as tiny beads. It is also used to describe various water-shedding hairs or scales in insects.

the feather and the oily sebum secreted by the skin. Ducks, loons, grebes, auks, penguins, and petrels all use this dry-suit strategy and are able to thrive in polar and cold temperate-zone seas. "Dry suits" can also help control body temperature where the climate is very warm, and lightweight versions are used by boobies, pelicans, and the petrels that breed on tropical islands.

A few birds use a "wetsuit" technique, in which water replaces the air close to the skin. Once the body warms the water, the feathers help prevent it from circulating to the outside. Not surprisingly, it is most appropriate for tropical species that do not go out onto the ocean, such as the anhinga and finfoot, but it can be effective in cold water. Cormorants use the wet-suit strategy and are able to range from Greenland to Antarctica. Their large, loose feathers trap water when they are submerged but they appear to maintain a thin layer of air next to the skin – a belt and suspenders approach. This hybrid insulation system is surprisingly efficient. Off the coast of British Columbia, Pelagic Cormorants spend much of the long, rainy winter submerged up to their ears in seawater that is colder than 7°C. Even on the occasional clear day, only a few of these birds perch in trees to dry their wings. The rest just loaf away the hours between meals, sitting on bell buoys and other convenient roosts. Although the cormorant's plumage looks saturated, its loose construction enables the bird to shed excess water quickly, and even submerged birds can take off and fly away without a great deal of extra effort.

As important as being insulated against the icy cold of seawater is being able to get rid of the excess heat generated by the muscles in strenuous flight. The dense, watertight plumage of typical seabirds insulates in both directions, and the accumulation of heat in the body could, in theory, interfere with normal physiology. The loose feathers of a frigatebird offer ample opportunities for heat to escape, but their black colour absorbs solar energy and the bird needs an elaborate system to direct heat transduction. The bones of its pectoral girdle are exceptionally large but their walls are extremely thin, and they are highly pneumatized. Excess heat from the flight muscles passes through the paperlike walls of the bones into extensions of the thoracic air sac, and is then carried out of the body as the bird exhales.

Other seabirds are able to maintain body temperature with less dramatic structural specialization. The bones of tropical petrels are not as inflated or as pneumatized as those of a frigatebird, but their flight style may also make intense demands on the muscles and generate excess heat in the process. Petrels can shed that heat through a thinly insulated area under the wing, and may enhance that area's effectiveness as a radiator with quick dips into the sea.

Auks have a much more vigorous flight style than petrels and must also generate a great deal of excess body heat in flight. They do not have thinly insulated areas on their wings and their bones are not pneumatized at all. They must depend on other cooling mechanisms to protect organs, such as the brain, that are highly sensitive to temperature change. One of the structures that keeps their brain cool is the *rete mirabile ophthalmicum*. A *rete* is literally a network, in this case a network of counterflowing blood vessels located near the nasal passages, where air movement increases the cooling effect of evaporation. Cool blood coming from surface of the nasal passages picks up heat from blood coming from areas deeper in the body and releases it into the exhaled air. Many birds have a similar but less elaborate network in the hind limbs, which makes particularly good use of radiation from naked legs and webbed feet. Usually the blood vessels entering a *rete* can be constricted to control heat loss in the cold or expanded to increase flow when the body is too warm.

Specialized anatomical structures may not always be adequate to cool seabirds that live in the tropics, and many of them fall back on simple behavioural tricks to cool off. Cormorants and other pelecaniform birds flutter their throats and cool themselves by increasing the rate of evapotranspiration. Tropical petrels escape the heat of the day by dozing in nesting burrows; other birds simply seek out patches of shade thrown by a shrub or rock. Cooling does not seem to be a problem for the auks and petrels that nest on the fog-shrouded islands of the temperate zone.

Suitable Prey

The physiological processes that maintain body temperature and stabilize osmotic balance consume energy, and seabirds must work hard to meet their daily energy budget. Food is rarely lying on the surface of the sea, just waiting to be picked up; neither is it evenly distributed either horizontally across the water's surface or vertically in the water column. Land birds can learn to concentrate their search for food on certain plants or in special habitats, but marine birds must search huge areas of apparently featureless sea for suitable prey, and then catch it as it attempts to escape deeper into its own natural environment.

The organisms that make up the prey of seabirds move about in the sea for their own purposes and may be available to air-breathing predators only at certain times of day or during certain parts of the tide cycle. Hidden features of sea-bottom topography such as seamounts, deep-sea canyons, or shelves in channels are often marked by concentrations of birds loafing on the surface. They may be waiting for regularly scheduled foraging opportunities created when underwater obstructions deflect tidal currents upward and carry prey items within range. There are places in the world, especially along continental shelves or over seamounts, where upwelling oceanic currents deliver a constant supply of food. The Humboldt Current, which passes up the Pacific coast of South America, and the Benguela Current off the coast of Southwest Africa carry food to millions of seabirds. When they fail, the effects are catastrophic. When the well-known El Niño/Southern Oscillation (ENSO) effect disrupts the normal currents in the southern Pacific Ocean, the effects can be seen in reduced breeding activity among seabirds in North America, as far north as Alaska.

It is not too surprising that the most obvious adaptations for life at sea tend to involve special features for locomotion both above and below the surface. Locomotion of any sort requires energy, and seabirds must invest heavily to search large areas for suitable foraging opportunities and then catch enough prey to recoup their energy expenditure. In the breeding season, they must also find enough food for their young. Surprisingly, some of the same characteristics of prey that help seabirds with osmotic regulation also help them balance a daily energy budget. Most prey species used by seabirds are mid-water organisms that maintain a preferred position in the water column by controlling buoyancy. Shrimp and other crustaceans and many small fish depend on oils that are slightly less dense than water. The oils are also an energy store that greatly increases the value of their owners as food for other animals. Mid-water crustaceans such as shrimp and euphausiids (krill), and various oily fish such as herring, anchovetta, and sand lance, often swarm in huge numbers, creating patches of energy-rich oil that whales, seabirds, and other marine predators compete for. The high caloric density of these animals (about 5,000 calories per gram) is particularly important to seabirds trying to meet the energetic demands of fast-growing nestlings.

It is no accident that the seabirds carefully select certain species of fish. Coastal seas are full of swarming fish similar to young rock cod, which use an air bladder for buoyancy instead of oil in their flesh. But air has no nutritive value at all, so these fish have a much lower caloric content than, for example, sand lance or herring. When fish with air bladders are delivered to the young of birds that typically exploit oily fish, it may be a precursor to widespread reproductive failure. If the shortage is not too severe, the nestlings may be able to fledge at a lower than normal weight, but often some die in the nest. The adults are usually unaffected but the surviving fledglings may not be strong enough to endure a winter at sea. When El Niño depletes the anchovetta off Peru, thousands of adult seabirds and other marine animals, including sharks and sea lions, simply die of starvation. Deep windrows of their bones pile up on the beaches and it is hard to believe that the ecosystem will ever recover.[4]

Oceanic Distances

The true masters of the sea are birds that have developed sophisticated forms of locomotion to overcome the issues of time and space. Their life strategies are built around the huge expanse of the sea and the long periods they must spend crossing it. When the ancestors of cranes and other large birds left the forest, they met the challenge of thinly scattered resources with increased flight range and efficiency. Ancestral seabirds faced challenges that were an order of magnitude greater when they first ventured out onto the oceans. It seems unlikely that the earliest marine pioneers were able to conquer the high seas. They probably stayed close to shore and foraged in shallow waters. It took a long time to evolve the special suite of flight techniques and foraging skills needed to capitalize on the opportunities in deep water.

---

[4] The ecological reach of a strong El Niño is surprising. In 1998, I was taken to see a hibernaculum for male vampire bats at Punta San Juan, Peru. At the beginning of the austral winter, the male bats migrate from montane forests across the coastal desert to feed on the nestlings of a penguin colony. That year, the birds did not breed and only one lonely vampire remained in the cave.

The biggest structural change that accompanied the shift from terrestrial to marine habitats was the conversion of the inner bird from a scrawny gnome to something more akin to a muscular troll. A few exceptional seabirds, such as the tiny storm-petrels and the predatory frigatebirds, maintain a low overall body density, but typical marine birds are much more robust and densely muscular than any terrestrial species. A great increase in the size of the flight muscles has been particularly important, especially among birds that use their wings for underwater propulsion. In typical forest birds, the flight muscles comprise between 10% and 15% of the body mass. The hummingbird is an exception because its unique flight technique requires a great deal of power; its flight muscles account for over 25% of its body weight. The flight of the auk is also based on the naked application of power, and its flight muscles account for about 20% of its body weight.

The size and weight of the flight muscles in marine birds have played a major role in the birds' interactions with people. The seabird's preference for remote habitats meant that scientists did not get a close look at their life histories until the 20th century, but aboriginal peoples and mariners knew them well. Rich, fatty breast muscles, and a habit of collecting in huge numbers at breeding colonies, made seabirds an important part of the diet of subsistence hunters and passing sailors. Consequently, much of the human history of seabirds has been a rather grim record, spotted with stories of wasteful exploitation, gross indifference, and local extermination. Exploitation was particularly intense in the 18th and 19th centuries, before ships had refrigeration, but in the 20th century, there were local crises that put pressure on seabird colonies. Birds of the Sea of Okhotsk were a · major source of protein during the Second World War, and later for Stalin's gulags. Many of the worst modern examples have been a case of "out of sight, out of mind," but most maritime nations actively participate in seabird conservation programs.[5]

Types of Seabirds

The Penguin

Before the 20th century, explorers and whalers in the south polar seas depended heavily on meat collected at penguin colonies. Penguins represent one end of a spectrum in seabirds that stretches from the greatest of aerial performers to birds that have given up flight altogether – at least aerial flight. Flightlessness in birds has come to represent loss of evolutionary opportunity and impending extinction. Most of the flightless birds in history have become extinct, and flightless seabirds are no exception. Many went with human help but others died out from natural

---

[5] This is a very recent phenomenon. In my career as a seabird conservation biologist, I saw business proposals to can shearwaters for pet food and import penguin skins to make high-fashion glove leather. Naval forces of many countries still use seabird colonies for artillery practice (especially during the breeding season), and in the 1980s, the French Air Force used penguin colonies to study the effect of fragmentation bombs. The shameful and unnecessary by-catch of albatrosses in long-line fisheries threatens the existence of some species and is a major international scandal.

causes. The giant, toothed *Hesperornis* of the Cretaceous disappeared with the dinosaurs, and the huge *Plotopterus* of the North Pacific may have been displaced by
newly evolving seals in the Miocene. The disappearance of the later mancallid
auks of California is more puzzling. They may simply have been unable to cope
with climate change or oceanic regime shifts. Perhaps sea-level variation reduced
their supply of secure nesting opportunities. The Great Auk survived into the 19th
century as the only large flightless bird of the North Atlantic. A century after it
became known to science, collectors were offering cash rewards for the skins of the
last survivors, and the Great Auk became an icon for extinction brought on by
greed and mindless exploitation. In the southern seas, penguins have survived, and
not just in the frozen wastes of Antarctica. With a little human help to control introduced mammalian predators, healthy colonies survive in New Zealand, Australia,
South Africa, Argentina, the Falkland Islands, Chile, and even along the equatorial
coasts of Peru and the Galapagos Islands. By any standard, penguins are a diverse
and successful group with as good a chance of continued survival as any other kind
of bird on our battered planet.

Considering that a great deal of ink has been used to compare the Great Auk to
the more familiar penguin, it is appropriate that the earliest European name for the
penguin is *sotilicário,* the vernacular name for the Great Auk in Portuguese. When
Vasco da Gama sailed into Sanbras on the coast of Africa on 25 November 1497,
his crew of Portuguese fishermen thought that they recognized the flightless birds
that bred there. They had eaten a great many Great Auks while fishing and whaling
across the North Atlantic, and they were happy to do the same at Sanbras.

Penguins are giants among birds. A few birds may stand taller but no other group
boasts such a densely packed array of muscles. Only ostriches and their relatives
are a great deal heavier. Over the years, I have banded ducks, geese, swans, loons,
albatrosses, and most of the larger auks, as well as dangerous birds like eagles and
large owls. I have never had much difficulty preventing them from injuring either
themselves or me. I had been warned that penguins were strong, but I could not
imagine a threat from anything so docile and friendly looking. My opinion was
sharply changed by an angry penguin at Phillips Island, Australia. It objected to
the application of a wing tag and quickly took meaty chunks out of my arm and
chest in spite of my best attempts to restrain it. It was a Little Blue, one of the
smallest species. When an opportunity arose to band the slightly larger Humboldt
Penguin a few years later, I left it to the muscular Peruvian park wardens, who had
already impressed me with their enjoyment of lunchtime football at 44°C.

The massive muscles, powerful flippers, and streamlined shape are all part of a
penguin's specialization as a diving bird. These birds spend most of their lives at
sea, and their size and weight makes them more at home underwater than on land.
Because they do not fly, they are free from the aerodynamic constraints that have
such a strong effect on the architecture of other birds. They have become exceptionally large and heavy, with a greater body density than any other living species

of bird. When a penguin prepares for a dive by depressing its feathers fully, it can raise its body density to 0.98 grams per cubic centimetre, very close to the density of the seawater (1.025 grams per cubic centimetre) in which it hunts.

Penguins are among the most specialized birds, and their adaptations for life underwater have interested scientists for over two centuries. Curiosity about their ability to dive to depths greater than 500 m has made them one of the most extensively studied and thoroughly documented birds. Deep diving in birds and mammals is still a mysterious process. Birds, in particular, have been able to overcome a series of biomechanical and physiological problems through adaptations that we are just beginning to understand.

One of the most obvious impacts of a deep dive is compression due to the great weight of water overhead. Diving birds use elaborate combinations of flexible bones and tough connective tissues to avoid pressure differentials that might distort sensitive organs or tear membranes. The most obvious example of such special protection is the arrangement of greatly elongated ribs that surround most of the lower abdomen in auks and penguins. They seem insubstantial and are often little more than easily bent bristles, but they are springy and quickly return to their original shape when pressure is released. They also have considerable strength parallel to the long axis of the body, preventing its wall from being pulled out of shape during a dive.

Human divers may suffer from a variety of problems associated with attempts to reach great depths. In both birds and mammals, some problems stem from the presence of nitrogen in air carried in the lungs. At high pressures, the nitrogen readily passes into the blood. This may cause nitrogen narcosis, which disorients the diver. If the diver rises to the surface too quickly, the nitrogen can form bubbles, which cause the bends. Not all problems are related to gases in the body. There are also poorly understood impacts of compression on the nervous system. In humans, the effects are known as "high pressure nervous syndrome." Somehow birds have found a way to avoid this problem, although no one has found any evidence of structural modification to their nervous systems [45].

Once underwater, air breathers are cut off from any source of fresh oxygen. Most animals must surface once they have exhausted the air in the lungs, but species that achieve extreme depths often exceed the aerobic diving limit. Surprisingly, these species also carry less air in their lungs than you might expect. In birds, the secret of success appears to be the myoglobin concentrations in the large muscles. Because myoglobin has a greater affinity for oxygen than hemoglobin, deep-diving birds are actually able to carry far more oxygen than you would expect from a simple measurement of their lung capacity.

In order to meet the challenges of their lifestyle, penguins have developed a unique skeletal architecture, a specialized physiology, and a suite of features that separate them from other birds, just as unique and highly specialized structures distinguish the dolphin from any terrestrial mammal. Penguins represent such a fundamental departure from the evolutionary trends in other groups of birds that they may be entitled to removal from the common mass and placed in a new sub-class of their own.[6]

Penguins are not just a flightless diving bird; they are a feathered alternative to dolphins and the differences between these two groups of marine animals offer an interesting contrast in the solution to biomechanical problems of life underwater. Where dolphins have abandoned hips and legs to develop a nearly boneless semi-fluid architecture, penguins have extended their skeleton to sheath much of the body. Long, flexible ribs protect the lower abdomen and a pair of platelike scapulas cover the back. The much smaller auks exhibit very similar ribs but have a more normal scapula (Figure 2.8). Throughout the penguin's skeleton, joints are mas-sively reinforced with connective tissue to absorb the stress of diving and high-speed underwater locomotion.

In most cases, penguins have modified their skeletons without sacrificing overall flexibility. The wings are the exception, if the highly specialized flippers of these wing-propelled divers can still be called wings. The flippers are flat plates of bone and connective tissue bound by a thin layer of "flesh" and feathers. There is little muscle tissue and only a minimal blood supply. Each major wing bone has become a broad flat plate (Figure 10.1). The humerus is flattened and arched as in other wing-propelled diving birds. The arch may help transmit bending stress towards the ends of the bone, reducing the risk of a break at the midpoint. Two bones that are tiny in most birds, a sesamoid of the elbow and the ulnare of the wrist, have become large flat plates in their own right. The alula (digit 1) has disappeared, and the leading edge of the flipper is a smooth, uninterrupted arc.

As you might expect, the bones of the pectoral girdle are massive in order to absorb the stress of vigorous wing-propelled swimming, but are generally similar in shape to those of other birds. The scapula, however, is a broad plate instead of the more familiar scimitar. It reaches far down the back, where its broad surface may help redistribute the pressures generated by muscle movement. Penguin flippers generate thrust during their upstroke, when they must work against the resistance of the water, and the large scapula serves to anchor dorsal muscles that assist the supracoracoideus in raising the flipper.

Seabirds that "Fly" in Two Media

Auks and some tube-nosed seabirds have learned to use their forelimbs for propul-sion underwater without losing their ability to fly in the air. Such a strategy presents these birds with some extremely difficult biomechanical challenges because water is some 300 times denser than air. Moving a wing vigorously through any

---

6 The specializations in penguins stand out because of the general uniformity of other birds. Pen-guins and sparrows both belong to the Neoaves, but the differences between the two are arguably much greater than the differences between a sparrow and a chicken (Galloanserinae) or even be-tween a sparrow and a tinamou (Paleognathae).

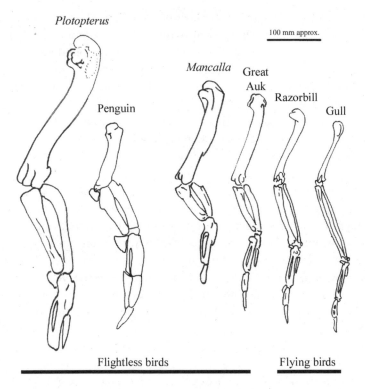

Figure 10.1 Comparison of wing skeletons in five wing-propelled diving birds and a non-diving gull (after R. Storer [200] and S.L. Olson and Y. Hasegawa [153]).

medium generates bending and twisting stresses in the bones and joints. In the air, the stresses are never great enough to threaten the structure of the wing. Underwater, however, the strains are much greater and the bones must be proportionately stronger. The flightless penguins have overcome the problem with brute force. The wing bones became greatly enlarged, but in the process, the limbs became so heavy and rigid that they lost the capability for aerial flight.

The stresses of both swimming and flight strokes are concentrated in the shoulder, so it is not surprising to find that it is the largest joint in the bird's body and is strongly supported by connective tissue. It is also not a simple ball-and-socket assembly. The head of the humerus is a broad bar that fits into a wide trough in the shoulder, and the joint works more like a robust hinge that bends in only one direction. In wing-propelled diving birds, the shoulder and the humerus may be responsible for absorbing most of the torsional stresses generated by locomotion. Adaptations in the outer bones and joints of the wing are subtler and less obvious.

Although the flying types of wing-propelled diving birds have retained small and light bones in their wings, there are some similarities to penguins. As in the penguin, but unlike in most other birds, the humerus of an auk or diving-petrel is the longest bone in the wing. It is also somewhat flattened and distinctly arched, presumably to cope with stresses from a resistant medium like water. Although the humerus of an auk is usually described as flattened, widened may be a more accurate description. Its profile is only slightly thinner than the round humerus of a comparable land bird but its internal structure is very different. In a land bird, the thickness of the walls of the humerus is more or less uniform; in an auk, the walls along the narrow edge are much thicker than the walls on the flat surface (Figure 2.1). This results in a box-beam with great resistance to strain against its edge but much less against its flat surface. Like a knife blade, it can twist or bend to a certain degree to take up strain without breaking [41].

Resistance to bending is usually a simple matter of bone strength, but twisting generates shearing forces that present animals with difficult biomechanical problems. Birds can respond by varying the cross-sectional shape of their bones, as in the auk's humerus, but torsion also puts stress on joints, where it can cause excess wear on the articulating faces or damage the supporting connective tissue. In a human athlete, twisting may result in knee injuries or torn rotator cuffs of the shoulder that are slow to heal. For a bird, comparable injuries to the wing are fatal.

In the air, a bird exposes the whole surface of the wings to stress when it flies with the wings fully extended. Forces on the joints are greatest during the powerful downstroke, but remain relatively modest because the medium is not very dense and full extension of the wings enables the bird to take advantage of the glide to recover between strokes. Underwater, the resistance of a dense medium works against the movement of a large surface such as the extended wing. Friction quickly terminates any glide. Birds have adopted the obvious solution, reducing the affected area by closing the wings slightly. They move underwater on partially folded wings. Whether by accident or design, the geometry of the partially folded wings greatly reduces torsional stress.

The wide head of the bird's humerus at the shoulder is somewhat reminiscent of the grip at the end of a canoe paddle, which fits the hand comfortably and helps it resist twisting in the power stroke. A well-built canoe paddle, however, is a rigid object carved from a single piece of wood and carefully balanced so that the surfaces of the blade have an equal area on both sides of the shaft. A bird's wing is not rigid. It has joints at the elbow and wrist. When partially folded, it assumes a "Z" shape that seems an eccentric choice for a balanced blade. Nonetheless, it is this particular response to moving the wing through a very dense medium that is surprisingly effective at reducing the amount of twisting.

By partially folding its forelimb into a "Z," the bird moves its elbow back from the leading edge of the wing and places it closer to the geometric centre of the forces acting on it. A membrane, the propatagium, stretches from the shoulder to the wrist in front of the elbow and makes up the actual leading edge of the wing, while the secondary and tertiary feathers fan out behind to make the trailing edge. Rotation of the limb is largely the responsibility of the shoulder, and the avian elbow is designed to bend in only one direction (at right angles to the action of the shoulder). It pronates (bends parallel to the sagittal plane), while rotation across the sagittal plane is limited by small bony structures in the joint. To protect the elbow from rotating or bending inappropriately under stress, the joint is heavily bound by connective tissue and supported externally by the layer of stiff secondary coverts. These feathers are unusually robust in auks and may absorb some of the strain on the skeleton of the wing.

The second bend of the "Z" occurs further out on the limb, where the hand, or manus, meets the radius and ulna. In birds, the wrist is designed to bend in two directions. It pronates parallel to the sagittal plane so that the outermost digit moves towards the elbow when the wing folds, and it rotates so that the wingtip can twist during flight. When the bird is using its wings, either underwater or in the air, the wrist rotates most during the upstroke. This action allows the loosely trailing primary feathers to drop down, reducing the drag on their surface. The primaries also separate slightly so that it is easier for the water to pass between them. In the downstroke, the primaries stack on top of each other for mutual support, to increase the stiffness of the wingtip.

Underwater, any twisting strain is ultimately transmitted towards the shoulder, but the overall geometry of the Z-shaped paddle helps reduce the forces on that

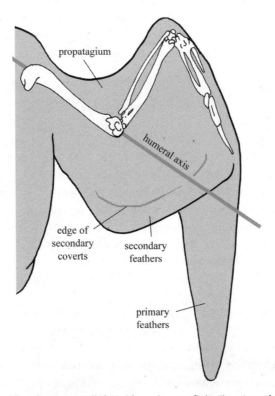

Figure 10.2   Outline of an auk wing, partially folded for underwater flight. The primary feathers would lie in front of the humeral axis if the wing were fully opened for aerial flight.

joint. If you draw a line along the humerus, as though it were the shaft of an imaginary canoe paddle, and extend that line through the rest of the partially folded wing, it will divide the wing's surface area into roughly two equal parts, half in front of the humerus and half behind (Figure 10.2). The wrist and the base of the primary feathers lie in front, while their trailing edge and the secondary feathers lie behind. When the wing is fully extended for aerial flight, the division is much less equal; about 80% of the wing's surface lies behind the line through the humerus. Air is such a low-density medium, however, that the shoulder can absorb the torsional forces that such an unequal division creates.

Underwater locomotion greatly increases the wear and tear on the plumage, and even the tough feathers of an auk must be replaced regularly. All wing-propelled diving birds undergo a synchronous moult, losing all their primary and secondary feathers at the same time. The featherless stub looks very much like a penguin's paddle, and the moulted birds are flightless until new feathers grow. This moult occurs after the birds have completed their annual set of parental duties and are no longer carrying food to the nestlings. All their foraging activity can be devoted to self-maintenance. Without flight feathers, they cannot search over large areas of ocean, but the stubbier wings probably generate less drag underwater and the birds may become more efficient as underwater predators. Moulting auks appear to enhance their similarity to penguins by extending the moulted stub of their wing more than usual while swimming underwater. Even fully feathered birds, however, will sometimes extend their wings as they twist and turn to chase fish.

Delicate Seabirds

Although most seabirds are robust and muscular, other designs have been successful, and surprisingly delicate birds can survive ocean storms. Terns, storm-petrels, and phalaropes demonstrate that there are opportunities for small, lightly built birds on the open ocean. They find ample food on the surface, without diving. The much larger frigatebird is also lightly and somewhat delicately built in spite of its size. It is one of the largest birds to depend on an exceptionally low overall body density (Table 10.1). It has huge wings that greatly reduce wing loading, and it seems even more lightly built than terrestrial aerial specialists such as swifts or kites. Low density has helped make it so comfortable on the wing that it can preen itself in flight; it is even reported to snooze without landing. It feeds entirely from the sea surface and never dives. Reports that it can swim on the surface are rare and poorly documented, but an ineffectual remnant of web between the bird's toes suggests that it had swimming ancestors. Tiny legs and feet make it almost impossible for a frigatebird to walk properly, and it is most frequently seen on the wing or sitting at a roost.

The aerial acrobatic capabilities of the frigatebird make it the most agile aerial predator among the birds. Its manoeuvrability depends on the long, pointed feathers that extend its wingspan and give its tail an exceptionally deep fork. It also has some unusual internal adaptations that carry weight reduction to an extreme. The major bones of the skeleton look much larger than those of other birds of similar

Table 10.1

Body and wing proportions of a frigatebird compared with those of two other marine birds and a terrestrial bird

Species	Weight (g)	Wing loading[a] (N per m²)	Wingspan (m)	Aspect ratio
American Robin	80	34.2	0.37	6.0
Marbled Murrelet	237	134.3	0.44	10.6
Royal Albatross	8,700	158.0	2.82	14.7
Magnificent Frigatebird	1,520	36.5	2.29	12.8

[a] Wing loading is measured in Newtons (N) per square metre (m²). 1 N = 100 g × the acceleration due to gravity.

size but they are little more than inflated shells. The entire skeleton adds up to no more than 5% of the total body weight. In fact, its feathers weigh more than its skeleton.

Although the frigatebird can perform unusual contortions as it manoeuvres in the air, many parts of its skeleton are rigidly fused together. The furcula and coracoids are exceptionally long and extend the length of the flight muscles. They frame a large open space that accommodates the bird's gular pouch, in which it carries food for later consumption. The greatly inflated shoulder bones are thoroughly fused to each other and to a surprisingly small sternum (Figure 10.3), implying that the flight muscles may be elongated and also unusually slim. Long, slim muscles can contract more than short ones but they cannot generate as much power as thick ones. They are exactly the kind of muscles you would expect to

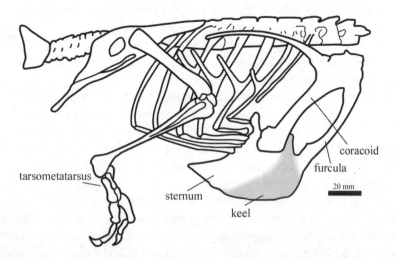

Figure 10.3 Thoracic skeleton of a frigatebird, showing the very large furcula and coracoids fused to each other and to the scapulas. The rather small keel on the sternum is well suited to very deep but slow wing beats.

find in a bird that flies with deep, slow wing strokes, like the frigatebird. The sternum is also carried in a peculiar semi-vertical position, far forward in the body. It is quite unlike the arrangement of any other bird and may also be related to the bird's unusual flight stroke.

There is one very well-preserved fossil of a frigatebird (*Limnofregata azygosteron*) and some additional fragments from the Lower Eocene Green River Formation of Wyoming. Peculiarly, many of its bones are less thoroughly pneumatized than in the modern forms. In an active flier like a modern frigatebird, pneumatized bones are important radiators of excess body heat, but *Limnofregata* may have been less active than its descendants. Various aspects of its structure suggest that it lived somewhat like a gull over the shallow seas that once covered central North America. The wings are shorter and the legs larger than those in a modern frigatebird, as though it spent more time swimming than the recent species.

Extreme specialization for life in the air does not seem to have made the descendants of *Limnofregata* particularly successful, but the frigatebird family has survived while the plotopterids, pseudodontorns, and others have not. The modern Family Fregatidae has only one genus, *Fregata*, which includes only five species. They all seem to fit the same very narrow ecological niche as predators of flying fish and kleptoparasites that steal food from less agile birds. Not surprisingly, all five species are strikingly similar in appearance and vary only in size.

The fossil of *Limnofregata* offers some important lessons in the interpretation of bird fossils. Many finds consist of solitary bones or fragments. Although these finds are intriguing, it may be misleading to assign them to a modern group or, worse, use them to define a new group. Storrs Olson points out that *Limnofregata* consists of an unusually large assemblage of bones but only the pelvis and sternum, and to some extent the remains of the skull and tarsometatarsus, show features of the Fregatidae [148]. The outer end of the tarsometatarsus might easily be confused with that of a booby (Sulidae), which is at least a pelecaniform bird, but the inner end of this bone is quite different from that in any living family. The other large bones of the wing, the hind leg, and even the pectoral girdle would probably be misidentified if found alone.

This observation raises the second problem. The identification of *Limnofregata* is based on the expertise of Storrs Olson. Although he is one of the most respected avian paleontologists, he based his decision on the procedures of classical comparative morphology. He did not put the characteristics in a cladistic context by comparing them with living representatives of the various lineages that may have contributed to the "pelecaniforms" [54]. Consequently, we can say only that the fossil is a "pelecaniform" bird that looks like a frigatebird. A full cladistic analysis would help ensure that it was not a convergent type from some other lineage.

The modern frigatebirds highlight one of the major discrepancies between molecular and morphological classifications. Until Sibley and Ahlquist published their biomolecular phylogeny in 1990 [192], ornithologists accepted a close relationship among frigatebirds, tropicbirds, pelicans, cormorants, anhingas, gannets, and boobies, and placed them all together in the order Pelecaniformes. They shared the totipalmate foot, in which all four toes are joined by a web, as well as some other structural and developmental features. Analyses of both nuclear DNA and mitochondrial DNA suggest a different relationship. Cormorants, anhingas, gannets, and boobies appear to be closely related, and Sibley and Ahlquist placed them together in the clade Sulida. Biomolecular analyses also suggest a close relationship between the Sulida and the tropicbirds, placing the Phaethontidae[7] in a sister clade. Sibley and Ahlquist could not place the pelicans among the other pelecaniform birds, however, and placed them among the storks (Ciconiidae), close to the peculiar Shoebill Stork. They also set up the frigatebirds as a sister clade to the stork/vulture/pelican group. In other words, comparative morphology indicated one large group of birds with similar features but biomolecular analyses indicate convergence between two rather distantly related lineages (see pages 189-90, in Chapter 6).

Frigatebirds are the only group of seabirds in which feathers play as large a structural role as the skeleton. Other kinds of seabirds are far more dependent on their skeletal architecture and have much smaller feathers. Two distinct evolutionary paths have led to the development of a range of skeletons that lies between the extreme airiness of frigatebirds and the monumental solidity of the penguin. The petrels, or tube-nosed seabirds, have a lightly built skeleton, without the inflated bones of the frigatebird. They travel at relatively modest speeds using a flight style characterized by fine control of lift and sophisticated conservation of effort. The auks are at the other end of the spectrum. They travel at very high speeds, apparently disdain aerial control, and expend vast amounts of energy on thrust. They have compact, solid bodies and robust skeletons whose joints are strongly reinforced with connective tissue. The design of tropicbirds, cormorants, boobies, and other "pelecaniform" seabirds is more reminiscent of the terrestrial types of birds, to which they may be related. It tends to fall between the extremes represented by petrels and auks (Table 10.2).

Two Evolutionary Strategies in Truly Oceanic Birds

Auks and petrels are the most successful and diverse kinds of seabirds. Except for the penguins, they are the only groups that regularly inhabit the high seas. Their success as truly oceanic birds is based on two very different evolutionary strategies. The differences are most apparent in their flight. The petrels (albatrosses, shearwaters, fulmars, prions, petrels, and storm-petrels) carefully conserve energy and are generally very efficient fliers that use sophisticated control to take advantage of air movements and increase their flight range. The auks (puffins, murres, murrelets, auklets, and guillemots) depend entirely on muscle power to generate

---

7 Recently assigned to the Metaves and further removed from other "pelecaniform" types by Fain and Houde [59].

Table 10.2

Effects of the trade-off between lift and thrust in the locomotory and reproductive strategies of different kinds of seabirds

Frigatebirds	Petrels	Auks
Structural features		
Small sternum and keel, elongated furcula and coracoids	Short coracoids, large furcula curved forward, moderately long sternum and keel	Short coracoids, large furcula curved forward, long keel and sternum extending rearward beyond keel
Low wing loads (36-49 N per m²)	Low to moderate wing loads (13-158 N per m²)	High wing loads (75-206 N per m²)
Low design speed (10.0 m per sec)	Low design speed (11.9 m per sec)	High design speed (18.4 m per sec)
Very high aspect ratio (14.7)	High aspect ratio (10.7)	High aspect ratio (9.3)
Very long tail	Very short tail	Very short tail
Related reproductive strategies		
Small egg (about 6% of body weight)	Large egg (up to 28% of body weight)	Large egg (up to 34% of body weight)
Rapid development of nestling	Slow development of nestling	Rapid development of nestling
Whole prey delivered to nestling	Stomach oil delivered to nestling (13%-24% of adult body weight)	Whole prey delivered to nestling (2%-11% of adult body weight)

speed from basic flapping flight. In addition, they are diving birds that are able to combine high speed in the air with high speed underwater. They are so effective underwater that they can compete head-on with marine mammals for the same prey. It may be difficult to see their brute application of power as a sophisticated aerial technique, but aerodynamic theory suggests that we should look at the flight styles of these two groups as divergent examples of very advanced aerial techniques.

Understanding Avian Aerodynamics

Although the ability to fly is largely dependent on the structure of the outer bird's plumage, the activity is powered and controlled by the muscles and brain of the inner animal. It is a very complex process, which physicists and engineers are just beginning to understand. It has taken the application of advanced technology in the form of wind tunnels, radar, and high-speed cinematography to observe the details of avian flight and build predictive models of the performance of different kinds of birds. Such models have importance beyond aerodynamic theory. Ecologists can use them to calculate precisely the amount of energy a bird needs for its daily routine. By looking at changes in the effort that a bird must make in order to

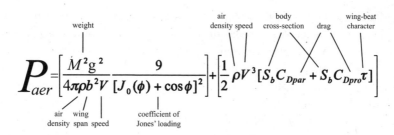

Figure 10.4 A mathematical expression of the power required for aerial flight, developed by J.M.V. Rayner [176]. See text for an explanation of the terms of the equation. P_{aer} = power required for aerial flight; M = mass; g = acceleration due to gravity; b = wingspan; V = velocity; ρ = air density; J_0 = coefficient for Jones' loading; S_b = cross-sectional body area; C_{Dpar} = coefficient of parasite drag; C_{Dpro} = coefficient of profile drag; τ = ratio of downstroke to upstroke.

survive, we can estimate the influence of seasonal weather patterns or the impact of human activities on the environment. Figure 10.4 shows the equation for one such model, developed by Jeremy Rayner to show the factors affecting the amount of power (P_{aer}) that birds require for aerial flight [176].

The equation is simpler than it looks. Basically, it says that the amount of power that a bird needs to get into the air is equal to the sum of two unrelated terms.

Term 1. The first term describes the components that need to be considered in overcoming gravity. The most important of these are the mass of the bird's body (M) and the effect of gravity (g) as a source of inertia. They are modified by two factors that describe the effectiveness of the bird's wings in generating lift: the size of the wings as indicated by their span (b) and the density of the air (ρ) that the wings are moving through. At lower elevations, air has a higher density than air at high altitudes. It therefore offers the bird more buoyancy close to the ground. In addition, air is not a uniform mixture of gases. The amount of water vapour is particularly variable, and water molecules are so large that very humid air is much denser than dry air. In fact, "air" as a medium is rarely a simple gas but a mixture of liquids and solids as well as gases. Birds often find themselves flying through a medium that includes microscopic water droplets (fog), cold water drops (rain), supercooled water drops (sleet), ice crystals (snow), and even ice pellets (hail). Birds captured in mist nets set over the sea are often wet to the touch and, on cold nights, they may be covered with tiny beads of moisture [107]. The large liquid content of this aerosol might be expected to greatly increase the energetic cost of flight, but small water droplets and fog may actually enhance the air's buoyant effect.

This first term also includes a factor for speed or velocity (V). Its presence in the denominator suggests that increased speed actually reduces that portion of the power output required to overcome the effects of gravity. The remaining factors in the first term are either constants (e.g., π) or mathematical coefficients (e.g., J_0) that account for various characteristics of an object moving through a fluid.

Term 2. The second term is concerned with characteristics of the bird's architecture and surface characteristics that affect its speed. The most important is air density (again). The rest are captured by the interaction of cross-sectional body area (S_b) with the effect of the bird's shape as reflected in air resistance, or drag (C). Basically, the greater the cross-sectional area, the greater the drag, but the two types of drag behave in different ways. Parasite or induced drag occurs mostly at wingtips and is caused by differences in pressure above and below the wing. It decreases with the speed of the bird. As the name suggests, profile drag is related to the shape of the wing. It can be decreased by streamlining but always increases with speed. The interaction of the two types of drag creates an optimum flight speed for each bird, at which drag is at a minimum. This is the so-called design speed.

Undoubtedly the most difficult feature for theoreticians to capture in a mathematical model of flight is the effect of the plumage. Not only do feathers form driving surfaces on the wing and the control surfaces on the tail but they also give the body a soft, flexible covering that is unlike any industrial material. The unusual porous and flexible surface of the plumage makes it particularly difficult to measure drag on an object as complex as a bird, but drag is important in flight and it appears as two separate mathematical terms, parasite drag (C_{Dpar}) and profile drag (C_{Dpro}), in the equation in Figure 10.4. At first, researchers attempted to measure drag by mounting dead birds on sticks and putting them in wind tunnels, but the results did not fit observations made on free-flying birds. Now researchers are able to put live birds in wind tunnels and watch them fly at different speeds. They have been able to test pigeons as though they were flying at speeds of 20 m per second (about 72 km per hour), but birds regularly fly much faster. The Marbled Murrelet, a small auk, has been clocked on radar at 154 km per hour in apparently level flight. At such speeds, it is moving so fast that the model may not reflect events accurately. At the speeds achieved by the murrelet, calculations from the equation suggest that it is flying with almost no drag at all.

Exploiting Muscle Power in Flight

The Marbled Murrelet[8] is a small auk and a good example of a flight style based on the exploitation of power and seemingly extravagant expenditures of energy. It

---

8 The Marbled Murrelet (*Brachyramphus marmoratus*) is a small (205 g), brown auk that lives in the eastern North Pacific. It is unlikely to be familiar to anyone who has not travelled in the area. An even more poorly known relative, the Long-billed Murrelet (*B. perdix*), lives along the North Pacific coast of Asia and shares many characteristics with its American cousin. Kittlitz's Murrelet (*B. brevirostris*) is a ground-nesting relative that inhabits treeless coasts in Alaska. The Marbled Murrelet has achieved considerable local notoriety because of its habit of nesting among the branches of the big-tree, old-growth forests that are the mainstay of the logging industry in western North America. In 1988, it was declared a threatened species, with a devastating impact on forestry in Oregon and serious repercussions in the adjoining states and in British Columbia. Its conservation remains a difficult issue that can be explored on a great many websites.

often flies for long periods at 70 km per hour. It achieves such high speeds by beating its wings at an exceptionally high rate, about 12.8 Hz (completed beats per second). It is able to sustain this for hours and often commutes more than 100 km between its nest and preferred feeding areas [97]. Not surprisingly, its flight muscles make up about 21% of its weight, but the entire body seems adapted for this vigorous lifestyle. In the hand, its body feels exceptionally solid, as though it consists entirely of dense masses of muscle. It has very little in common with the airy, low-density bodies of forest birds, such as starlings, or large, soaring birds, such as vultures. Large muscles require a robust skeleton and the bones of the murrelet's wings and shoulders are exceptionally large and strong. The wing skeleton is particularly strong, with three tendons controlling movement of the wingtip [8]. Within the chest, long hypapophyses that extend inward from the thoracic vertebrae support and protect the lungs and other vital organs. Comparable skeletal structures appear in the falcons, which sometimes hunt murrelets, but the seabird's bones are much more robust and dense than those of the predator. None of the bones are pneumatized (connected to the respiratory system).

To withstand the wear and tear of the murrelet's vigorous lifestyle, the flight feathers must be exceptionally durable and stiff. They are much heavier for their size than similar feathers on terrestrial birds [46]. The body plumage is also unusual, and even the small contour feathers are unusually thick and densely packed all over the body.

Murrelets typically fly in long, straight lines, a metre or two above the water. You might expect such a fast bird to fly at a higher elevation, where the air is less dense. Low-level flight should require more effort from the bird because the denser air offers more resistance to the wings and greater thrust is required to overcome drag. Air density (ρ) appears twice in the equation in Figure 10.4. In the second term on the right, which deals with drag and body diameter, air density appears in the numerator, where increased values increase the power required for flight. Air density also appears in the denominator, however, in the first term on the right, which deals with body mass. In that term, high air density decreases the cost of moving a heavy object, suggesting that the buoyancy of the air has a significant effect on power requirements. In that case, flight might be considered as the forceful displacement of air molecules – the more molecules moved out of the way during a given wing stroke, the more lift or thrust generated by the bird. Perhaps the buoying effect of the water droplets is so strong that it overrides the effect of increased drag, especially at the speeds achieved by murrelets.

If buoyancy is the factor that encourages a murrelet to fly low over the water, the bird might be taking advantage of special characteristics of air as a mixed medium to generate greater amounts of thrust per stroke. The surface temperature of seawater in the murrelet's habitat is usually less than 7°C, and may fall below 0°C in winter in the protected inlets of British Columbia's coast. The density of the air increases as its gaseous components are chilled by the water's proximity and it becomes saturated with relatively large and heavy water molecules. The cold, dense layer of air also includes suspended droplets of liquid water. When there are enough of these droplets, the mixture becomes visible as fog, but even without visible fog, there may be enough droplets in this aerosol to affect a bird's flight. The droplets of water are small enough to act like molecules, but they are huge in comparison and their great weight increases the medium's overall density (ρ). The greater the weight of the medium displaced by the movement of the bird's wing, the more energy is transferred into thrust and the faster the bird can fly.

The ability to fly through a mixture of air and water droplets might be a very useful characteristic for a bird that encounters periods of fog, drizzle, rain, sleet, or snow at some point every day. Other seabirds face similar conditions and the presence of water droplets in the air may affect their selection of a preferred altitude for cruising flight. The transition from saturated air mixed with water droplets to ordinary humid air, at 10-20 m above the sea surface, is not smooth. There are layers of air with specific characteristics separated by boundary zones or discontinuities. When a cloud of water droplets becomes visible as a fog, the discontinuities give it a distinct top several metres in the air and a bottom that lies a metre or so over the water. Murrelets and other auks tend to fly through the zone that becomes fog, while cormorants fly below it and loons, grebes, and albatrosses apparently fly above it.

The wing stroke of the murrelet is so vigorous that you might even expect cavitation as water droplets vaporize during sudden drops in pressure across the murrelet's wing. The act of vaporization consumes energy that would otherwise be available for generating thrust. Cavitation is a frequent problem with poorly designed boat propellers that move too quickly through the water, and it can cause a significant loss of power. If cavitation is an issue for murrelets, they may be protected by their choice of habitats. Vaporization is less likely at temperatures near freezing simply because cold droplets are less likely to turn into a gas.

The cormorant is another type of seabird that frequently flies close to the water's surface, but its exceptionally slow flight has little in common with the high-speed performance of the Marbled Murrelet. All the fast or efficient seabirds have densely packed feathers and a waterproof plumage that probably sheds water droplets without much additional expenditure of energy. A cormorant has loose feathers that easily get wet, so it might be expected to avoid layers of air laden with water droplets. Nonetheless, the cormorant often flies within a wing's length of the sea. This exceptionally low elevation enables it to take advantage of ground effect. The vortices from its slow-moving wings bounce off the sea's surface and provide extra lift.

It is not clear whether the albatross flies through a high-density mixture of air and water droplets or above it. It is not easy to measure air densities over large ocean swells. Because albatrosses do not flap their wings as frequently as other birds, the effects of buoyancy may not be useful to them, and suspended water droplets may merely cause increased drag. Whatever the effect of the droplets, it seems certain that albatrosses choose an elevation at which the medium enables them to operate with the greatest efficiency.

It is also unclear why fast-flying grebes and loons choose higher elevations than murrelets. The most obvious difference in flight style between loons and murrelets is the depth of the wing stroke. Murrelets have a very deep and rapid stroke, while the wings of loons vibrate through a very narrow arc. The short stroke displaces a much smaller volume of air than the deep one. How the differences between these birds are related to the air layers might provide an interesting study in blade theory, but it would be extremely difficult to collect data rigorously enough to test any hypotheses.

Exploiting Efficient Design of the Wing

The flight of petrels contrasts sharply with that of auks. Petrels are the great masters of controlled flight and energy conservation. Their elegant use of highly efficient wings and their almost effortless sailing over the sea have made them darlings of aeronautical engineers and ornithologists alike. Extremely fine control of the extremities has given petrels sufficient speed and range in the air to explore remote sectors of the ocean. Like auks, petrels have a complete suite of physiological adaptations that enable them to remain at sea for extended periods. They return to land only to nest.

None of the petrels offer better examples of the group's lifestyle than the albatrosses. They appear to capture or reject tiny amounts of energy almost instantaneously from the movement of air past their bodies. It has been very difficult to work out the mechanisms behind their spectacular sailing flight. There are many misleading clues and several things are going on at the same time. Most observers are certain that there must be some special significance to the albatross' habit of making great wheeling pull-ups as it flies, and we know that many soaring seabirds can make some use of winds deflected upward from the waves. The birds do not have to flap their wings so often because the updraft compensates for their tendency to sink through the air. They appear to take advantage of a tendency in wind speed to increase with height over the water. As a bird pulls up against the wind, it gains height without a great loss of air speed. Once at the top of a pull-up, it can increase air speed by falling downwind until it begins another pull-up.

C.J. Pennycuick, who has spent a lifetime observing birds on the wing and developing aerodynamic models of their flight, has found a discrepancy in this "wind gradient" theory [161]. Wind gradient might explain why albatrosses reach 3 m (about one wingspan) in a pull-up but not why they regularly pull up to 15 or 20 m above the waves, far beyond the likely effect of wind gradient. He believes that albatrosses take advantage of a separation bubble of confused air that forms behind a wave. As an albatross crosses the crest of the wave, it meets a gust of wind from that bubble. If it flew into it head on, the energy of the separation bubble would simply slow the bird down. The albatross wheels upward, however, and the energy catches the bird from underneath, lifting it higher into the air. From that height, it can gain speed by falling in a glide towards the next bubble.

To capture the most energy, the albatross must have exactly the right orientation to the gust. The sensitivity needed for such control is far beyond any mechanical device, and the process of control in the albatross is not known. Pennycuick suggests that the unique pair of tubes that hold the nostrils of the albatross might be the key because the nostrils of petrels are a particularly important part of their sensory system. Petrels are one of the few birds that find their food through a sense of smell [76]. Pennycuick goes on to propose that the separated nostrils of the albatross might house an undescribed sensory apparatus capable of detecting subtle differences in the pressure of air coming in from one side or the other [161].

So far the evidence for such a structure is circumstantial and based on the behaviour of and absence of appropriate nostril characteristics from birds that do not fly like an albatross. The smaller relatives of the albatross and "pelecaniform" seabirds such as gannets and boobies have appropriately shaped wings and often soar in the stable updrafts that form near sea cliffs. For level flight, however, they resort to relatively simple flapping flight. Even when strong winds permit them to soar over the waves at sea, they do not make wheeling pull-ups like an albatross. They may be too small to take advantage of separation bubbles but they also lack the equivalent of the albatross' double-tube structure behind the opening of the nostril. Although they are closely related to albatrosses, the smaller petrel (genus *Pterodroma*) and the shearwaters (genus *Puffinus*) have nostril openings that open into a single tube. There is no space for the hypothetical organ with which albatrosses may perceive subtle clues about the atmosphere. There is no possibility of any such structure in the gannets and boobies, whose nostrils are covered by the edge of the beak.

Whatever the explanation, the flight of albatrosses is exceptional and appears to depend on their large size and unusually long wings. It should be no surprise, then, that they exhibit skeletal novelty in the form of unique bony structures. Completely new bones are a rare phenomenon among birds, but one appears in the wings of the albatross and other long-winged, tube-nosed seabirds.

The albatross wing is a masterful combination of plumage and skeleton, but it is not merely a scaling up of a less extreme design. It includes some unique structures. Between the wrist and the shoulder, the leading edge of the wing is formed by the patagium, a thin flap of skin supported by a tendon. In the unusually long wing of the albatross, there is a risk that the patagium will begin to flap and reduce the overall efficiency of the wing. Yacht crews call it "luffing" when it happens to sails. The albatross has solved the problem by developing a free-floating bony strut that crosses the patagium at its widest point [17, 224] (Figure 10.5). It sits on a small ossicle that extends from the humerus just in front of the elbow, and it is quite visible in the wing of a flying bird when viewed against the sky. Some smaller petrels have a comparable structure but nothing quite like it appears in any other groups.

The albatross must also overcome the strain of holding its wings out for long periods of time, and has responded with another skeletal adaptation. Instead of developing large, heavy shoulder muscles, it has produced an elegant and lightweight

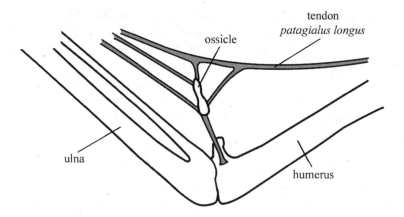

Figure 10.5   The patagial ossicle, or spreader bone, that appears in the wings of albatrosses and some of their relatives. It helps stiffen the thin layer of skin that stretches from the shoulder to the wrist and forms the leading edge of the inner wing.

lock in the shoulder joint. The head of the humerus slides into a notch that holds the wing out from the body without muscular effort. A special muscle releases the lock by sliding the head of the humerus back into the correct position for folding the wing.

Power versus Efficiency

It is difficult to imagine a flight technique further from the elegant performance of the albatross than the high-speed flapping flight of an auk. It is such a demanding technique that adaptations to sustain elevated rates of energy expenditure permeate all aspects of the bird's biology. Nonetheless, biologists were dismissive of its sophistication for many years. Its flagrant consumption of energy and apparent lack of control appeared to contradict ideas about fitness, optimization, and efficiency. It turns out, however, that the biologists were looking at too small a picture and simply did not appreciate the physics of flight. The auk might be expending huge amounts of energy at any given moment but that expenditure is well within its total daily budget and there are unexpected advantages to flight at great speeds. Modern mathematical models show that auks are actually exploiting vortex generation and blade theory in a way that we are only beginning to understand (Figure 10.4).

The differences in flight style have had a strong effect on the reproductive strategies in the two groups. Albatrosses and other petrels use their efficient flight styles to achieve range and weight-carrying capability. They may travel thousands of kilometres over the ocean in search of food for their young, and they are able to make deliveries that represent 15% of their body weight. During the long trip, they absorb nutrients from the water-soluble portion of their prey and convert the remainder into refined and calorie-rich oil for the nestling. Needless to say, the nestling must be able to wait long periods between meals. This seemingly leisurely schedule of parental care in petrels contrasts sharply with the frenetic activity of the parent songbird, which makes many tiny deliveries each day.

In the northern seas, the auks maintain a nestling-care schedule that is busier than the petrel's but not nearly as busy as the songbird's. Auks do not range as far from the nest as an albatross – only tens of kilometres, not thousands – and they deliver whole, undigested food. Some species make several deliveries each day while others return to the nest only once each night, with a single large delivery of food. They may be making use of darkness to avoid some of the avian predators and thieves that lurk near breeding colonies.

The Marbled Murrelet is one of those auk species that make only one or two visits to the nest each day. It is a bird that flies very fast, and birds that fly very fast have a skewed energy allocation. They put a great deal of effort into thrust and just enough energy into lift to get airborne. Theoretically, thrust produces only forward motion and the amount of energy it requires is not related to the to-tal weight of the bird. Consequently, in a bird that is expending most of its energy on thrust, additional loads such as a heavy egg or a relatively large fish add only a little extra cost to the total amount of energy already committed to flight. A Mar-bled Murrelet carrying an egg to a nest 800 m up the side of a mountain uses only a little more energy flying uphill than when it flies over the sea without an egg.

Murrelets have been clocked on sea-level radar at speeds up to 154 km per hour, but those birds might have been taking advantage of momentum during dives from high-elevation nests. The average speed in level flight is closer to 70 km per hour, with frequent bursts up to 130 km per hour. The average foraging trip at such a high speed requires the murrelet to maintain a continuous wing beat fre-quency between 12 and 13 Hz for an hour or more. Similar wing beat frequencies can be found in the Giant Hummingbird (*Patagona gigas*), but the hummingbird is only a tenth the size of a murrelet and it cannot not fly for such long periods with-out a rest. Like the hummingbird, the murrelet depends on calorie-rich food items. The hummingbird gets its energy from the sugar in nectar; the murrelet preys on small, oily fish whose energy content is about 5,000 calories per gram.

Familiar forest birds, such as the sparrow, demonstrate the high cost that many birds pay for flying slowly. In the air, they expend roughly equal amounts of energy on thrust and lift. Any extra load adds significantly to the total cost of flight be-cause it increases the effort going into lift. It is this distribution of costs that forces slow-flying songbirds to make hundreds of deliveries of tiny meals to their nest-lings. Even large terrestrial birds tend to carry small loads to the nest. Ospreys and fishing eagles, in particular, take surprisingly small fish. In spite of their large wings, ospreys often struggle to lift their prey into the air to fly a few metres into a treetop.

Most of the murrelet's relatives fly more slowly, and the additional cost of a lower speed may have a significant effect on their choice of nest sites. All auks must spend several weeks every year carrying heavy food loads to their nestlings. Faster species, such as the murre, may nest on cliffs hundreds of metres above the sea, but puffins, guillemots, and small auklets fly more slowly and typically nest just a few metres above the tide line. The Pigeon Guillemot, one of the slow auks, is one of the few that must habitually run across the water to take off. Most other auks can spring into the air on the opening wing stroke, even when they are carrying a fish, although I have seen heavily loaded puffins laboriously taxiing back and forth in a dead calm.

The speedy flight style of the Marbled Murrelet depends entirely on muscle power and follows a strategy exactly the opposite of the flight technique used by the albatross. The murrelet's flight has four outstanding features: high speed, long range, great carrying capacity, and high cost. Because of its special conservation status, the murrelet happens to be one of the few auks whose flight has been examined in any detail. During the nesting season, a murrelet may fly over 70 km each way between its nest in the forest and foraging areas in waters over the continental shelf. The nest is rarely at sea level and is often between 500 and 1,000 m up the steep sides of a coastal fiord. Early in the breeding cycle, the female may make several trips to the nest carrying a single 30 g egg that is equivalent to 15% of her body weight. Later, both parents repeat the trip at least once each day, carrying a 5-10 g fish for the nestling. Such a heavy workload is possible, and even made easier, because Marbled Murrelets can achieve very high speeds.

The murrelet beats its wings very vigorously. If you watch other waterbirds in flight, such as ducks or loons, you will see that the wings appear to vibrate as they travel through a short arc. The murrelet's wings are little more than a blur, but on video records you can see the murrelet making very deep strokes even though its wings are beating at a very high rate. At normal cruising speeds, the wings cycle through an arc of about 120° at about 12.6 Hz, just within the 15 Hz resolution of digital video recorders operating at 29.5 frames per second [55]. In loons, the arc of the stroke is usually less than 40° and may be as little as 20°. Their wing beat frequency is about 8-10 Hz.

The Marbled Murrelet exhibits many of the same structural features as other birds that depend on muscle power to achieve very high speed. The furcula is strongly curved forward so that the anterior section of the pectoral muscle can pull the wing strongly forward during the downstroke. The flight muscles, especially the supracoracoideus muscles, are very large. The wings are long and narrow, like those of a falcon or shearwater, and the aspect ratio is very high at 10.6.

Most of the structural adaptations needed for fast flight in the Marbled Murrelet are quantitative – more muscle there and bigger tendons here. There is, however, one distinctive structural feature that suggests that the wing of the Marbled Murrelet needs exceptional strength. In most birds, the wingtip folds when pulled on by one or two tendons arising from the muscle *Musculus tensor patagii brevis* in the forearm. Marbled Murrelets have three [8, p. 361]. To accomplish its aerial

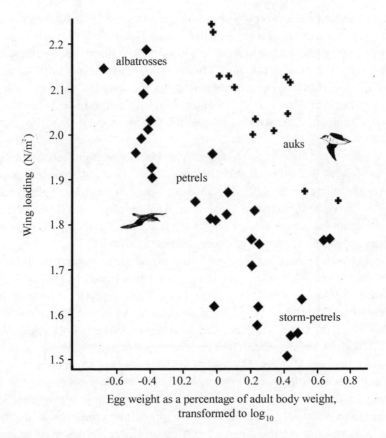

Figure 10.6   Relative egg size compared with wing loading in two important groups of seabirds [66, 224, and museum specimens].

feats, the murrelet has also developed a specialized physiology. Where most birds invest in large reserves of fat to fuel expensive high-speed flights, the murrelet remains near the same weight throughout the year, even when it is making long flights to deliver food to its nestling. It uses a "cash flow" strategy that depends on the presence of enough food in the sea to recoup each day's expenditure [97]. It is a strategy that could make the murrelet very vulnerable to human interference or competition for shared resources. We might not eat the energy-rich sand lance (*Ammodytes hexapterus*) or other small fish that the murrelet catches, but we do exploit them as a source of protein for fish farms and pet food.

The phenomenon of high-speed flight affects other aspects of life history. One of the predictions made from mathematical models of bird flight is that birds with lower wing loading and higher flight efficiency should have larger eggs. This is certainly true within a particular flight style or clade of birds. Figure 10.6 shows that the larger species of albatrosses (with relatively high wing loading) have smaller

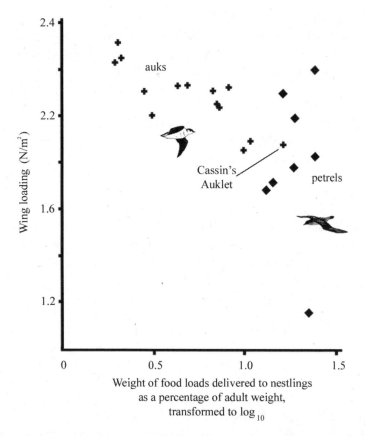

Figure 10.7 The weight of food loads delivered to nestlings compared with wing loading in two important groups of seabirds [66, 224].

eggs for their body size than the tiny storm-petrels (with exceptionally low wing loading). Large eggs enable the nestling to be more fully developed at hatch.

Unexpectedly, we see in the same figure that all the auks have larger eggs than petrels of similar size. The mathematical models of bird flight suggest that auks do not have to work very much harder than petrels to carry such eggs around. Interestingly, the auks do not attempt to maintain such a high level of effort during the whole breeding season. Figure 10.7 shows that the auks deliver smaller loads of food to their nestlings than petrels. Significantly, the only exception is the Cassin's Auklet, which carries provisions for its nestling within a gular pouch, whereas petrels carry their supply of provisions for the nestling in their stomach. Perhaps the issue is not the weight of the provisions but its volume. A Marble Murrelet can lift a 20 g herring out of the water, but a fish carried crosswise in the beak greatly increases the cross-sectional area of the bird's body. Perhaps the murrelet is unwilling or unable to pay the energetic cost of the increased drag.

Birds and Aircraft

One way of understanding the various flight styles of birds is to compare them to modern aircraft. Hummingbirds are energetically expensive analogs of helicopters. Sparrows are analogs of single-seater aircraft that fly slowly and carry small loads. Albatrosses are somewhat similar in design to advanced sailplanes. Loons and swans are heavy lifters, with speed and range like a modern jet transport. To carry this line of thinking to its logical conclusion, its speed and carrying capacity make the auk comparable to the modern fighter-bomber. Unlike the fighter-bomber, however, the auk is also a high-speed submarine.

Diving

Flight speed and flight efficiency are merely tools that help seabirds locate food and meet the logistical demands of nestling care. Other adaptations are needed to actually capture the prey [224]. In the three-dimensional world underwater, there is no "higher ground"; a predator must simply be bigger, faster, and more manoeuvrable than its prey. Learning to manipulate buoyancy is the secret of success in all diving birds. In the layers near the surface, a bird will pop to the top if it stops swimming downward. Lower down, there is a layer of negative buoyancy in which the bird tends to sink if it stops swimming, and must actively swim upward to reach the surface. Between the two is a narrow layer of neutral buoyancy, in which all the muscular energy of a bird can be used for speed and manoeuvrability. It is in this zone that a seabird may find it easiest to outswim and catch single fish. A seabird may locate a school of fish in the zone of neutral buoyancy and use its superior speed and agility to drive it to the surface. On the surface, it does not need to expend more energy than it takes to dive 1 or 2 m, and it can take frequent breaths without giving the school time to disperse.

In the past few years, scientists have been attaching tiny depth gauges to diving birds in an attempt to understand the secrets of their underwater behaviour. Nearly all the work has been done with nesting birds that must return to the colonies each night to feed their young. Scientists simply recover the gauge when the bird returns to its nest. The work began with penguins, which proved to be prodigious divers. A very large species like the Emperor Penguin can reach depths of 550 m. Flying seabirds are also good divers. The murre can dive to more than 70 m and smaller auklets will dive to 40 m. Shearwaters would seem to be unlikely divers. They migrate from one hemisphere to another on long, slender wings, but they can use those same wings to push themselves down to 70 m [225]. None of these birds, however, appears to spend much time near the limits of their diving depths. Most dives take them only 15-40 m below the surface.

The average dive probably takes the bird into its zone of neutral buoyancy. Within that zone, the bird's body neither sinks to the bottom nor rises to the surface without a push, and all the swimming effort can go into forward movement in pursuit of prey. The tiny dipper or water ouzel is one of the few diving birds that can never reach its zone of neutral buoyancy. It is a songbird related to the wren that forages in shallow creeks and streams and must always struggle against buoyancy. Once submerged, it clings to the bottom with long, sharp claws and forages by walking about as though on a forest floor.

The dry-suit approach to diving, which is favoured by most seabirds, traps an insulating layer of air beneath the feathers. The air increases the bird's buoyancy, but the effect is somewhat reduced because a compact body shape keeps the volume of that layer to a minimum. When the bird flattens its feathers against its body, the volume of air is further reduced. The cormorant, a wet-suit diver, reduces its buoyancy by getting wet. Only a little air is trapped in the feathers and the birds can swim comfortably with only the head above water, like the periscope of a submarine. Regardless of the insulating system, buoyancy makes all diving birds work to get below the water's surface. Grebes, loons, and penguins often leap out of the water to gain momentum for the start of a dive, but most wing-propelled birds just push themselves under with powerful thrusts of their wings.

Many seabirds follow a daily pattern suggesting that the zone of neutral buoyancy plays a very important role in their ecological strategy. Unless they are feeding at every opportunity to build eggs or cope with a food shortage, they concentrate their foraging activity close to first light in the morning and near dusk in the evening. During the day, they just loaf, preen, and socialize unless an unusually rich feeding opportunity comes along. Their prey species often exhibit a complementary daily cycle. They migrate deeper beneath the surface for the daylight hours, probably just out of the effective foraging range of the birds. As dusk approaches, they move closer to the surface. Twice each day, this migration cycle carries the prey species through the layer of neutral buoyancy, where birds are most effective as predators. The close linkage between prey availability and bird behaviour may cause problems for marine ornithologists. Seabird observers have had a hard time relating concentrations of birds sitting on the water to concentrations of prey detected by depth sounders. The problem may be that the observers are able to see the birds only in daylight. By the time it is light enough for observers to see clearly, the birds may have drifted to loafing areas in one direction while deep-sea currents have carried the prey off in another.

Foot-propelled diving birds and wing-propelled diving birds are two entirely different kinds of animals. Their special suites of adaptations for life underwater have had a strong effect on the shape of their bodies. All diving birds are streamlined to help them overcome the increased drag of moving in a much denser medium than air. Foot-propelled divers, whether wet-suit (cormorants, anhingas, and finfoots) or dry-suit (ducks, loons, and grebes), stretch out into a long, slim conformation underwater. The wing-propelled divers, such as auks, diving-petrels, and penguins, use the dry-suit technique. They depend on speed underwater, and achieve a streamlined shape by pulling their heads in against their shoulders and neatly tucking away their feet to reduce drag. The gangly shearwaters do not appear to depend on speed underwater. In fact, they do not look particularly well designed for any underwater activity. Video recordings show that they use all four limbs more or less independently [224]. It's hard to see how such uncoordinated activity allows them to make any headway, but the film footage was shot in relatively shallow water (possibly in the zone of neutral buoyancy); more coordinated behaviour may help deeper dives that require speedier swimming. The locomotory behaviour of species dependent on only one type of locomotion – foot-propelled or wing-propelled – is always fully coordinated.

Wing-Propelled Diving

Use of the wings underwater is perhaps the ultimate form of locomotion achieved by birds. It has evolved independently in dippers (Passeriformes), auks (Charadriiformes), shearwaters and diving-petrels (Procellariiformes), penguins (Spheniscidae), and some extinct groups of seabirds such as the giant, flightless plotopterids of the North Pacific, whose relationship, if any, to modern lineages is poorly understood. The basis of its success may be the coarse similarity between this kind of diving and aerial flight. In fact, it is often referred to inaccurately as "underwater flight." Unlike aerial flight, underwater wing-propelled locomotion is entirely a matter of thrust. There is no significant opportunity for gliding and the bird does not need to generate lift. Using only one large mass of muscle to drive the wings both underwater and in the air offers obvious structural economies that are not available to foot-propelled divers.

The large central muscle mass in auks and other wing-propelled divers includes both of the major muscles responsible for aerial flight. The pectoral is the largest, as it is in all flying birds. It runs from the sternum to the lower surface of the humerus. When it contracts, it pulls the wing down to propel the bird through the water. The recovery stroke is the responsibility of the supracoracoideus muscle, which lies between the pectoral muscle and the sternum. Its tendon loops through the triosseal gap at the top of the coracoid and onto the upper surface of the humerus. In aerial flight, air rushing by the body helps the supracoracoideus to open the wings; consequently, the muscle is relatively small in all terrestrial birds except hummingbirds, which need a large supracoracoideus to help them hover. Underwater, the recovery stroke must be achieved against the drag of a much denser medium, and it requires considerable effort. As a result, wing-propelled divers have a massively developed supracoracoideus that is three or four times larger than in a typical forest bird (Table 10.3).

Penguins have such a massively developed supracoracoideus that they are able to generate thrust during the upstroke. This has not been demonstrated in auks, which also use their wings to fly in the air. An auk's wings are probably too flexible underwater, and the swimming stroke is not quite the same as the flying stroke. Underwater, auks pull the wings close to the breast as though they were giving themselves a hug. The action squeezes water between the wing and the chest, and the birds may be jetting forward just like a squid or octopus contracting its mantle. The upstroke that follows is visibly slower than the downstroke, and the birds seem to push through the water in a series of short pulses. When the bird decides

Table 10.3

Weights of the pectoral and supracoracoideus muscles compared with the total weight of birds

Species	Total weight (g)	Pectoral weight (g)	Supracoracoideus weight (g)	Ratio of pectoral to supracoracoideus
Ruby-throated Hummingbird[a]	2.9	0.51	0.23	2.2:1
Common Murre	870.0	141.80	42.20	3.4:1
Marbled Murrelet	237.0	32.20	8.10	4.0:1
American Robin[a]	72.5	10.00	1.15	8.7:1

[a] [184]. Other measurements by the author.

to pursue a fish, the flap rate increases sharply. The forward movement becomes continuous as the bird accelerates.

The large central muscle mass of a wing-propelled diver helps it maintain a usefully compact, spherical profile. In any medium, a spherical shape is ideal if you want to change directions quickly, because the cross-sectional area is the same from all aspects and there is much less drag in a sharp turn. Wing-propelled divers decrease the profile of the front end by folding the neck to the head snugly against the shoulders, and none of them has a long tail that would drag in a tight turn. The hind limbs have relatively light locomotory responsibilities, so their muscles are fairly small. Making short runs for take-off and digging nest burrows are probably their most onerous duties. The knees are specially designed so that the legs can be folded tightly against the body. The small, webbed feet are used mostly for slow swimming on the surface or steering underwater.

Many ornithologists have been dismissive of the aerial capabilities of wing-propelled divers, and have described them as mere waypoints on the road to a more "natural," penguin-like condition. If that conversion always had strong benefits, however, examples should have arisen more frequently than suggested by the fossil record of the last 65 million years. In addition to the penguins and the recently extinct Great Auk, only the Mancallidae and Plotopteridae were completely flightless. Wing-propelled divers can be efficient and successful flying birds. As we saw earlier, some of them stretch the envelope of aerodynamic design. It is only logical that birds using the same set of wings for locomotion in the air and in a medium 300 times as dense should be capable of spectacular aerial performance.

Foot-Propelled Diving

The body plan of a foot-propelled diver has much in common with the design of the old sternwheeler steamboats. All the propulsive force is behind the bird and there is no turbulence to increase drag along the forward part of the body. The feet drive and steer, but because they are located in the rear, turns tend to be broad and sweeping. Broad, sweeping curves are not ideal for catching erratically darting fish, but the elongated neck helps cormorants increase their overall agility. Neither the neck nor the head is wide enough to extend beyond the body's cross-section so they do not increase drag. The head can act as a small rudder because it is placed far forward; small movements can have a large effect on direction and the cormorant is literally led through the water by its beak. When prey comes close enough to the head, the low profile of the cormorant's beak and the thinness of the neck enable the bird to make a lightning-swift strike in any direction.

The long neck presents the cormorant with a more difficult problem when it comes to holding a large fish. The hooked beak can grip the prey securely, but the jaws are not strong enough to prevent it from thrashing around. As a result, cormorants usually try to swallow their meal almost as soon as it is caught, and it can often be seen wriggling as it slips down the birds' throat. The struggles of a snagged fish must put considerable stress on the neck joints initially, especially as cormorants often take rather large fish (>100 g). To protect themselves, cormorants have strong, muscular necks and massive neck tendons. In addition, large species such as the Double-crested Cormorant have an unusual structure on the back of their skulls. There is a large boss (Figure 1.7) above the foramen magnum that looks a lot like an accessory occipital condyle. It does not articulate with any bone, however, and is merely a massive attachment site for neck ligaments.

The snatch-and-swallow technique of cormorants and many other foot-propelled divers is very different from that of wing-propelled divers, and may reflect their inability to generate high speeds underwater. Penguins and auks use underwater speed to hit fish hard and stun them so that they do not struggle. Their short beaks have a powerful grip, and the overlap of the upper jaw over the lower bends the fish sharply. Either they crush the back of the head and the gill covers or they dislocate the fish's spine.

Large birds, such as the Rhinoceros Auklet or Tufted Puffin, often return to their nest with nothing but the head of a herring. The bird's bite is so strong that the body of the fish falls off. Herring are flat and soft, unlike other species taken by auks. The sand lance is also a favourite food. With a thick body, it is very strongly built and never falls apart. Either on the water or at the colony, a returning bird may display its prey, held crossways in its beak.

All foot-propelled divers have an elongated body. The slender profile is usually described as an adaptation for avoiding snags and tangles in weeds, but it simply reflects the bird's internal allocation of space. Foot-propelled divers need room for two large muscle masses. One powers the wings in the air; the other drives the legs underwater. Drag is proportional to cross-section, so an in-line arrangement is the most logical way to keep drag to a minimum. The long body also creates convenient external space for stowing large folded wings. Large wings have a lower wing loading and reduce the overall cost of aerial flight, but foot-propelled divers are often so large and heavy that they need long taxiing runs in order to take off. Anhingas have such large wings that they can indulge in thermal soaring and circle high over their tropical wetlands with vultures and eagles. The flight of cormorants is generally more laboured or ponderous. In spite of large wings, they may need to take advantage of "ground effect." To exploit this effect, the vortices leaving the wing must be reflected upward in a way that increases lift. The effect is useful only at moderate speeds and the bird needs to be less than a wing's length from the ground. Cormorants may use ground effect regularly, especially when they are trying to fly while carrying a heavy load of fish, and it may be the main reason that cormorants often fly very close to the water's surface. They will also fly at higher elevations but some species, such as Brandt's and Pelagic Cormorants, rarely fly over land, where turbulence could make stable flight more difficult.

Cormorants and anhingas are among a small number of waterbirds with a long tail. At low landing speeds, it can be fanned out to provide extra lift and prevent stalling, and many species may depend on its action to nest on narrow cliff ledges or tree branches. The tail does not increase drag when the bird is underwater because it remains folded and does not increase the bird's cross-sectional area. It may even assist by helping to confine the turbulence generated by the feet to the lower surface.

Loons have very short tails and grebes have virtually no tail at all. There thus seems to be little to help them reduce any drag created by turbulence from the feet while swimming. In the case of loons and grebes, the extreme rearward position of the legs may reduce the effectiveness of any structure, and turbulence may be directed outward and away from the body by their peculiar synchronous swimming stroke. Most ducks also have short tails but longer tails appear in several diving species. One of the deepest divers in the group, the Long-tailed Duck (formerly Oldsquaw) has unusually long central tail feathers, which may help to dampen turbulence by leading vortices backward into the bird's wake.

Loons and grebes also differ from cormorants by being much heavier and more muscular. They have relatively narrow wings that are too small to take advantage of ground effect or to enable their owners to soar. These birds depend entirely on large flight muscles to generate enough power for flight. They make long taxiing runs and use deep strokes to get enough speed for take-off; once airborne, they use rapid, shallow wing beats. Like many birds with a diet of calorie-rich fish, they expend whatever energy is necessary to fly at high speed.

Many waterbirds with short tails fan out their webbed toes to increase lift and delay stalling at low air speeds. Once on land, however, it is the legs and feet that cause the most difficulty. The legs are placed as far back as possible for efficient swimming, and walking may be difficult or even impossible. Cormorants are merely clumsy walking around their nests, but loons and grebes must slide on their bellies and rarely attempt to walk. The upper portions of their legs are bound to the body by connective tissue, and any walking depends on coordinating the long tarsometatarsi with the very large feet.

It is this inability to walk or run that makes loons and grebes obligatory waterbirds and prevents them from flying off a solid surface. They can accelerate to take-off speed by running across the water, but they cannot run on solid ground.

Glistening surfaces such as wet roadways or frozen lakes are death traps. They must come to land to lay eggs, but their nests are invariably built next to the shore or on a floating platform. In fact, the sight of a loon or grebe standing upright on land is a clear sign of a local disaster. To get erect, they point their feet out to the side, spreading their toes as much as possible to balance vertically on top of the upright tarsometatarsi. The posture looks ridiculous, and I have seen it only in oiled birds that have landed on a beach as a last resort for survival.

The rearward location of the legs has enabled the swimming techniques of foot-propelled divers to become more advanced and efficient than the paddling stroke of dabbling ducks. A duck's feet push backward against the drag of the water on the body, but only in the same line as the direction of travel. This style of locomotion is said to be "drag-based." The foot often travels half again as fast as the body, but the speed of the body can never exceed the speed of the foot. If a dabbling duck tries to achieve speed by paddling rapidly, the force from the webbed foot lifts the body out of the water, generating more splash than forward motion. Quite a few ducks use this noisy form of locomotion as a display to intimidate rivals, and some grebes have adapted it for elaborate courtship rituals. It is not particularly effective for generating thrust underwater, however. Divers that want to generate thrust with their hind legs for deep diving or speed underwater use a peculiar sprawled posture. At rest, they look a bit like a swimming frog. The legs are rotated slightly and splayed sideways so that the animal is very stable when resting on the surface. The knees lie fairly high along the side of the body. The ankles rest closer to the midline underneath and the feet extend outward to the side, like oars. Their thrust is directed backward and to the sides without lifting the tail out of the water. Underwater, mergansers and other ducks use alternate strokes of the feet, but the very fast foot-propelled divers such as loons and grebes use a synchronous stroke.

The Skeletons of Diving Birds

Although diving birds live very different lives from their terrestrial relatives, and presumably from the equally terrestrial ancestors of all birds, their skeletons are surprisingly generalized. Victorians would have described them as quite reptilian. Only a few bones, such as the humerus of the auk or the tarsometatarsus of the loon, are greatly modified for life in the water. The remaining bones are usually robust and unfused to their neighbours. Even the penguin has a generalized avian skeleton, except for its highly specialized flipper (Figure 10.1). The other obvious skeletal specialization in the of diving birds is the flexible ribcage with its greatly elongated but very thin ribs (Figure 2.8). Elongated ribs are not unique to marine diving birds but can be seen in loons and grebes as well as in ancient diving birds such as *Hesperornis*.

Avian Hydrofoil Propulsion

Grebes may have developed a unique and extremely sophisticated form of foot-propelled swimming that is exceptionally efficient [103]. All parts of their lower leg appear to be designed for a special purpose. The tibiotarsus is flattened and its covering of long scales makes it almost bladelike. The bone itself is a flat box with shallow grooves on the edges to accommodate massive tendons that travel to the toes. Each toe has lobes of toughened skin along both margins that are joined by a narrow web near the base. There is also a single flat lobe on the hallux. On each of the main toes, the outer lobe is much broader than the inner one. The lobes open flat during the power stroke and collapse together during recovery. The only other diving birds with comparable feet are coots, phalaropes, finfoots, and the long-extinct *Hesperornis*. Perhaps significantly, the finfoots (Heliornithidae) are a family within the Order Gruiformes that may be closely related to the grebes (Podicipedi-formes). There are no known relatives of *Hesperornis*. Unlike the coots (which occasionally perch on branches) and all diving birds with webbed toes, grebes lack a tendon locking mechanism to keep the toes in a flexed position.[9]

For years, ornithologists have assumed that the lobed toes of the grebe func-tioned like the webbed foot of a duck or gull. Observations of a grebe swimming under laboratory conditions, however, have led Johansson and Norberg to the con-clusion that a grebe's foot follows a peculiar circular path quite unlike any other bird's. A grebe's foot moves upward and outward, across the swimming direction of the body so that the locomotion is more like a form of flight, called "lift-based" locomotion. In lift-based swimming, the force generated by the foot moves in a different direction from that of the body. Johansson and Norberg compare the grebe's swimming action to the highly efficient stroke of a kayaker's paddle, as op-posed to the drag-based stroke of a typical canoe paddle [103]. Much of the force generated by the stroke is outward from a kayak, creating a very large vortex, whereas the stroke along the side of a canoe generates a relatively small vortex. The hydrodynamics are complex but the key factor appears to be that "a large vortex has a lower core velocity, which is much more efficient" [173].

In the air, slots in the wing help increase lift at low speeds, and the long, lobed toes of the grebe may have a similar function. Slots in a wing are most effective when they are staggered across the direction of flow, as in the wingtips of soaring eagles, where the lift-to-drag ratio can be improved by over 100%. The three toes of grebes are ideally spaced and shaped for such a function underwater and prob-ably control the direction of the propulsive force generated by the foot. The asym-metrical lobes on the toes may serve the same purpose as the asymmetrical vanes of flight feathers and help stabilize the limb during the power stroke. The foot movement in grebes is rotary, comparable to the wing movement used by kestrels and other large birds for a form of hovering.[10] Where a kestrel remains stationary against the force of gravity, however, a grebe is propelled forward. In effect, the pe-culiar flat, fringed toes act as a "self-stabilizing, multi-slotted hydrofoil" during the

---

9 This feature is also missing in flamingos and several other members of the Metaves, but not all groups have been examined [59, 162].

10 Hummingbirds use a different kind of stroke.

power stroke. The role of the extremely flat tarsometatarsus has not been examined. Given the rotary motion of the feet in grebes, it would not be too surprising if the tarsometatarsus were able to contribute significantly to lift in its own right, although the shape may merely reduce drag. The tarsometatarsus is also flat in loons and coots, even though loons cannot generate a hydrofoil effect with webbed feet and coots paddle like ducks.

Seabirds on Land

In contrast to loons or grebes, auks and other wing-propelled diving birds are often as at home on dry land as they are in the water. Penguins, auks, and petrels walk around their breeding colonies to find likely nesting sites.

Although their legs appear to be as far back as those of a loon, penguins are comfortable in a semi-vertical posture and will march great distances and climb sea cliffs to reach their breeding colonies. Penguin legs look short but they are really as long as those of many other birds. The long bones are merely folded at the knee and tucked away under the skin; only the tarsometatarsus is unusually short. The penguin's body hangs down from knees that sit up along the belly, and it moves with what athletes call a "duck walk." The legs are swung forward as the bird rocks from side to side. It looks awkward and running is difficult, but penguins have evolved in the absence of terrestrial predators and the comic rolling walk meets their needs.

The peculiar posture of penguins also enables some species to achieve a peculiar form of aerial locomotion. Rock-hopper Penguins spring up rock faces and over other obstacles on their way to and from the nest. The massive leg muscles are attached to exceptionally large tendons. When they flex, the toes kick downward with great force and the penguin is launched into the air. The flippers are probably too small to contribute any control.

In the Northern Hemisphere, the auks that walk around in the forest need to be nimble on land. On the Pacific coast of Canada, Bald Eagles gather regularly at seabird colonies to catch unwary Rhinoceros Auklets during their nightly visits [105]. Midnight at latitude 54°40'N may seem very dark to humans but there is enough light for eagles. Some just lurk in low shrubs and drop on passing auklets, but others prowl about like latter-day dinosaurs. The seabirds are not easy to catch. They can scamper across the forest floor quite quickly, and readily hurl themselves over hummocks and into narrow gullies, where the eagles cannot follow.

Auks also have an amazing skill that narrows the eagle's window of opportunity for an attack. Their memory for minute geographic details is very highly developed. No matter how dark or complex the nesting site is, an auklet can unerringly find one hole among tens of thousands of holes, and often plummets out of the sky to land at the doorstep of its burrow. The grand master of this trick is the solitary-nesting Marbled Murrelet, which nests in the forest canopy instead of underground. In absolute darkness and far from the sea, it can find its nest on the right branch in a forest of branches.

Conclusion

Success on and under the sea may be the single greatest accomplishment of birds. In the wing-propelled divers, it is the ultimate expression of centralization and the inner-bird strategy, which distinguishes birds from all other kinds of animals. It crowns the spread of birds into all the world's major habitats and demonstrates the flexibility and adaptability of their design. More than any other achievement, it sets the accomplishments of birds far beyond those of any of their ancestors among the dinosaurs and matches those of any mammal. We may disappear, like the dinosaurs, in some future cataclysm, but it is not too hard to believe that birds will find a way to survive, yet again.

Further Reading

After a lifetime of research, John Warham has produced two books on the Order Procellariiformes that touch on every aspect of the life of petrels and make them one of the most thoroughly documented groups of birds:

Warham, J. 1990. The petrels: Their ecology and breeding systems. Academic Press, London.
Warham, J. 1996. The behaviour, population biology and physiology of petrels. Academic Press, London.

Tony Gaston and Ian Jones have produced a comparable volume on the ecology of auks, and Michael Harris has produced an interesting account of his research on the Atlantic Puffin:

Gaston, A.J., and I.L. Jones. 1998. The auks: Alcidae. Bird families of the world. Oxford University Press, Oxford.
Gaston, A.J. 1992. The Ancient Murrelet. T. and A.D. Poyser, London.
Harris, M.P. 1984. The Puffin. T. and A.D. Poyser, Calton, UK.

Conclusion

Until late in the 20th century, there were only three Mesozoic bird fossils of consequence: *Archaeopteryx, Hesperornis,* and *Ichthyornis.* In 1985, Storrs Olson reviewed the entire Mesozoic avifauna in only 11 pages [151]. Today a comparable effort takes hundreds of pages: 589 in the Ostrom symposium (2001) [see 70], 472 in *Mesozoic birds* (2002) [see 25], and 342 more in *Feathered dragons* (2004) [41].

When we knew only three important fossil birds and a comparably small number of dinosaurs, it was reasonable to connect them in a story that involved the loss of fingers and teeth, a shortening of the tail, and rotation of the pubis. Today we know hundreds of dinosaurs, some with zero, one, or two fingers (e.g., *Unenlagia* and *Tyrannosaurus rex*) and many different forms with beaks instead of teeth. We even know of some with short tails whose tip ends in a fused pygostyle (e.g., *Chirostenotes* and *Nomingia*). We also know that many different lineages experimented with pubic bones of different shapes and angles in relation to the spine. The changing shape and size of dinosaurs in the Creataceous appears to have required an arrangement of the hipbones that redistributed the mass of their internal organs. Many widened their pelvic region and some, like birds, shortened their tail [156, Appendix A]. All these changes occurred long after the days of *Archaeopteryx,* in dinosaur lineages that lived alongside modern types of birds, and they happened solely because they conferred an adaptive benefit on the dinosaur involved. For some unknown reason, the bipedal, parasagittal posture of early, long-tailed dinosaurs was improved by a similar suite of changes that occurred in many later, unrelated lineages.

Birds were just one of those lineages, and apparently just one of several lineages of dinosaur covered with feathers. *Archaeopteryx* looked very much like other small dinosaurs, and it seems appropriate that its special features had parallels among them. Flapping flight may have been its only unique legacy. It is certainly the one with the clearest link to later, more birdlike animals, such as *Confuciusornis.* Unlike *Archaeopteryx, Confuciusornis* actually had specialized bones whose shape implied that their owner was more committed to active flight than any previous animal. It is the earliest known bird whose tail ends in a pygostyle, but the heavy pubic bones, long femur, and other features of its skeleton still show a clear relationship to more typical dinosaurs.

We lack the data to demonstrate a cause-and-effect relationship between changes in architecture and the development of enhanced flight capabilities in later birds, but we can suggest a degree of connectivity because a cause-and-effect relationship is more probable than any other explanation. We can link the possession of a furcula with long arms to stronger flight because the long arms were associated with the concurrent development of elongated coracoids and a hypocleideum that extended onto the breast. Vigorous flight requires large muscles, and large muscles need secure points of attachment. Those attachments were poorly developed in early birds like *Archaeopteryx* and *Confuciusornis,* but are an important part of the skeleton in every later bird.

There is a similar link between the bird's abandonment of the pubic ring and its gradual development of wider hips. Like the characteristic wishbone shape of the furcula, the broad vault of the avian abdomen has no parallel among the dinosaurs. It is a unique development in birds that is directly linked to the production of small clutches of large eggs in modern birds. All of the bird's elaborate courtship ritual and deep commitment to parental care stems from that reproductive strategy.

Liaoxiornis was a modern type of bird that lived in China long before the extinction of the dinosaurs. Except for a large keel on its sternum, it had most of the skeletal features of a modern bird and it probably looked and lived like some modern species. *Liaoxiornis* was a bird in the common sense of the word, no longer an animal with significant similarities to a dinosaur.

From that point on, the changes in birds were those of degree, not kind. Some bones are longer or more elaborate than ancestral shapes, and others are smaller or have disappeared altogether, but there are no important new additions to the skeleton or major changes in the way it functions. From that point on, there is also no clear evolutionary story for the birds. Only a few kinds of birds survived the great extinction event at the end of the Cretaceous, and they all share a very similar body plan. They all have the inner-bird type of architecture, with its highly centralized muscle masses, long tendons, and sticklike legs. The outstanding success of birds with just that one plan is what makes them so interesting and rewarding to study.

APPENDIX 1

Birds in Relation to Other Vertebrate Animals

In the following chart, the names of groups that include birds among their members are underlined.

<u>Chordates.</u> Animals with a spinal cord.
   <u>Vertebrates.</u> Chordates with backbones.
      <u>Tetrapods.</u> Vertebrates with four limbs.
         Anapsids. Vertebrates with no passage through the skull for jaw muscles.
            Chelonids. Turtles, the only living anapsids.
         Synapsids. Vertebrates with one passage through the skull for jaw muscles.
            Therapsids. An extinct lineage of "reptiles."
            Mammals. The only living synapsids.
         Euryapsids. Vertebrates with one passage through the skull for jaw muscles (extinct).
         <u>Diapsids.</u> Vertebrates with two passages through the skull for jaw muscles.
            <u>Archosaurs.</u> A major lineage of diapsids.
               Lepidosaurs. A lineage of archosaurs now represented only by lizards, snakes, and the tuatara.
               Pterosaurs. A lineage of flying archosaurs (extinct).
               Crocodiles. A living lineage of archosaurs.
               <u>Dinosaurs.</u> A diverse lineage of archosaurs.
                  Ornithischia. A major lineage of dinosaurs with birdlike hips (extinct).
                  <u>Saurischia.</u> A major lineage of dinosaurs with lizard-like hips.
                     <u>Theropods.</u> A diverse lineage of predatory dinosaurs.
                        <u>Maniraptors.</u> A lineage of advanced theropods that included feathered dinosaurs.
                           <u>Dromaeosaurs.</u> A lineage of theropods that shared many characters with birds (extinct).
                           Avialae (birds in the common sense). Descendants of the last ancestor of both *Archaeopteryx* and the Class Aves.
                              Archaeopteryx. The earliest known fossil bird (extinct).
                              Aves or Pygostylia. Birds with a pygostyle at the end of their tail.
                                 Confuciusornithes. "Confucius birds" (an extinct lineage of birds).
                                 Enantiornithes. "Ball-shouldered birds" (an extinct lineage of birds).
                               Neornithes. Modern or "socket-shouldered" birds (living).
                                 "Odontornithes." Birds with teeth; not a natural grouping (extinct).
                               Paleognaths. Birds with the paleognathous palate.
                                 Lithornithids. A lineage of flying paleognaths (extinct).
                                 Ratites. Ostriches, kiwis, and their relatives (living).
                                 Tinamous. A lineage of flying paleognaths (living).
                               Neognaths. Modern birds with a neognathous palate (living).
                                 Galloanserines. Chickens, ducks, and their relatives (living).
                                 Plethaves. The great bulk of modern birds (living).
                                   Metaves. A lineage within Plethaves suggested by genetic information.
                                   Coronaves. A sister lineage of the Metaves that contains most of the world's species.

Geological Time Scale

Era	Period	Epoch (and age in Mesozoic)	Start date (millions of years)	Important dinosaurs and other tetrapods (all extinct)	Fossil and molecular evidence for important birds (* = extinct)
CENOZOIC	NEOGENE (formerly Quaternary)	Holocene	0.01		
		Pleistocene	1.8		
		Pleiocene	5.3		
		Miocene	23.0		Podicipediformes, Columbiformes, *Plotopterus**, Phorusrhacidae*
	PALEOGENE (formerly Tertiary)	Oligocene	33.9		Passeriform birds arrive in Europe by the end of this epoch; Trochilidae; modern type Piciformes
		Eocene	55.8		*Paleotis** (Paleognathae); *Diatryma**, *Foro panarum**, *Presbyornis**, and *Anatalavis** (Anserae); Gallinuloides* and Paraortygoides* (Gallinae); *Prophaeton** (Procellariiformes); Falconiformes; *Messelornis** (Gruiformes); Psittaciformes, Strigiformes, Caprimulgiformes, Apodiformes; Sandcoleidae* (Coliiformes)
		Paleocene	65.5		*Lithornis** (Paleognathae), possible Sphenisciformes

The "K-T Event" ended the Mesozoic Era 65.5 million years ago

Era	Period	Epoch (and age in Mesozoic)	Start date (millions of years)	Important dinosaurs and other tetrapods (all extinct)	Fossil and molecular evidence for important birds (* = extinct)
MESOZOIC	CRETACEOUS	Maastrichian	70.6	Advanced theropods	A probably paraphyletic group called graculavids*; equivocal evidence for Charadriiformes, Procellariiformes, Gaviiformes, Pelecaniformes, and Psittaciformes; sound evidence for *Vegavis*, an anseriform
		Upper Campanian	83.5		*Apsaravis?**, *Patagopteryx**, *Palintropus** (Galliformes?); *Presbyornis** (Anseriformes)
		Santonian	85.8		
		Coniacan	89.3		
		Turconian	93.5		Galliform-anseriform split (90 million years ago)
		Cenomanian	99.6		
		Albian	112.0	Xiagou Formation, China 115-105 million years	*Enaliornis* (Hesperornithes)* *Gansus**
		Aptian	125.0	Upper limit of Jiufotang formation, China, at 110 million years. First oviraptor, first tyrannosaur?	*Sinornis?* (Enantiornithes)*; *Ambiortus** (Paleognathae); *Plotopterus**, *Yanornis**, *Yixianornis**
		Lower Barremian	130.0	First dromaeosaur	*Confuciusornis**, *Iberomesornis**, *Liaoxiornis**
		Hauterivian	136.4	Ornithomimisauria	
		Valanginian	140.2	Fossils of feathered dinosaurs from Yixian formation, China, are younger than 139.4 million years.	
		Berriassian	145.5		*Noguerornis?*[1]*
	JURASSIC	Upper	161.2	*Compsognathus*, possibly first troodontid	*Archaeopteryx**
		Middle	175.6	*Ornitholestes*	
		Lower	199.6	First therizinosaur	
	TRIASSIC	Upper	228.0	First theropod fossils: *Eoraptor*, *Megalancosaurus*, sphenosuchians	*Protavis**
		Middle	245.0	*Lagerpeton*, *Quianosuchus*	
		Lower	251.0	*Euparkeria*	

A great extinction at the end of the Permian ended the Paleozoic Era 251 million years ago

PALEOZOIC			542.0		Molecular evidence suggests that avian lineage (Diapsida) split from mammalian lineage (Synapsida) 310 million years ago.
PRECAMBRIAN			3,600+		

[1] Could be as recent as Lower Barremian.

Glossary of Ornithological Terms

Although not all of the following terms are used in this book, they appear regularly in avian literature.

A

Acetabulum	Joint in the hip that receives the head of the femur.
Acrocoracoid process	The point on the dorsal surface of the humerus for the attachment of the tendon from the supracoracoideus muscle, whose other end is anchored to the sternum.
Acromion	A large process on the forward edge of the scapula that creates part of the triosseal gap. Its location guides the passage of the tendon through the triosseal gap and allows the supracoracoideus muscle to raise the wing (see *biceps tubercle* and *triosseal gap*).
Albatross	Largest member of the petrel group, the Procellariiformes.
Alcid	An auk; a member of the Family Alcidae in the Order Charadriiformes.
Alectoromorphae	The group of forest birds with an ambiens muscle.
Allometric	Referring to a situation where one variable increases exponentially (i.e., non-linearly) in relation to changes in another (see *isometric*).
Alula	The small feathered digit on the leading edge of the bird's wing. It is believed to function as an anti-stalling device, as in the slotted wings of aircraft. It is absent from penguins. Its embryological origin as digit 1 or digit 2 is significant in discussion of the origin of birds from dinosaurs.
Ambiens	A small muscle that is supposed to help birds grip a perch. It is often used as a taxonomic feature.
Anapsida	Tetrapod animals without an opening in the dermal skull roof (e.g., turtles). The muscles that operate the jaw pass over the surface of the skull.
Anhinga	A tropical snakebird or darter. A member of the Pelecaniformes or the Sulida of Sibley and Ahlquist [192].
Anomalogonatae	A group of birds without an ambiens muscle, including the Piciformes [54].
Anseriformes	The order of birds containing ducks, mergansers, geese, swans, and screamers.

Apodiformes	The order of birds containing swifts (Apodidae or Chaeturidae) and hummingbirds (Trochilidae). Equivalent to the Superorder Apodimorphae in Sibley and Ahlquist [192].
Archaeopteryx	A Jurassic fossil, generally accepted as the oldest known bird.
Ardeiformes	The order of birds containing herons and bitterns.
Apteria	Areas of skin between feather tracts (pterylae), without feather follicles.
Archosauria	A group of diapsid animals that includes crocodilians, pterosaurs, dinosaurs, and birds.
Aspect ratio	The ratio of the lifting surface (i.e., both wings plus the area of body between them) to a circular disc whose diameter is equivalent to the wingspan.
Atlas	The first vertebra in the neck.
Auk	A member of the Family Alcidae, Order Charadriiformes; same as alcid.
Avian	An adjective referring to characteristics of members of the Class Aves.
Axis	The second vertebra in the neck.

B

Basipterygoid processes	A pair of processes on either side of the rostrum of the parasphenoid bone, which lies on the midline of the palate. Each process articulates with an end of the pterygoid bone and is the point of action when the pterygoid is pushed by the quadrate to raise the upper jaw. Similar structures are found in Ratitae, Galloanserinae, many Charadriiformes, many Cathartinae, some Trogoniformes, some Strigiformes, and some Caprimulgiformes, but each may be an example of convergence rather than an indication of a relationship among those birds.
Biceps tubercle	In *Archaeopteryx*, the equivalent of the acrocoracoid, found on the ventral surface of the humerus and the shoulder bones. It does not form a triosseal gap. Contraction of *Archaeopteryx*'s supracoracoideus muscle might have pulled the wing down, not up. Considering that *Archaeopteryx* may have had no more than a short cartilaginous sternum without a keel, the contraction of this muscle would probably not have generated a great deal of force [13].
Bittern	A small member of the Order Ardeiformes.
Booby	A group of tropical plunge-divers in the Order Pelecaniformes or the Parvorder Sulida of Sibley and Ahlquist [192].
Brachial index	The ratio of the humerus to the ulna.
Brachyramphus marmoratus	Marbled Murrelet, a small auk (Family Alcidae).

C

Calcaneum	A small ankle bone fused into the top of the tarsometatarsus, facing tailward.
Caprimulgiformes	The order of nocturnal aerial insectivores, including nightjars, frogmouths, Oilbirds, poor-wills, potoos, nighthawks, etc.
Carinatae	Birds with a keel on the sternum.

Charadriiformes	The order of birds containing sandpipers, plovers, avocets, stilts, oystercatchers, jacanas, sandgrouse, seedsnipe, Plains Wanderer, sheathbills, thick-knees, pratincoles, terns, gulls, skuas, jaegers, and auks.
Caudo-femoral muscle	A muscle between the tail and the femur. It is important in the locomotion of dinosaurs and crocodiles, in which it contracts to pull the hind limb back. In birds, it is small.
Centrum	The drum-shaped central portion of a vertebra, whose front and back faces articulate with its neighbours.
Ciconiiformes	The order of birds containing storks and New World vultures; it may also include pelicans.
Cladistics	A method of determining the degree of similarity among groups of organisms and organizing the similarities according to a phylogeny based on the presence of shared derived features.
Cnemial crest	A bony process in front of the knee for the attachment of muscles, especially in swimming birds.
Codon	A group of three nucleotides that usually specifies the use of a particular amino acid during protein synthesis.
Coliiformes	The order of birds containing African collies or mousebirds.
Columbiformes	The order of birds containing pigeons, doves, and the extinct dodos and solitaires.
Condyle	A large, smooth dome of bone in a joint whose surface slides on the close-fitting face of the cotyle on the adjacent bone.
Confuciusornithes or Confucius birds	A fossil subclass of birds from the Cretaceous of China that may be closely related to *Microraptor* and other dromaeosaurs. Known for its unusually large humerus. The most primitive member of the Pygostylia.
Convergence	The development of superficial similarity in unrelated organisms.
Coot	A small, semi-aquatic member of the Gruiformes.
Coraciiformes	The order of birds containing jacamars, puffbirds, hornbills, hoopoes, rollers, motmots, kingfishers, and bee-eaters. In some classifications, it includes trogons (Trogoniformes).
Coronaves	A subclass of birds defined by characteristics of the fibrinogen gene (see *Metaves*) [59].
Cotyle	Depression in a joint that fits the condyle on an adjacent bone.
Crane	Archetypal member of the Order Gruiformes.
Crop	A thin-walled sac for temporary food storage, located between the esophagus and the proventriculus in the upper digestive tract of birds.
Crossbill	A small seed-eating member of the Order Passeriformes.
Cuculiformes	The order of birds containing cuckoos, coucals, roadrunners, and occasionally the Hoatzin.

D

Dentary bone	A bone in the lower jaw that often supports teeth in primitive archosaurs. It forms the whole lower jaw of mammals.
Diapsida	Animals with two holes in the dermal skull for the passage of muscles to operate the lower jaw. The group includes the archosaurs and lepidosaurs, as well as all the living reptiles except turtles. It also

includes birds. Turtles are members of the Anapsida and mammals are members of the Synapsida.

Diastataxy
: The absence of a fifth secondary feather in the wing in some groups of birds that are considered to be more "reptilian" than others.

Digitigrade
: Walking on the fingers.

Diving-petrel
: Auk-like member of the Order Procellariiformes.

Dove
: A small member of the Order Columbiformes.

E

Enantiornithes, opposite birds, or ball-shouldered birds
: An extinct subclass of birds whose fossils are recognized by certain characters in the shoulders and legs that are constructed in a manner opposite to that found in the Neornithes (modern birds).

Enantiornithean
: An adjective describing the characteristics of members of the Enantiornithes (see *Ornithes* and *ornithean*).

Eoaves
: A subclass of Aves that includes ratites, tinamous, and the Galloanserinae in Sibley and Ahlquist [192] (see *Paleognathae* and *Neoaves*).

Epaxial muscles
: A long series of muscles that lie along the backbone. In primitive animals, the series is unbroken from head to tail. In birds, the series is broken at the interface between the thorax and the synsacrum.

Eutaxy
: The presence of a fifth secondary feather in the wing (see *diastataxy*).

F

Falconides
: An infraorder in the Sibley and Ahlquist biomolecular phylogeny [192], equivalent to the Order Falconiformes in other classifications.

Falconiformes
: The order of birds containing hawks, eagles, kites, harriers, secretary birds, and Old World vultures.

Feather tracts
: See *pterylae.*

Femur
: The uppermost or proximal leg bone.

Fibula
: A small splint-like bone in a bird's leg that lies along the upper section of the tibiotarsus.

Finfoot
: A member of the tropical Family Heliornithidae in the Order Gruiformes, often called a sun-grebe.

Finch
: A member of the Order Passeriformes; a small seed eater.

Flamingo
: A member of the Order Phoenicopteriformes.

Foramen
: An opening in a bone or other structure (see *pneumatic foramen*).

Foramen magnum
: The opening at the base of the skull for the passage of the spinal cord into the vertebral column.

Fossa
: A deep trench in a bone, often leading to a foramen.

Frigatebird
: A group of pelecaniform seabirds, famous as highly accomplished fliers, belonging to the Family Fregatidae or Order Fregatiformes.

Fulmar
: A mid-sized, albatross-like petrel in the Order Procellariiformes.

G

Galliformes
: The order of birds containing domestic chickens, jungle fowl, grouse, ptarmigan, pheasants, peafowl, etc.

Gallinule
: A small, semi-aquatic member of the Order Gruiformes.

Galloanserinae
: A superorder that includes both Galliformes and Anseriformes in the Sibley and Ahlquist biomolecular phylogeny [192].

Gannet	A plunge-diving member of the Order Pelecaniformes or the Parvorder Sulida of Sibley and Ahlquist [192].
Gansus yumenensis	A fossil web-footed bird from Lower Cretaceous deposits in China.
Gastralia	Long, thin bones that lie across the abdomen of crocodiles and dinosaurs. They function as free-floating ribs.
Gaviiformes	Loons, an order of large aquatic birds.
Glycogen body	A poorly understood structure in the spinal cord near the hips, constructed mainly of glycogen. It may actually be an organ of balance [71, 138].
Goose	A group of large herbivorous and generally terrestrial members of the Order Anseriformes.
Graminivore	A seed eater.
Greater trochanter	An extension of the femur that limits rotation at the hip joint.
Grebe	A member of the Order Podicipediformes.
Gruiformes	An order of birds containing cranes (Gruidae), rails (Rallidae), coots, gallinules, and tropical finfoots or sungrebes (Heliornithidae).
Gular pouch	A sac of loose skin near the top of the throat of all pelecaniform seabirds and some small auks, in which they can store food.
Gull	One of the larger members of the Order Charadriiformes.

H

Hallux	Digit 1 of the foot (see *pollex*).
Heron	One of the larger members of the Order Ardeiformes.
Herbivore	Plant eater.
Hesperornis	A very large flightless, toothed diving bird from the Cretaceous, belonging to the Order Hesperornithiformes.
Heterocoelous centrum	Centrum with saddle-shaped face.
Hoatzin	*Opisthocomus hoazin,* a tropical leaf-eating bird of uncertain relation-ships [94, 95, 155].
Homalogonatae	A group that includes parrots, cuckoos, and turacos because of the presence of an ambiens muscle [65].
Homobatrachotoxin	A poison that is present in the feathers of the songbirds *Pitohui* and *Ifrita* apparently as a result of their consumption of poisonous beetles.
Hornbill	A large member of the Order Coraciiformes.
Humerus	The bone in the forelimb that is closest to the body.
Hummingbird	A family (Trochilidae) of small, colourful nectivorous birds (Order Apodiformes).
Hyoid apparatus	A collection of small bones that support the tongue and help the bird swallow its food.
Hypocleideum	A flat piece of bone that grows across the joint between the two clavicles in the furcula (wishbone).

I

Ibis	A member of the Order Threskiornithiformes.
Ifrita kowaldi	A member of the Order Passeriformes from Papua Niugini, noted for its toxic plumage.

Ilium	The largest and most anterior of the large bones in the pelvic girdle.
Ischium	The most posterior bone in the pelvic girdle.
Isometric	Referring to a situation where one variable increases linearly with changes in another (see *allometric*).

J

Jaeger	A member of the Order Charadriiformes; also called a skua.

K

K-T boundary	Jargon for the division between the Cretaceous and the Tertiary periods.
Keratin	Tough, proteinaceous material found in feathers, claws, beaks, etc.
Kingfisher	A member of the Order Coraciiformes.

L

Ligament	Connective tissue linking bone to bone.
LINE	In the avian chromosome, a long insertion of nucleotide elements.
Lithornithidae	Extinct, rail-like paleognath birds that could fly. Originally thought to be related to vultures.
Loon	A member of the Order Gaviiformes.

M

Maniraptora	A group believed to include dromaeosaurs and birds.
Manus	The hand.
Maxilla	A large tooth-bearing bone in the upper jaw. It may have been lost in birds along with the teeth.
Mesokinesis	Condition in which the hinge between the upper jaw and the braincase lies above the eye (see *prokinesis*).
Metaves	A major lineage of birds defined by characteristics of the fibrinogen gene [59]. Includes the Hoatzin, grebes, tropicbirds, doves, swifts, hummingbirds, and several other groups.
Microraptor	A small, feathered theropod that probably achieved a form of gliding flight (see *Tetrapteryx*).
Monophyletic	Descended from a single ancestor.
Mousebird	A member of the Order Coliiformes.
mtDNA	Mitochondrial DNA; found only in the mitochondria and transmitted only in the female line because the egg contains some mitochondria whereas the sperm contributes only its nuclear DNA during fertilization.
Murrelet, Marbled	*Brachyramphus marmoratus*, an alcid found on the west coast of North America and a member of the Order Charadriiformes.
Musophagiformes	An order of uncertain relationships that contains only the African turacos or plantain eaters.

N

nDNA	Nuclear DNA, found in the chromosomes.
Neoaves	The subclass of Aves that includes all living birds except the Eoaves, in the Sibley and Ahlquist phylogeny [192].

Neornithean	An adjective that describes characteristics of the members of the Neornithes; more correct linguistically than "neornithine."
Neornithine	An adjective widely used to describe members of the Neornithes; linguistically, however, it erroneously implies the existence of a group named "Neornithinae."
New World vultures	Members of the Subfamily Cathartinae. The subfamily was formerly in the Order Falconiformes, but biomolecular and morphological evidence tends to support a new relationship within the Order Ciconiiformes.
Nidicolous	Birds whose nestlings stay in the nest until ready to fledge.
Nidifugous	Birds whose nestlings leave the nest shortly after hatching, long before they develop flight feathers.
Nightjar	A member of the Order Caprimulgiformes.
Notarium	A ridge on the spinal column formed by the fusion or near fusion of the dorsal spines.

O

Obturator foramen	An opening in the hip bones to the rear of the joint with the hind limb.
Obturator process	A vertical process on the top of the pubis.
Obturator ridge	A ridge on the femur for the attachment of the obturator muscle, which arises on the edge of the obturator foramen.
Occipital condyle	An articulating surface on the base of the skull that meets the cotylar surfaces of the atlas (the first cervical vertebra).
Odontornithes	Birds with teeth.
Old World vulture	A large scavenger belonging to the Order Falconiformes.
Opisthocomus hoazin	The Hoatzin (Opisthocomidae, Opisthocomiformes).
Ornithurae	Birds with a "birdlike" tail; very close in meaning and content to Pygostylia.
Ornithurean	An adjective describing the characteristics of a member of the Ornithurae.
Os cuneatum	One of the sesamoid bones between the toes and the tarsometatarsus.
Os pisiforme	A specialized sesamoid bone in the wrist.
Osteology	The study of bones.
Outgroup	In biomolecular and cladistic analyses, a very distantly related group that is chosen as a contrast. Crocodiles are the preferred outgroup for biomolecular studies of birds.
Owl	A member of the Order Strigiformes.

P

Paleognathae	A subclass of birds with an unusual palate. Includes Ratitae (ostriches, etc.), tinamous, and the extinct Lithornithidae (Eoaves).
Paraphyletic	Not descended from a single ancestor.
Parasagittal stance	A posture in which the limbs are mounted at right angles to the sagittal plane, the plane that divides humans into a front and a back but other animals into a top and a bottom.
Parrot	A member of the Order Psittaciformes, a colourful group concentrated in the Southern Hemisphere.

Passeriformes	The order of birds containing the songbirds. About half of all bird species belong to this order.
Patella	Kneecap, a cartilage that helps control movement in the knee.
Pelecaniformes	A traditional order of waterbirds based on morphological similarities between boobies, gannets, cormorants, anhingas, pelicans, frigatebirds, and tropicbirds. It may be artificial (see Sulida, Fregatiformes, Phaethontiformes, Ciconiiformes [192]).
Pelican	Relationships uncertain; molecular evidence suggests the Order Ciconiiformes [192], morphological evidence suggests the Order Pelecaniformes.
Penguin	A member of the Order Sphenisciformes.
Petrel	A member of the Order Procellariiformes.
Phalarope	A swimming sandpiper with lobed toes like a coot or grebe; a member of the Order Charadriiformes.
Phoenicopteriformes	The order of birds containing only the flamingos.
Piciformes	The order of birds containing woodpeckers, piculets, barbets, toucans, and toucanets.
Pigeon (or dove)	A member of the Order Columbiformes.
Piscivore	Fish eater.
Pitohui	A member of the Order Passeriformes from Papua Niugini, noted for its toxic plumage.
Plantar tendons	Tendons in the foot that move the toes. Their varied arrangement has been used as a taxonomic feature.
Plethaves	All the living groups of birds not included in Paleognathae or Galloanserinae.
Pneumatic foramen	A passageway through which the air sac passes into the interior of a bone.
Plantigrade	Walking on the flat of the foot.
Plover	A member of the Order Charadriiformes.
Podicipediformes	The order of birds containing grebes.
Pollex	Digit 1 of the hand (see *hallux*).
Polytomy	Situation in a dendrogram where more than two branches originate at one point; more than a dichotomy.
Premaxillary bone	A bone that lies anterior to the maxilla.
Prion	A member of the Order Procellariiformes.
Procellariiformes	The order of birds containing petrels, storm-petrels, diving-petrels, prions, shearwaters, fulmars, and albatrosses.
Prokinesis	Condition in which the hinge between the upper jaw and the braincase lies in front of the eye (see *mesokinesis*).
Protavis texensis	The name assigned to a controversial collection of fossils from Texas that might represent a bird 75 million years older than *Archaeopteryx* [24].
Psittaciformes	The order of birds containing parrots, budgerigars, macaws, cockatiels, cockatoos, etc.
Ptarmigan	A northern member of the Order Galliformes.
Pterygoid bone	One of the independent bones of the skull that participates in the movement of the upper jaw.
Pterylae	Specific tracts in which feather follicles are organized, varying from family to family.

Pubic apron	A name sometimes applied to the pubic symphysis or area of fusion at the ends of the pubic bones in dinosaurs and *Archaeopteryx*.
Pubic bone	The third bone of the pelvic girdle. It meets the other two adjacent to the hip joint.
Pubic foot	An expansion of the tip of the pubic bone that dinosaurs may have used as a prop while resting.
Pubic spoon	A concave area on the rear surface of the pubic symphysis in some dinosaurs and *Archaeopteryx*.
Puffin	A large type of auk (Family Alcidae, Order Charadriiformes).
Pygostyle	A plough-shaped bone at the end of a bird's tail.
Pygostylia	A monophyletic group of birds, all of which have a pygostyle [25]. Excludes only *Archaeopteryx, Rahonavis,* and their near relatives.

Q

Quadrate bone	A large independent bone of the bird skull. Muscles attached to it cause the movement of the pterygoid and quadratojugal bones, which lack muscles of their own.
Quadratojugal bone	One of the independent bones of the skull that participates in the movement of the upper jaw.

R

Rachis	The central supporting rib of a feather.
Rail	A very small, semi-aquatic member of the Order Gruiformes.
Ramphotheca	The horny or chitinous part of the beak.
Ratitae	Belonging to the Paleognathae, a diverse group of flightless birds from the Southern Hemisphere, all of which lack a keel on the sternum.
Rectrices	The large tail feathers.
Remiges	The large wing feathers, or primaries, attached to the hand (manus).
Reptilia	No longer considered to be a monophyletic taxonomic grouping. Living representatives have been placed in the Anapsida or the Diapsida according to their skull structure.
Roadrunner	A member of the Order Cuculiformes.
Rotular groove	A groove for the passage of tendons between the condyles of the lower tibiotarsus.

S

Sacrum; sacral	Hip; of the hip (see *synsacrum*)
Sacral girdle	The synsacrum and the associated hipbones (ilium, ischium, and pubis).
Salt glands	Large glands above the eye in seabirds that excrete salt. Believed to be modified tear glands.
Sandpiper	A member of the Order Charadriiformes.
SAPE	Society for Avian Paleontology and Evolution.
Sapeornis	The most primitive bird with a hypocleideum on its furcula. Known from Cretaceous fossils in China.
Sauriurae	Birds with a reptilian, frondlike tail, such as *Archaeopteryx*.
Scleral ossicles	A ring of small bony plates on the surface of the eye in birds and dinosaurs. In owls, they form a tube to give the eye a unique shape.

Sclerotic tissue	A tough protective layer created by the walls of dead skin cells.
Scopiformes	An order of birds containing only the Hammerhead Stork.
Sebum	A mixture of organic chemicals that maintains the flexibility of the keratin in feathers.
Septum	A plate of bone separating two areas; especially the septum of the nose, which separates the two nostrils.
Shearwater	A member of the Order Procellariiformes.
SINE	In the avian chromosome, a short insertion of nucleotide elements.
Sinusoidal movement	The S-shaped wriggle of snakes that can be seen in many tetrapod animals.
Skua	A large type of jaeger (Order Charadriiformes).
SORA	Searchable Ornithological Research Archive (http://www.elibrary.unm.edu/sora/index.php).
Sphenisciformes	An order of birds containing only the penguins.
Stork	A member of the Family Ciconiidae, Order Ciconiiformes.
Storm-petrel	A member of the Order Procellariiformes.
Strigiformes	An order of birds containing the owls.
Struthioniformes	An order of birds containing ostriches, emus, rheas, etc. (Ratitae, Paleognathae).
Sulida	A biomolecular superfamily proposed by Sibley and Ahlquist [192] for boobies, gannets, cormorants, and anhingas.
Supra-orbital glands	See *salt glands*.
Swift	A member of the Order Apodiformes (Family Chaeturidae or Apodidae).
Synapomorphy	A derived feature that is shared by all members of the group in question.
Synapsida	A group of tetrapods with only one hole in the dermal skull. Its only living members are mammals.
Synsacrum	Structure in birds consisting of the fused sacral vertebrae. The original string of two or three sacral vertebrae may acquire up to nine postsacral vertebrae from the tail and often several thoracic vertebrae, lengthening the whole hip area.
Syrinx	The voice box in birds. It is often used in taxonomy.

T

Tarsal	An ankle bone.
Tarsometatarsus	Bones of the foot and ankle (tarsals) that are fused in birds and function as the lower section of the leg.
Tendon	Connective tissue linking muscle to bone.
Tern	Lightly built member of the Order Charadriiformes.
Tetrapteryx	An imaginary four-winged animal proposed by William Beebe in 1915 as an intermediary stage between early tetrapods and birds (see *Microraptor*).
Thecodont	A group of primitive reptiles that has been considered ancestral to birds. It has no unifying features, however, and is now considered to be an artificial assemblage.
Tetrapod	An animal with four legs.
Theropoda	A group of predatory dinosaurs.

Threskiornithiformes   An order of birds containing only the ibises.

Tibiotarsus   The bone in the middle section of the bird's leg. It is formed by the fusion of the tibia with some of the ankle bones (tarsals).

Tinamiformes   An order of birds containing only the tinamous (Paleognathae).

Tinamou   A chicken-like flying bird found only in South and Central America (Paleognathae). A member of the Order Tinamiformes.

Toucan   A member of the Order Piciformes.

Triosseal gap   A prominent structure in the glenoid of the shoulder that is critical to the flight of birds. It is basically a smooth tube formed by the conjunction of special features at the tips of the coracoid, the furcula, and the base of the scapula. The two large flight muscles are attached to the sternum. The tendon from the large pectoral muscle is attached to the underside of the humerus to pull the wing down when its muscle contracts. The tendon from the supracoracoideus muscle passes up through the triosseal gap, like a rope through a pulley. It could have been attached anywhere on the humerus and still raise the wing, but it is attached in the most logical position on the dorsal surface, to the acrocoracoid.

Tropicbirds   A group of tropical, white seabirds with only three species (Order Phaethontiformes or Family Phaethontidae). Not included in the Order Pelecaniformes by Sibley and Ahlquist [192].

Tube-nosed birds   Petrels. Members of the Order Procellariiformes.

Tubinares   Former synonym for Procellariiformes.

V

Vulture   See *New World vultures* and *Old World vultures*.

W

Woodpecker   A member of the Order Piciformes.

Y

Yixianornis grabaui   An ornithurean or modern bird from the Lower Cretaceous of China, with a modern-looking keel and pectoral girdle. Its pelvic girdle looks rather primitive [31].

Literature Cited

1 Allman, J. 1999. Evolving brains. Scientific American Library, New York.

2 Alonso, P.D., A.C. Millner, R.A. Ketcham, M.J. Cookson, and T.B. Rowe. 2004. The avian nature of the brain and inner ear of *Archaeopteryx*. Nature 430:666-69.

3 Alvarengo, H.M.F., and E. Höfling. 2003. Systematic review of Phorusrhacidae (Aves, Galliformes). Papéis Avulsos de Zoologia 43:55-91.

4 Avise, J.C., W.S. Nelson, and C.G. Sibley. 1994. Why one-kilo-base sequences from mito-chondrial DNA fail to solve the Hoatzin phylogenetic enigma. Molecular Phylogenetics and Evolution 3:175-84.

5 Bailey, J.P., and M.E. DeMont. 1991. The function of the wishbone. Canadian Journal of Zoology 69:2751-58.

6 Bakker, R.T. 2004. Introduction: Dinosaurs acting like birds, and vice versa – An homage to Reverend Edward Hitchcock, first director of the Massachusetts Geological Survey. Pages 1-11 *in* P.J. Currie, E.B. Koppelhus, M.A. Shugar, and J.L. Wright (eds.). Feathered dragons: Studies in the transition from dinosaurs to birds. Indiana University Press, Bloomington.

7 Baumel, J.J., and L.M. Witmer. 1993. Handbook of avian anatomy: Nomina Anatomica Avium, 2nd ed. Pages 45-132 *in* J.J. Baumel, A.S. King, J.E. Breazile, H.E. Evans, J.C. Vanden Berge (eds.). Publication of the Nuttall Ornithological Club 23. Nuttall Ornithologi-cal Club, Cambridge, MA.

8 Beddard, F.E. 1898. The structure and classification of birds. Longmans, Green, London.

9 Beebe, C.W. 1915. A *Tetrapteryx* stage in the ancestry of birds. Zoologica 2:38-52.

10 Bigu del Blanco, J., and C. Romero-Sierra. 1975. The properties of bird feathers as converse piezoelectric transducers and as receptors of microwave radiation. II. Bird feathers as dielec-tric receptors of microwave radiation. Biotelemetry 2:354-64.

11 Bock, W.J. 1965. Experimental analysis of the avian passive perching system. American Zoologist 5:681.

12 Bock, W.J. 1974. The avian skeletomuscular system. Pages 119-257 *in* D.S. Farner and J.R. King (eds.). Avian biology, vol. 4. Academic Press, London.

13 Bock, W.J. 1986. The arboreal origin of flight. Pages 57-72 *in* K. Padian (ed.). The origin of birds and the evolution of flight. Memoirs of the California Academy of Sciences 8. Califor-nia Academy of Sciences, San Francisco.

14 Brochu, C.A., and M.A. Norell. 2001. Time and trees: A quantitative assessment of tempo-ral congruence in the bird origins debate. Pages 511-35 *in* J. Gauthier and L.F. Gall (eds.). New perspectives on the origin and early evolution of birds: Proceedings of the international

symposium honoring John H. Ostrom. Peabody Museum of Natural History, Yale University, New Haven, CT.

15 Brodkorb, P. 1971. Origin and evolution of birds. Pages 19-55 *in* D.S. Farner and J.S. King (eds.). Avian biology, vol. 4. Academic Press, London.

16 Brooke, M. de L., S. Hanley, and S.B. Laughlin. 1999. The scaling of eye size with body mass in birds. Proceedings of the Royal Society of London B 266:403-12.

17 Brooks, A. 1937. The patagial fan in the Tubinares. Condor 39:82-83.

18 Brown, R.E., and A.C. Cogley. 1996. Contributions of the propatagium to avian flight. Journal of Experimental Biology 276:112-24.

19 Buffetaut, E., G. Grellet-Tinner, V. Suteethorn, G. Cuny, H. Tong, A. Kosir, L. Cavin, S. Chitsing, P.J. Griffiths, J. Tabouell, J. Le Loeuf. 2005. Minute theropod eggs and embryo from the Lower Cretaceous of Thailand and the dinosaur-bird transition. Naturwissenschaften 92:477-82.

20 Bühler, P. 1988. Light bones in birds. Pages 385-93 *in* K.E. Campbell (ed.). Papers in avian paleontology honoring Pierce Brodkorb. No. 36, Science Series. Natural History Museum of Los Angeles County, CA.

21 Carrier, D.R., D.V. Lee, and R.M. Walter 2001. Influence of rotational inertia on the turning performance of theropod dinosaurs: Clues from humans with increased rotational inertia. Journal of Experimental Biology 204:3917-26.

22 Chamberlain, F.W. 1943. Atlas of avian anatomy: Osteology, arthrology, myology. Michigan State College, East Lansing, MI.

23 Li Chun, Wu Xiao-chun, Cheng Yen-nien, T. Sato, and Wang Liting. 2006. An unusual archosauran from the marine Triassic of China. Naturwissenschaften 93:200-6.

24 Chatterjee, S. 1997. The rise of birds: 225 million years of evolution. Johns Hopkins University Press, Baltimore, MD.

25 Chiappe, L.M. 2002. Basal bird phylogeny: Problems and solutions. Pages 448-72 *in* L.M. Chiappe and L.M. Witmer (eds.). Mesozoic birds: Above the heads of dinosaurs. University of California Press, Berkeley.

26 Chubb, A.L. 2004. New nuclear evidence for the oldest divergence among neognath birds: The phylogenetic utility of ZENK(i). Molecular Phylogenetics and Evolution 30:140-51.

27 Clark, J.M., M.A. Norell, and T. Rowe. 2002. Cranial anatomy of *Citipati osmolskae* (Theropoda, Oviraptosauria), and a reinterpretation of the holotype of *Oviraptor philoceratops*. Novitates 3364:1-24. American Museum of Natural History, New York.

28 Clarke, J.A. 2004. Morphology, phylogenetic taxonomy, and systematics of *Ichthyornis* and *Apatornis* (Avialae: Ornithurae). Bulletin of the American Museum of Natural History 286. American Museum of Natural History, New York.

29 Clarke, J.A., and M.A. Norell. 2002. The morphology and phylogenetic position of *Apsaravis ukhaana* from the late Cretaceous of Mongolia. Novitates 3387. American Museum of Natural History, New York.

30 Clarke, J.A., C.P. Tambussi, J.I. Noriega, G.M. Erickson, and R.A. Ketcham. 2005. Definitive fossil evidence for the extant avian radiation in the Cretaceous. Nature 433:305-8.

31 Clarke, J.A., Zhou Zhonghe, and Zhang Fuchang. 2006. Insight into the evolution of avian flight from a new clade of early Cretaceous ornithurines from China and the morphology of *Yixianornis grabaui*. Journal of Anatomy 208:287-308.

32 Cobb, S. 1959. On the angle of the cerebral axis in the American Woodcock. Auk 76:55-59.

33 Cottam, P.A. 1957. The pelecaniform characteristics of the skeleton of the Shoebill Stork, *Balaeniceps rex*. Bulletin of the British Museum (Natural History) Zoology 5:49-72.

34 Coues, E. 1884. Key to North American birds, 4th ed. Estes and Lauriat, Boston.

35 Cracraft, J. 1981. Toward a phylogenetic classification of the recent birds of the world (Class Aves). Auk 98:681-714.

36 Cracraft, J. 1982. Phylogenetic relationships and monophyly of loons, grebes, and hesperornithiform birds, with comments on the early history of birds. Systematic Zoology 31:35-56.

37 Cracraft, J. 2001. The basal clades of modern birds. Pages 143-56 *in* J. Gauthier and L.F. Gall (eds.). New perspectives on the origin and early evolution of birds: Proceedings of the international symposium in honor of John H. Ostrom. Peabody Museum of Natural History, Yale University, New Haven, CT.

38 Cubo, J., and A. Casinos. 2000. Incidence and mechanical significance of pneumatization in the long bones of birds. Zoological Journal of the Linnean Society 130: 499-510.

39 Currey, J.D. 2002. Bones: Structure and mechanics. Princeton University Press, Princeton, NJ.

40 Currey, J.D., and R. McN. Alexander. 1985. The thickness of the walls of tubular bones. Journal of Zoology, London 206A:453-68.

41 Currie, P.J., E.B. Koppelhus, M.A. Shugar, and J.L. Wright (eds.). 2004. Feathered dragons: Studies in the transition from dinosaurs to birds. Indiana University Press, Bloomington.

42 Czerkas, S.J. (ed.). 2002. Feathered dinosaurs and the origin of flight. Dinosaur Museum, Blanding, UT.

43 Dacke, C.G., S. Arkle, D.J. Cooke, I.M. Wormstone, S. Jones, M. Zaidi, and Z.A. Bascal. 1993. Medullary bone and avian calcium regulation. Journal of Experimental Biology 184: 63-88.

44 Darwin, C. 1839. Voyage of the *Beagle* (journal of the researches into the natural history and geology of the countries visited during the voyage of HMS *Beagle* round the world). Reprint, 2001. Modern Library, New York.

45 Davis, R.W., L. Polasek, R. Watson, A. Fuson, T.M. Williams, and S.B. Kanatous. 2004. The diving paradox: New insights into the role of the dive response in air-breathing vertebrates. Comparative Biochemistry and Physiology138:263-68.

46 Dawson, A. 2005. The scaling of primary flight feathers and mass in relation to wing shape, function and habitat. Ibis 147:283-92.

47 De Gennaro, L.D. 1982. The glycogen body. Pages 341-71 *in* D.S. Farner, J.R. King, and K.C. Parkes (eds.). Avian biology, vol. 6. Academic Press, New York.

48 De Quieroz, K., and D.A. Good. 1988. The scleral ossicles of *Opisthocomus* and their phylogenetic significance. Auk 105:29-35.

49 Dial, K.P. 2003. Evolution of avian locomotion: Correlates of flight style, locomotor modules, nesting biology, body size, development, and the origin of flapping flight. Auk 120:941-52.

50 Dingus, L., and T. Rowe. 1998. The mistaken extinction: Dinosaur evolution and the origin of birds. W.H. Freeman, New York.

51 Dumbacher, J.P., A. Wako, S.R. Derrickson, A. Samuelson, T.F. Spandle, and J.W. Daly. 2004. Melyrid beetles (*Choresine*): A putative source for the batrachotoxin alkaloids found in poison-dart frogs and toxic passerine birds. Proceedings of the National Academy of Sciences 101:15857-60.

52 Dunning, J.B. 1983. CRC handbook of avian body masses. CRC Press, Boca Raton, FL.

53 Dyke, G. 2001. The evolutionary radiation of modern birds: Systematics and patterns of diversification. Geological Journal 35:304-15.

54 Dyke, G.J., and M. van Tuinen. 2004. The evolutionary radiation of modern birds (Neornithes): Reconciling molecules, morphology and the fossil record. Zoological Journal of the Linnean Society 141:153-77.

55 Elliott, K.H., M. Hewett, G.W. Kaiser, and R.W. Blake. 2004. Flight energetics of the Marbled Murrelet, *Brachyramphus marmoratus*. Canadian Journal of Zoology 82:644-52.

56 Elzanowski, A. 2002. Archaeopterygidae (Upper Jurassic of Germany). Pages 129-59 *in* L.M. Chiappe and L.M. Witmer (eds.). Mesozoic birds: Above the heads of dinosaurs. University of California Press, Berkeley.

57 Engels, W.L. 1938. Cursorial adaptations in birds. Limb proportions in the skeleton of *Geococcyx*. Journal of Morphology 63:207-17.

58 Ericson, Per G.P. 1997. Systematic relationships of the palaeogene family Presbyornithidae (Aves: Anseriformes). Zoological Journal of the Linnean Society 121:429-83.

59 Fain, M.G., and P. Houde. 2004. Parallel radiations in the primary clades of birds. Evolution 58:2558-73.

60 Feduccia, A. 1996. The origin and evolution of birds. Yale University Press, New Haven, CT.

61 Fisher, H.I. 1946. Adaptations and comparative anatomy of the locomotive apparatus of New World vultures. American Midland Naturalist 35:545-727.

62 Fürbringer, M. 1888. Untersuchungen zur Morphologie und Systematik der Vögel, zugleich ein Beitrag zur Anatomie der Stüz – und Bewegungsorgane. Van Holkema, Amsterdam, Netherlands.

63 Gadow, H.F. 1933. The evolution of the vertebral column: A contribution to the study of vertebrate phylogeny. Cambridge University Press, Cambridge, UK.

64 Galis, F., M. Kundrat, and. J.A.J. Metz. 2005. Hox genes, digit identities, and the theropod/bird transition. Journal of Experimental Zoology Part B Molecular and Developmental Evolution 304B:91-106.

65 Garrod, A.H. 1873. On certain muscles of the thigh of birds and of their value in classification. Part I. Proceedings of the Zoological Society of London 1873:626-44.

66 Gaston, A.J., and I.L. Jones. 1998. The auks. Oxford University Press, Oxford.

67 Gatesy, S.M., and K.P. Dial. 1996. Locomotory modules and the evolution of avian flight. Evolution 50:331-40.

68 Gatesy, S.M., and K.M. Middleton. 1997. Bipedalism, flight, and the evolution of theropod locomotor diversity. Journal of Vertebrate Paleontology 17:308-29.

69 Gauthier, J. 1986. Saurischian monophyly and the origin of birds. Pages 1-55 *in* K. Padian (ed.). The origin of birds and the evolution of flight. Memoirs of the California Academy of Sciences 8. California Academy of Sciences, San Francisco.

70 Gauthier, J., and K. de Quieroz. 2001. Feathered dinosaurs, flying dinosaurs, crown dinosaurs, and the name "Aves." Pages 7-46 *in* J. Gauthier and L.F. Gall (eds.). New perspectives on the origin and early evolution of birds: Proceedings of the international symposium in honor of John H. Ostrom. Peabody Museum of Natural History, Yale University, New Haven, CT.

71 Giffin, E.B. 1995. Postcranial paleoneurology of the Diapsida. Journal of the Zoological Society of London 235:389-410.

72 Gingerich, P.D. 1973. The skull of *Hesperornis* and the early evolution of birds. Nature 243:448-62.

73 Grant, P.J. 1986. Ecology and evolution of Darwin's finches. Princeton University Press, Princeton, NJ.

74 Gregory, J.T. 1952. The jaws of the Cretaceous birds *Ichthyornis* and *Hesperornis*. Condor 54:73-88.

75 Grimmer, J.L. 1962. Strange little world of the Hoatzin. National Geographic 122:390-401.

76 Grubb, T.C. Jr. 1972. Smell and foraging in shearwaters and petrels. Nature 237:404-5.

77 Gussekloo, S.W., M.G. Vosselmann, and R.G. Bout. 2001. Three-dimensional kinematics of skeletal elements in avian prokinetic and rhynchokinetic skulls determined by Roentgen stereophotogrammetry. Journal of Experimental Biology 204:1735-44.

78 Gussekloo, S.W.S., and G.A. Zweers. 1999. The paleognathous pterygoid-palatinum complex. A true character? Netherlands Journal of Zoology 49:29-43.

79 Harrison, C.J.O. 1977. The limb osteology of the diving petrels and the Little Auk as evidence of the retention of characters in morphologically convergent species. Ardea 65:43-52.

80 Hebert, P.D.N., M.Y. Stoeckle, T.S. Zemlak, and C.M. Francis. 2004. Identification of birds through DNA barcodes. Public Library of Science Biology 2 DOI: 10.1371/journal.pbio. 0020312 (available on line at http://biology.plosjournals.org/perlserv?request=index-html).

81 Hedenstrom, A. 2002. Aerodynamics, evolution, and ecology of avian flight. Trends in Ecology and Evolution 17:415-22.

82 Heilmann, G. 1926. The origin of birds. Witherby, London; 1927, Appleton, New York.

83 Henderson, D.M. 2002. The eyes have it: The sizes, shapes, and orientations of theropod orbits as indicators of skull strength and bite force. Journal of Vertebrate Paleontology 22:766-78.

84 Hennig, W. 1966. Phylogenetic systematics. University of Illinois Press, Urbana.

85 Hone, D.W.E., and M.J. Benton. 2005. The evolution of large size: How does Cope's Rule work? Trends in Ecology and Evolution 20:4-6.

86 Hope, S. 2002. The Mesozoic radiation of the Neornithes. Pages 339-88 *in* L.M. Chiappe and L.M. Witmer (eds.). Mesozoic birds: Above the heads of dinosaurs. University of California Press, Berkeley.

87 Hopp, T.P., and M.J. Orsen. 2004. Dinosaur behavior and the origin of feathers. Pages 234-50 *in* P.J. Currie, E.B. Koppelhus, M.A. Shugar, and J.L. Wright (eds.). Feathered dragons: Studies in the transition from dinosaurs to birds. Indiana University Press, Bloomington.

88 Hou Lianhai, L.M. Chiappe, Zhang Fucheng, and Chuong Cheng-Ming. 2004. New early Cretaceous fossil from China documents a novel trophic specialization for Mesozoic birds. Naturwissenschaften 91:22-25.

89 Hou Lianhai, L.D. Martin, Zhou Zhonghe, A. Feduccia, and Zhang Fucheng. 1999. A diapsid skull in a new species of the primitive bird *Confuciusornis.* Nature 399:679-82.

90 Houde, P. 1986. Ostrich ancestors found in the Northern Hemisphere suggest new hypothesis of ratite origins. Nature 324:563-65.

91 Houde, P. 1987. Histological evidence for the systematic position of *Hesperornis* (Odontornithes: Hesperornithiformes). Auk 104:125-29.

92 Houde, P. 1988. Paleognathous birds from the early Tertiary of the Northern Hemisphere. Publications of the Nuttall Ornithological Club 22. Nuttall Ornithological Club, Cambridge, MA.

93 Houde, P., and S.L. Olson. 1981. Paleognathous carinate birds from the early Tertiary of North America. Science 214:1236-37.

94 Hughes, J.M. 2000. Monophyly and phylogeny of cuckoos (Aves, Cuculidae) inferred from osteological characters. Zoological Journal of the Linnean Society 130:263-307.

95 Hughes, J.M., and A.J. Baker. 1999. Phylogenetic relationships of the enigmatic Hoatzin (*Opisthocomus hoazin*) resolved using mitochondrial and nuclear gene sequences. Molecular Biology and Evolution 16:1300-7.

96 Hui, C.A. 2002. Avian furcula morphology may indicate relationships of flight requirements among birds. Journal of Morphology 251:284-93.

97 Hull, C.L., G.W. Kaiser, C. Lougheed, L. Lougheed, S. Boyd, and F. Cooke. 2001. Variation in commuting distance of Marbled Murrelets (*Brachyramphus marmoratus*): Ecological and energetic consequences of nesting further inland. Auk 118:1036-46.

98 International Chicken Genome Sequencing Consortium. 2004. Sequence and comparative analysis of the chicken genome provide unique perspectives on vertebrate evolution. Nature 432:695-716.

99 Jacob, J., and V. Ziswiler. 1982. The uropygial gland. Pages 199-324 *in* D.S. Farner, J.R. King, and K.C. Parkes (eds.). Avian biology, vol. 6. Academic Press, New York.

100 Jenkins, F.A. Jr., K.P. Dial, and G.E. Goslow Jr. 1988. A cineradiographic analysis of bird flight: The wishbone in starlings is a spring. Science 241:1495-98.

101 Ji Qiang, P.J. Currie, M.A. Norell, and Ji Shu-An. 1998. Two feathered dinosaurs from north-eastern China. Nature 393:753-61.

102 Ji Qiang, Ji S., You H., Zhang J., Yuan C., Ji X., and Li Y. 2002. Discovery of an Avialae bird *Shenzhouraptor sinensis gen. et sp. nov.* from China. Geological Bulletin of China 21:363-69.

103 Johansson, L.C., and U.M.L. Norberg. 2001. Lift based paddling in diving grebe. Journal of Experimental Biology 204:1687-96.

104 Johnson, K.P., S.M. Goodman, and S.M. Lanyon. 2000. A phylogenetic study of the Malagasy Couas with insights into cuckoo relationships. Molecular Phylogenetics and Evolution 14:436-44.

105 Kaiser, G.W. 1989. Nightly concentrations of Bald Eagles at an auklet colony. Northwestern Naturalist 70:12-13.

106 Kaiser, G.W. 2000. Alternative origins for flight in birds. Neues Jahrbuch für Geologie und Paläontologie 217:27-39.

107 Kaiser, G.W., A.E. Derocher, S. Crawford, M.J. Gill, and I. Manley. 1995. A capture technique for Marbled Murrelets in coastal waters. Journal of Field Ornithology 68:321-33.

108 Kemp, A.C. 2001. Bucerotidae (hornbills). Pages 436-523 *in* J. del Hoyo, A. Elliott, and J. Sargatal (eds.). Handbook of birds of the world. Birdlife International and Lynx Edicions, Barcelona.

109 Korzun, L.P., C. Erard, J.-P. Gasc, and F.J. Dzerzhinsky. 2003. Biomechanical features of the bill and jaw apparatus of cuckoos, turacos and the Hoatzin in relation to food acquisition and processing. Ostrich 74:48-57.

110 Kurochkin, E.N. 2000. Mesozoic birds of Mongolia and the former USSR. Pages 533-59 *in* M.J. Benton, M.A. Shishkin, D.M. Unwin, and E.N. Kurochkin (eds.). The age of dinosaurs in Russia and Mongolia. Cambridge University Press, Cambridge. (*Paeornis sharovi*, Fig. 27.9, p. 553)

111 Kurochkin, E.N., and C.A. Walker. 1999. What were the Enantiornithes? Proceedings of the International Ornithological Congress (Durban, South Africa) 22:3219-22.

112 Kuroda, N.H. 1967. Morpho-anatomical analysis of parallel evolution between Diving Petrel and Ancient Auk; with comparative osteological data of other species. Miscellaneous Reports of the Yamashina Institute for Ornithology 5:111-37.

113 Lerner, H.R.L., and D.P. Mindell. 2005. Phylogeny of eagles, Old World vultures, and other Accipitridae based on nuclear and mitochondrial DNA. Molecular Phylogenetics and Evolution 37:327-46.

114 Ligon, J.D. 1967. Relationships of the cathartid vultures. Occasional Papers 651, University of Michigan Museum of Zoology. University of Michigan, Ann Arbor.

115 Lind, L.R. (trans.). 1963. Aldrovandi on chickens. The ornithology of Ulisse Aldrovandi (1600), vol. 2, book 14. University of Oklahoma Press, Norman.

116 Livezey, B.C. 1997. A phylogenetic analysis of basal Anseriformes, the fossil *Presbyornis*, and the interordinal relationships of waterfowl. Zoological Journal of the Linnean Society 121:361-428.

117 Livezey, B.C., and R.L. Zusi. 2001. Higher-order phylogenetics of modern Aves based on comparative morphology. Netherlands Journal of Zoology 51:179-205.

118 Makovicky, P.J., S. Apesteguia, and F.L. Angolin. 2005. The earliest dromaeosaurid theropod from South America. Nature 437:1007-11.

119 Manegold, A., G. Mayr, and C. Mourer-Chauviré. 2004. Miocene songbirds and the composition of the European passeriform avifauna. Auk 121:1155-60.

120 Marsh, O.C. 1880. Odontornithes: A monograph on the extinct toothed birds of North America. Report of the US Geological Exploration of the Fortieth Parallel, No. 7. US Geological Survey, Washington, DC.

121 Martin, G.R. 1985. The eye. Pages 311-73 *in* A.S. King and J. McLelland (eds.). Form and function in birds, vol. 3. Academic Press, London and New York.

122 Martin, G., L. Marina Rojas, Y. Ramírez, and R. McNeil. 2004. The eyes of Oilbirds (*Steatornis caripensis*): Pushing at the limits of sensitivity. Naturwissenschaften 91:26-29.

123 Maurer, B.A. 1998. The evolution of body size in birds. I. Evidence for non-random diversification. Evolutionary Ecology 12:925-34.

124 Maybury, W.J., and J.M.V. Rayner. 2001. The avian tail reduces body parasitic drag by controlling flow separation and vortex shedding. Proceedings of the Royal Society of London B 268:1405-10.

125 Maynard-Smith, J. 1952. The importance of the nervous system in the evolution of flight. Evolution 6:127-29.

126 Mayr, E. 2001. What evolution is. Basic Books, New York.

127 Mayr, G. 2000. Tiny hoopoe-like birds from the middle Eocene of Messel (Germany). Auk 117:964-70.

128 Mayr, G. 2004. Old World fossil record of modern-type hummingbirds. Science 304:861-64.

129 Mayr, G. 2004. Morphological evidence for sister group relationship between flamingos (Aves: Phoenicopteridae) and grebes (Podicipedidae). Zoological Journal of the Linnean Society 140:157-69.

130 Mayr, G. 2005. Tertiary plotopterids (Aves, Plotopteridae) and a novel hypothesis on the phylogenetic relationships of penguins (Spheniscidae). Journal of Zoological Systematics and Evolutionary Research 43:61-71.

131 Mayr, G., and J. Clarke. 2003. The deep divergences of neornithine birds: A phylogenetic analysis of morphological characters. Cladistics 19:527-53.

132 Mayr, G., and A. Manegold. 2004. The oldest European fossil songbird from early Oligocene of Germany. Naturwissenschaften 91:173-77.

133 Mayr, G., B. Pohl, and D.S. Peters. 2005. A well-preserved *Archaeopteryx* specimen with theropod features. Science 310(5753):1483-86.

134 Mearns, B., and R. Mearns. 1988. Audubon to Xantus; 1992. Biographies of bird watchers; and 1998. The bird collectors. Academic Press, San Diego and London.

135 Merrem, B. 1813. *Tentamen systematis naturalis avium.* Abh. Königel. (Preussische) Akadem. Wiss. Berlin 1812-1813 (1816) (Physikal.):237-59.

136 Mourer-Chauviré, C. 1995. Dynamics of avifauna during the Paleogene and early Neogene of France. Settling of the recent fauna. Acta Zoologica Cracoviensis 38:325-42.

137 Murphy, M.E., T.G. Taruscio, J.R. King, and S.G. Truitt. 1992. Do molting birds renovate their skeletons as well as their plumages? Osteoporosis during the annual molt in sparrows. Canadian Journal of Zoology 70: 1109-13.

138 Necker, R. 1994. Sensorimotor aspects of flight control in birds: Specializations in the spinal cord. European Journal of Morphology 32:207-11.

139 Norberg, U.M. 1990. Vertebrate flight. Springer-Verlag, Berlin.

140 Noriega, J.I., and C.P. Tambussi. 1995. A late Cretaceous Presbyornithidae (Aves: Anseriformes) from Vega Island, Antarctic Peninsula: Paleogeographic implications. Ameghiniana 32:57-61.

141 Norell, M.A., and J.A. Clarke. 2001. Fossil fills a critical gap in avian evolution. Nature 409:181-84.

142 Norell, M.A., J.M. Clark, L.M. Chiappe, and D. Dasheveg. 1995. A nesting dinosaur. Nature 378:774-76.

143 Norell, M.A., J.M. Clark, and P.J. Makovicky. 2001. Phylogenetic relationships among coelurosaurian theropods. Pages 49-67 *in* J. Gauthier and L.F. Gall (eds.). New perspectives on the origin and early evolution of birds: Proceedings of the international symposium in honor of John H. Ostrom. Peabody Museum of Natural History, Yale University, New Haven, CT.

144 Norell, M.A., and Xu Xing. 2005. Feathered dinosaurs. Annual Review of Earth and Planetary Science 33:277-99.

145 Novas, F.E. 2004. Avian traits in the ilium of *Unenlagia comahuensis* (Maniraptora: Avialae). Pages 150-66 *in* P.J. Currie, E.B. Koppelhus, M.A. Shugar, and J.L. Wright (eds.). Feathered dragons: Studies in the transition from dinosaurs to birds. Indiana University Press, Bloomington.

146 Nudds, R.L., G.J. Dyke, and J.M.V. Rayner. 2004. Forelimb proportions and the evolutionary radiation of the Neornithes. Proceedings of the Royal Society of London Series B, Biology Letters 271: S324-27.

147 O'Connor, P.M., and L.P.A.M. Claessens. 2005. Basic avian pulmonary design and flow-through ventilation in non-avian theropod dinosaurs. Nature 436:253-56.

148 Olson, S.L. 1977. A Lower Eocene frigatebird from the Green River Formation of Wyoming (Pelecaniformes: Fregatidae). Smithsonian Contributions to Paleontology 35:1-33.

149 Olson, S.L. 1979. Multiple origins of the Ciconiiformes. Proceedings of the Colonial Waterbird Group 1978:165-70.

150 Olson, S.L. 1982. A critique of Cracraft's classification of birds. Auk 99:733-39.

151 Olson, S.L. 1985. The fossil record of birds. Pages 79-238 *in* D.S. Farner, J.R. King, and K.C. Parkes (eds.). Avian biology, vol. 8. Academic Press, Orlando, FL.

152 Olson, S.L., and A. Feduccia. 1979. Flight capability of the pectoral girdle of *Archaeopteryx*. Nature 278:247-48.

153 Olson, S.L., and Y. Hasegawa. 1979. Fossil counterparts of giant penguins from the North Pacific. Science 206:688-89.

154 Ostrom, J.H. 1969. Osteology of *Deinonychus antirropus,* an unusual theropod from the Lower Cretaceous of Montana. Peabody Museum of Natural History Bulletin 30:1-165.

155 Parker, W.K. 1891. On the morphology of a reptilian bird, *Opisthocomus cristatus.* Transactions of the Zoological Society of London 13:43-89.

156 Paul, G.S. (ed.). 2000. The Scientific American book of dinosaurs. St. Martin's Griffin, New York.

157 Paul, G.S. 2001. Were the respiratory complexes of predatory dinosaurs like crocodilians or birds? Pages 463-82 *in* J. Gauthier and L.F. Gall (eds.). New perspectives on the origin of birds: Proceedings of the international symposium in honor of John H. Ostrom. Peabody Museum of Natural History, Yale University, New Haven, CT.

158 Paul, G.S. 2002. Dinosaurs of the air. Johns Hopkins University Press, Baltimore and London.

159 Penny, D., and D.J. Phillips. 2004. The rise of birds and mammals: Are microevolutionary processes sufficient for macroevolution? Trends in Ecology and Evolution 19:516-22.

160 Pennycuick, C.J. 1989. Bird flight performance: A practical calculation manual. Oxford University Press, Oxford.

161 Pennycuick, C.J. 2002. Gust soaring as a basis for flight of petrels and albatross (Procellariiformes). Avian Science 2:1-12.

162 Perry, S.F., and M. Sander. 2004. Reconstructing the evolution of respiratory apparatus in tetrapods. Respiratory Physiology and Neurobiology 144:125-39.

163 Peters, J.L. 1934. Check-list of birds of the world. Harvard University Press, Cambridge, MA.

164 Pisani, D., A.M. Yates, M.C. Langer, and M.J. Benton. 2002. A genus-level supertree of the Dinosauria. Proceedings of the Royal Society of London Series B 269:915-21.

165 Pond, C.M. 1977. The significance of lactation in the evolution of mammals. Evolution 31:177-99.

166 Proctor, N.S., and P.J. Lynch. 1993. Manual of ornithology: Avian structure and function. Yale University Press, New Haven, CT.

167 Prum, R.O., and S. Williamson. 2001. Theory of the growth and evolution of feather shape. Journal of Experimental Zoology Part B Molecular and Developmental Evolution B291:30-57.

168 Pycraft, W.P. 1910. A history of birds. Methuen, London.

169 Quinn, T.H., and J.J. Baumel. 1990. The digital tendon locking mechanism of the avian foot (Aves). Zoomorphology 109:281-93.

170 Quinn, T.W. 1997. Molecular evolution of the mitochondrial genome. Pages 4-28 *in* D.P. Mindell (ed.). Avian molecular evolution and systematics. Academic Press, San Diego.

171 Raikow, R.J. 1985. Problems in avian classification. Current Ornithology 2:187-212.

172 Rayner, J.M.V. 1979. A new approach to animal flight mechanics. Journal of Experimental Biology 80:17-54.

173 Rayner, J.M.V. 1979. A vortex theory of animal flight. II. The forward flight of birds. Journal of Fluid Mechanics 112:97-125.

174 Rayner, J.M.V. 1985. Vertebrate flight: A bibliography to 1985. University of Bristol Press, Bristol, UK.

175 Rayner, J.M.V. 1988. Form and function in avian flight. Pages 1-66 *in* R.F. Johnston (ed.). Current ornithology, vol. 5. Plenum Press, New York.

176 Rayner, J.M.V. 1993. On aerodynamics and the energetics of vertebrate flapping flight. Contemporary Mathematics 141:351-97.

177 Rayner, J.M.V. 2001. On the origin and evolution of flapping flight aerodynamics in birds. Pages 363-85 *in* J. Gauthier and L.F. Gall (eds.). New perspectives on the origin and early evolution of birds: Proceedings of the international symposium honoring John H. Ostrom. Peabody Museum of Natural History, Yale University, New Haven, CT.

178 Rietschel, S. 1985. Feathers and wings of *Archaeopteryx,* and the question of her flight ability. Pages 251-60 *in* M.K. Hecht, J.H. Ostrom, G. Viohl, and P. Wellnhoffer (eds.). The beginnings of birds. Freunde des Jura-Museums, Eichstätt.

179 Romer, A.S. 1966. Vertebrate paleontology, 3rd ed. University of Chicago Press, Chicago.

180 Rowe, T., R.A. Ketcham, C. Denison, M. Colbert, X. Xu, and P.J. Currie. 2001. Forensic paleontology: The *Archaeoraptor* forgery. Nature 410:539-40.

181 Ruben, J.A., C. Dal Sasso, N.R. Geist, W.J. Hillenius, T.D. Jones, and M. Signore. 1999. Pulmonary function and metabolic physiology of theropod dinosaurs. Science 283:514-16.

182 Ruben, J.A., T.D. Jones, N.R. Geist, and W.J. Hillenius. 1997. Lung structure and ventilation in theropod dinosaurs and early birds. Science 278:1267-70.

183 Ruxton, G.D., and D.C. Houston. 2004. Obligate vertebrate scavengers must be large soaring fliers. Journal of Theoretical Biology 228:431-36.

184 Saville, D.B.O. 1950. Flight mechanisms in swifts and hummingbirds. Auk 67:499-504.

185 Saville, D.B.O. 1957. Adaptive evolution in the avian wing. Evolution 11:212-24.

186 Schweitzer, M.H., J.A. Watt, R. Avci, L. Knapp, L. Chiappe, M. Norell, and M. Marshall. 1999. Beta-keratin specific immunological reactivity in feather-like structures of the Cretaceous alvarezsaurid *Shuvuuia deserti*. Journal of Experimental Zoology Part B Molecular and Developmental Evolution 285B:146-57.

187 Schweitzer, M.H., J.L. Wittmeyer, and J.R. Horner. 2005. Gender-specific reproductive tissue in ratites and *Tyrannosaurus rex.* Science 308:1456-60.

188 Seebacher, F. 2003. Dinosaur body temperatures: The occurrence of endothermy and ectothermy. Paleobiology 29:105-22.

189 Seibold, I., and A.J. Helbig. 1995. Evolutionary history of New and Old World vultures inferred from nucleotide sequences of the mitochondrial cytochrome *b* gene. Philosophical Transactions of the Royal Society of London B Biological Sciences 350(1332):163-78.

190 Seki, Y., M.S. Schneider, and M.A. Meyers. 2005. Structure and behavior of a toucan beak. Acta Materialia 53:5281-96.

191 Shipman, P. 1998. Taking wing. Simon and Schuster, New York.

192 Sibley, C.G., and J.E. Ahlquist. 1990. Phylogeny and classification of birds: A study in molecular evolution. Yale University Press, New Haven, CT.

193 Siegel-Causey, D. 1997. Phylogeny of the Pelecaniformes: Molecular systematics of a primitive group. Pages 159-72 *in* D.P. Mindell (ed.). Avian molecular evolution and systematics. Academic Press, San Diego.

194 Smith, V.S. 2001. Avian louse phylogeny (Phthiraptera: Ischnocera): A cladistic study based on phylogeny. Zoological Journal of the Linnean Society 132:81-144.

195 Sorenson, M.D., E. Oneal, J. Garcia-Moreno, and D.P. Mindell. 2003. More taxa, more characters: The Hoatzin problem is still unresolved. Molecular Biology and Evolution 20:1484-99.

196 Spearman, R.I.C., and J.A. Hardy. 1985. Integument. Pages 1-56 *in* A.S. King and J. McLelland (eds.). Form and function in birds, vol. 3. Academic Press, London and New York.

197 Stegmann, B.C. 1978. Relationships of the superorders Alectoromorphae and Charadriomorphae (Aves): A comparative study of the avian hand. Publications of the Nuttall Ornithological Club 17. Nuttall Ornithological Club, Cambridge, MA.

198 Stettenheim, P. 1972. The integument of birds. Pages 2-63 *in* D.S. Farner and J.R. King (eds.). Avian biology, vol. 2. Academic Press, New York.

199 Stokstad. E. 2001. Exquisite Chinese fossils add new pages to book of life. Science 291: 232-36.

200 Storer, R.W. 1960. Evolution in the diving birds. International Ornithological Congress 12:694-707.

201 Storer, R.W. 1971. Classification of birds. Pages 1-18 *in* D.S. Farner and J.R. King (eds.). Avian biology, vol. 1. Academic Press, New York.

202 Stresemann, E. 1951 [English trans. 1975]. Ornithology from Aristotle to the present. H.J. Epstein and C. Epstein (trans.), W.G. Cottrell (ed.). Harvard University Press, Cambridge, MA.

203 Stresemann, E. 1959. The status of avian systematics and its unresolved problems. Auk 76:269-80.

204 Suhel, Q., K. Isvaran, R.E. Hale, B.G. Miner, and N.E. Seavy. 2004. Nonlinear relationships and phylogenetically independent contrasts. Journal of Evolutionary Biology 17(3):709-15.

205 Swartz, S.M., M.B. Bennett, and D.R. Carrier. 1992. Wing bone stresses in free flying bats and the evolution of skeletal design for flight. Nature 359:726-29.

206 Sy, M. 1936. Funktionell-anatomische Untersuchungen am Vogelflügel. Journal für Ornithologie 84:199-296.

207 Tanner, J.A. 1967. Effects of microwave radiation on birds. Nature 210:636.

208 Tanner, J.A., C. Romero-Sierra, and S.J. Davie. 1967. Non-thermal effects of microwave radiation on birds. Nature 216:1139.

209 Tennekes, H. 1998. The simple science of flight: From insects to jumbo jets. MIT Press, Cambridge, MA.

210 Tucker, V.A. 2000. Gliding flight: Drag and torque of a hawk and falcon with straight and turned heads, and a lower value for the parasitic drag co-efficient. Journal of Experimental Biology 203:3733-44.

211 Tucker, V.A., A.E. Tucker, K. Akers, J.H. Enderson. 2000. Curved flight paths and sideways vision in peregrine falcons (*Falco peregrinus*). Journal of Experimental Biology 203:3755-63.

212 van der Leeuw, A.H.J., R.G. Bout, and G.A. Zweers. 2001. Evolutionary morphology in the neck system of ratites and waterfowl. Netherlands Journal of Zoology 51:243-62.

213 van Tuinen, M., C.G. Sibley, and S.B. Hedges. 2000. The early history of modern birds inferred from DNA sequences of nuclear and mitochondrial ribosomal genes. Molecular Biology and Evolution 17:451-57.

214 Vargas, A.O., and J.F. Fallon. 2004. Birds have dinosaur wings: The molecular evidence. Journal of Experimental Zoology Part B Molecular and Developmental Evolution 304B:1-5.

215 Vasquez, R.J. 1994. The automating skeletal and muscular mechanisms of the avian wing (Aves). Zoomorphology 114:59-71.

216 Verheyen, R. 1958. Contribution a la Systématique des Alciformes. Bull. Inst. Roy. Sci. Nat. Belgique 34:1-22.

217 Verstappen, M., P. Aerts, and R. van Damme. 2000. Terrestrial locomotion in the Black-billed Magpie: Kinematic analysis of walking, running, and out-of-phase hopping. Journal of Experimental Biology 203:2159-70.

218 Vidiera Marceliano, M.L. 1996. Estudo osteológico e miológoco do crânio de *Opisthocomus hoazin* (Müller, 1776) (Aves: Opisthocomidae), comparado com algumas espécies de Cracidae, Musophagidae, e Cuculidae. Boletim de Museu Paraense Emílio Goeldi, Série Zoologia 12(2).

219 Villarreal, L.P. 2005. Viruses and the evolution of life. ASM Press, Washington, DC.

220 Wagner, G.P., and J.A. Gauthier. 1999. 1,2,3 = 2,3,4: A solution to the problem of the homology of the digits in the avian hand. Proceedings of the National Academy of Sciences 96:5111-16.

221 Walker, C.A. 1981. New subclass of birds from the Cretaceous of South America. Nature 292:51-53.

222 Walters, M. 2003. A concise history of ornithology. Yale University Press, New Haven, CT.

223 Wappler, T., V.S. Smith, R.C. Dalgleish. 2004. Scratching an ancient itch: An Eocene bird louse fossil. Proceedings of the Royal Society of London B (Suppl.), Biology Letters. DOI 10.1098/rsbl.2003.0158. FirstCite e-publishing.

224 Warham, J. 1996. The behaviour, population biology, and physiology of the petrels. Academic Press, London.

225 Weimerskirch, H., and P.M. Sagar. 1996. Diving depths of Sooty Shearwaters *Puffinus griseus*. Ibis 138:786-94.

226 Welten, M.C.M., F.J. Verbeek, A.H. Meijer, and M.K. Richardson. 2005. Gene expression and digital homology in the chicken wing. Evolution and Development 7:18-28.

227 Whetstone, K.N. 1983. Braincase of Mesozoic birds. I. New preparation of the "London" *Archaeopteryx*. Journal of Vertebrate Paleontology 2:439-52.

228 White, A.T. 1999. Vertebrate Notes Cladogram 5.0: Aves. http://www.home.houston.rr.com/vnotes/clad5.html#Avesmore (no longer active; last accessed 20 November 2001).

229 Wilson, R.P., and C.A.R. Bain. 1984. An inexpensive depth gauge for penguins. Journal of Wildlife Management 48:1077-84.

230 Wink, M. 1995. Phylogeny of Old and New World vultures (Aves: Accipitridae and Cathartidae) inferred from nucleotide sequences of the mitochondrial cytochrome *b* gene. Zeitschrift für Naturforschung C-A Journal of Biosciences 50:11-12.

231 Witmer, L.M. 2001. Nostril position in dinosaurs and other vertebrates and its significance for nasal function. Science 293:850-53.

232 Woodbury, C.J. 1998. Two spinal cords in birds: Novel insights into early avian evolution. Proceedings of the Royal Society of London B 265:1721-29.

233 Wright, J.L. 2004. Bird-like features of dinosaur footprints. Pages 167-81 *in* P.J. Currie, E.B. Koppelhus, M.A. Shugar, and J.L. Wright (eds.). Feathered dragons: Studies in the transition from dinosaurs to birds. Indiana University Press, Bloomington.

234 Wyles, J.S., J.G. Kunkel, and A.C. Wilson. 1983. Birds, behavior, and anatomical evolution. Proceedings of the National Academy of Sciences 80:4394-97.

235 Yalden, D. 1971. The ability to fly of *Archaeopteryx*. Ibis 113:349-56.

236 Yalden, D. 1985. Forelimb function in *Archaeopteryx*. Pages 91-97 *in* M.K. Hecht, J.H. Ostrom, G. Viohl, and P. Wellnhoffer (eds.). The beginnings of birds. Freunde des Jura-Museums, Eichstätt.

237 Yanega, G.M., and M.A. Rubega. 2004. Hummingbird jaw bends to aid insect capture. Nature 428:615.

238 You Hai-lu, M.C. Lamanna, J.D. Harris, L.M. Chiappe, J. O'Connor, Ji Shu-an, Lü Jun-chang, Yuan Chong-ix, Li Da-qing, Zhang Xing, K.J. Locavara, P. Dodson, Qiang Ji. 2006. A nearly modern amphibious bird from the Early Cretaceous of northwestern China. Science 312:1640-43.

239 Zhou Zhonghe and J.O. Farlow. 2001. Flight capability and habits of *Confuciusornis*. Pages 237-54 *in* J. Gauthier and L.F. Gall (eds.). New perspectives on the origin and early evolution of birds: Proceedings of the international symposium honoring John H. Ostrom. Peabody Museum of Natural History, Yale University, New Haven, CT.

240 Zhou Zhonghe and Zhang Fucheng. 2002. A long-tailed, seed-eating bird from the early Cretaceous of China. Nature 418:405-9.

241 Zhou Zhonghe and Zhang Fucheng. 2003. Anatomy of the primitive bird *Sapeornis chaoyangensis* from the early Cretaceous of Liaoning, China. Canadian Journal of Earth Sciences 40:731-47.

242 Zhou Zhonghe and Hou Lianhai. 2002. Mesozoic birds in China. Pages 160-83 *in* L.M. Chiappe and L.M. Witmer (eds.). Mesozoic birds: Above the heads of dinosaurs. University of California Press, Berkeley, CA.

Index

DATE DUE

	9-2008		